AF353105

Recent Advances in OMICs Technologies and Application for Ensuring Meat Quality, Safety and Authenticity

Recent Advances in OMICs Technologies and Application for Ensuring Meat Quality, Safety and Authenticity

Editors

Mohammed Gagaoua
Brigitte Picard

MDPI • Basel • Beijing • Wuhan • Barcelona • Belgrade • Manchester • Tokyo • Cluj • Tianjin

Editors
Mohammed Gagaoua
Teagasc Ashtown Food
Research Centre
Ireland

Brigitte Picard
UMR Herbivores
France

Editorial Office
MDPI
St. Alban-Anlage 66
4052 Basel, Switzerland

This is a reprint of articles from the Special Issue published online in the open access journal *Foods* (ISSN 2304-8158) (available at: https://www.mdpi.com/journal/foods/special_issues/omics_meat).

For citation purposes, cite each article independently as indicated on the article page online and as indicated below:

LastName, A.A.; LastName, B.B.; LastName, C.C. Article Title. *Journal Name* **Year**, *Volume Number*, Page Range.

ISBN 978-3-0365-5665-9 (Hbk)
ISBN 978-3-0365-5666-6 (PDF)

Contents

About the Editors

Mohammed Gagaoua

Dr. Mohammed Gagaoua is a meat scientist at PEGASE research unit, INRAE, France who develops research activities on pork meat quality and production systems. Mohammed has expertise on muscle biology, meat biochemistry and the eating qualities of meat and meat products along their evaluation by consumers. He aims by his research a better understanding of the mechanisms underpinning meat quality development and determination. Mohammed has expertise in foodomics, mainly proteomics, and statistical approaches that consider environmental, production systems, and post-slaughter harvesting practices. He has respected knowledge in data integration in the context of farm-to-fork continuum and identification of the main drivers of the final eating qualities of meat. He is expert in MeatOmics and their use for the discovery of biomarkers of eating qualities of meat. Mohammed was runner-up in 2015 for the International Meat Secretariat Award; he received the ICoMST2018 award by Robin Shorthouse for outstanding contribution to advancing Meat Science and the ICoMST2018 Meat Science journal award for best presentation in Australia. He has published more than 100 peer-reviewed papers in international journals, edited 2 books; has 18 book chapters, 50 proceedings, along with over 150 contributions in national and international conferences. Dr. Mohammed Gagaoua is scientific editor and/or member of editorial boards of several journals including *Meat Science*, *Meat Technology*, *Scientific African*, *Chemical and Biological Technologies in Agriculture*, *Peer Community in Animal Science*, *Foods*, *Frontiers in Animal Science*.

Brigitte Picard

Dr. Brigitte Picard is a Director of Research at the National Research Institute for Agriculture, Food and Environment (INRAE) in France. After a bachelor at the University of Clermont-Ferrand and a thesis on the baking quality of hybrid wheat, she was recruited by INRA in 1991. Her research topic at the Joint Unit of Research on Herbivores concerned skeletal muscle development and ruminants' meat quality. Her skills include skeletal muscle physiology, analysis of proteins (electrophoresis, Western blot, ELISA, and immunohistochemistry), primary cells cultures (myoblasts and satellite cells), and proteomic analysis of skeletal muscle. Her work focuses on the prediction of beef tenderness from proteomic biomarkers and the link between rearing practices and beef sensory qualities. Brigitte, with an h-index of 46, is the author of more than 230 peer-reviewed papers and several communications at national and international conferences. From 2004 to 2009, she was the French correspondent of the COST 925 action "The importance of prenatal events for postnatal muscle growth in relation to the quality of muscle-based foods". Then, she was a member of the COST FA1002 "Advances in Farm Animal Proteomics". From 1992 to 2016, she was the head of several INRA research teams. Since 2017, she has been co-director of the regional Ruminant Meat Consortium, comprising professionals in the meat sector and researchers.

Preface to "Recent Advances in OMICs Technologies and Application for Ensuring Meat Quality, Safety and Authenticity"

OMICs-based approaches, also known as foodomics, are used for the in-depth characterization and better management of numerous food products, including muscle foods. In meat research, foodomics can be described as a discipline combining food and nutrition studies through the implementation of OMICs technologies to improve the quality and ensure the safety of fresh meat and processed meat products, with the objective of guaranteeing the health and well-being of consumers. OMICs technologies are further used to increase our knowledge of the pathways and biological mechanisms underlying the development and determination of intrinsic quality traits of muscle foods, hence advancing our understanding regarding variations in meat quality and the origin of certain meat quality defects and adulterations. Genomics, transcriptomics, proteomics, peptidomics, metabolomics and lipidomics, among other OMICs-based technologies, allow for comprehensive and high-throughput information to be obtained regarding the composition, safety, nutritional value, etc., of muscle foods thanks to the large amount of data that can be yielded before applying advanced statistical analyses and bioinformatics. This Special Issue aimed to gather the current advances in and applications of OMICs technologies to guarantee the quality, safety and authenticity of muscle foods.

Mohammed Gagaoua and Brigitte Picard

Editors

Editorial

Recent Advances in OMICs Technologies and Application for Ensuring Meat Quality, Safety and Authenticity

Mohammed Gagaoua

Food Quality and Sensory Science Department, Teagasc Food Research Centre, Ashtown, D15 KN3K Dublin, Ireland; gmber2001@yahoo.fr or mohammed.gagaoua@teagasc.ie

Citation: Gagaoua, M. Recent Advances in OMICs Technologies and Application for Ensuring Meat Quality, Safety and Authenticity. *Foods* **2022**, *11*, 2532. https://doi.org/ 10.3390/foods11162532

Received: 13 August 2022
Accepted: 17 August 2022
Published: 22 August 2022

Publisher's Note: MDPI stays neutral with regard to jurisdictional claims in published maps and institutional affiliations.

Consumers and stakeholders are increasingly demanding that the meat industry guarantees high-quality meat products with stable and acceptable sensory and safety properties. To achieve this lofty goal, it is prerequisite for meat researchers to address current meat quality issues and consider certain important goals. First, it is essential to decipher the unknowns concerning the underlying mechanisms of meat quality determination and development. Second, we need a better understanding of the biochemical pathways behind the conversion of muscle into fresh meat and those related to the manufacturing steps and their impact on processed meat products. Third, it is more than necessary to refine our knowledge on the impact of pre- and post-harvest procedures on both the molecular aspects of muscle foods and the final quality and safety of meat products in order to develop management and decision tools. Over the last two decades, sophisticated OMICs technologies—genomics, transcriptomics, proteomics, peptidomics, metabolomics and lipidomics, also known as foodomics—have been powerful approaches that extended the scope of traditional methods and opened up impressive possibilities to explore the above objectives in significant ways [1–6]. Foodomics are used for in-depth characterization and better management of numerous food products including for muscle foods. Overall, these techniques aimed to study in a comprehensive manner the dynamic link(s) between the genome and the quality traits of the meat we eat compared to the traditional methods, hence improving both the accuracy and sensitivity thanks to the large quantities of data that can be generated. Accordingly, this Special Issue focused on cutting-edge research applications of OMICs tools to characterize or manage the quality of muscle foods. Eleven published papers applied transcriptomics, targeted and untargeted proteomics, metabolomics, and genomics, among others, to evaluate meat quality, to determine the molecular profiles of meat and meat products, to discover and/or evaluate biomarkers of meat quality traits, and to characterize the safety, the adulteration, and the authenticity of meat and meat products.

In the frame of the discovery and evaluation of beef quality biomarkers, González-Blanco et al. [7] assessed different extraction methods of the sarcoplasmic and myofibrillar sub-proteomes of the *Longissimus thoracis et lumborum* (LTL) to evaluate the most reliable protocol for the identification of biomarkers of dark-cutting beef condition, also known as dark, firm, and dry (DFD) meat [8]. By means of one-dimensional sodium dodecyl-sulfate polyacrylamide gel electrophoresis (SDS-PAGE), the authors investigated the protein fractions of each extraction protocol. Within the sarcoplasmic sub-proteome, the extraction buffers that contain Triton X-100 led to a higher protein extractability, while TES buffer containing Tris, EDTA, and Sucrose was effective to distinguish differences in the protein pattern between the normal and DFD meat. Within the myofibrillar sub-proteome, the non-denaturing buffer allowed higher intensity protein bands while the lysis buffer increased protein extractability with more sensitivity in the differences between the treatments. In a following paper, Sierra et al. [9] focused on the myofibrillar sub-proteome to explore the effects of production systems (intensive versus extensive) and transport and lairage (mixing versus non-mixing with unfamiliar animals) and the post-mortem time ageing (rate and extent of tenderization) of LTL muscle of yearling bulls. Twenty-one proteins were differentially abundant due to any of the factors considered: farm, transport and lairage, and post-mortem time ageing. The proteins were from three major and interconnected pathways, such as muscle structure and associated proteins, energy metabolism and associated pathways, and heat shock proteins, of which several were known biomarkers of beef tenderness [4,10]. The study by Zhu et al. [11] applied a shotgun proteomics approach to identify biomarkers of beef tenderness evaluated using Warner–Bratzler shear force on young Limousin-sired bulls reared under an Irish production system. The authors revealed 34 putative protein biomarkers discriminating between the tender and tough meat groups. These proteins belong to biological pathways related to muscle structure, heat shock proteins, energy metabolism, response to oxidative stress, and apoptosis, from which 23 belong to the previous list of biomarkers of beef tenderness gathered by Gagaoua and co-workers in one repertoire by means of an integromics data mining approach [4]. Furthermore, Zhu et al. proposed a regression model using three proteins (Myozenin 3, Bridging Integrator-1, and Mimecan) that yielded a predictive power of 79%. Another study by Gagaoua et al. [12] aimed to evaluate, by means of Reverse Phase Protein Array (RPPA) quantification (a quantitative microformat Dot-Blot approach), a list of 20 protein biomarkers previously shortlisted to explain and predict both tenderness (evaluated by WBSF) and marbling (intramuscular fat (IMF) content) of 188 Protected Designation of Origin (PDO) Maine-Anjou cows. Using three statistical methods, namely, correlations analyses, clustering of WBSF and marbling into three quality clusters, and Partial Least Squares regressions (PLS-R), several biomarkers were selected. Whatever the statistical method, seven putative biomarkers for both WBSF values and marbling were qualified as being robust, hence allowing the authors to move forward in the pipeline of biomarker discovery for beef eating qualities. In this study, 10 and 9 proteins were qualified using a large database as significantly related to the determination of beef tenderness and marbling, respectively, in PDO Maine-Anjou cows.

In lamb research, two papers evaluated the variation in color [13] using proteomics and tenderness using a combination of Iso-seq, RNA-seq, and CTCF ChIP-seq data [14]. The first study by Gao et al. [13] investigated the sarcoplasmic and myofibrillar sub-proteomes of *Longissimus lumborum* (LL, color-stable) and *Psoas major* (PM, colour-labile) from Small-tailed Han sheep in relation to color stability during post-mortem storage (1, 3, and 5 days). The study revealed that the main differentially abundant proteins were from the glycolysis, others belong to the energy metabolism enzymes, chaperones and heat shock proteins, and proteins of structure. Thanks to correlation analyses, proteins such as adenylate kinase isoenzyme 1 (AK1), Pyruvate kinase (PKM), Carbonic anhydrase 3 (CA3), and Creatine kinase M-type (CKM) were significantly related to color stability in agreement with the available proteome repertoire of meat color [5]. This study allowed to validate predictors of color discoloration in sheep meat during storage. The second paper (a communication) by Yuan et al. [14] performed an experiment on sheep from two crossbred populations, Dorper x Hu x Hu (DHH) and Dorper x Dorper x Hu (DDH), with divergent meat tenderness. The authors aimed to identify key isoforms associated with this important quality and to better understand the underlying mechanisms of alternative splicing regulations leading to the production of isoforms. The authors revealed in this preliminary study 624 differentially expressed isoforms between DDH and DHH.

Meat and processed meat products have high nutritional value and economic importance, which makes them appealing commodities for fraudulent activities. Fraud activities associated with meat and meat products include addition (allergic proteins, preservatives), dilution (addition of water for yield increase and cost reduction), substitution, and mislabeling or misdescription, which are critical issues for economic, health, and religious reasons. Therefore, meat authentication is an important concern to protect consumers from illegal and unwanted ingredients. Accordingly, three papers dealing with meat authenticity, origin, and detection of meat adulteration using OMICs methods were published [15–17]. Cai et al. [15] proposed a simple and reliable single-tube septuple PCR assay based on mitochondrial DNA to simultaneously recognize seven meat species from pig, beef, sheep, chicken, turkey, goose, and duck. Furthermore, the authors validated the method in terms of sensitivity, specificity, robustness, and low costs for broad application to detect the origin of meat in foodstuffs with suspected adulteration. Another interesting study by Dobrovolny et al. [16] consisted of a collaborative work among 15 laboratories (inter-laboratory ring trial) that aimed to harmonize an analytical method based on DNA metabarcoding assay to detect adulteration from poultry and mammalian species. In this European study, each research team received and analyzed 16 anonymously labeled samples (8 samples, 2 subsamples each) containing six mixtures of DNA extract, one DNA extract from a model sausage, and another from maize, considered in this trial as a negative control. The evaluation parameters of the method allowed the researchers to confirm the reliability of the DNA metabarcoding approach for meat species authentication in routine analysis. The third study by Chen et al. [17] developed a duck genomic reference material by means of digital PCR platforms to detect meat adulteration through the detection of the duck *interleukin 2* (*IL2*) gene. Similarly, eight independent laboratories proceeded with the validation and certification of the proposed method

Two other research papers aiming to evaluate the freshness in gilthead sea bream (*Sparus aurata*) using metabolomics [18] and a better understanding of wooden breast myopathy in commercial broilers using proteomics [19] were published in this Special Issue.

In summary, the content of this Special Issue fits in with the current trend toward the use of foodomics to ensuring the quality, safety, and authenticity of meat and meat products. We hope that this Special Issue will attract the interest of the community of meat scientists, as well as students and scholars, by inspiring them to undertake more research in this emerging and important area of research towards the development of methods and decision tools to ensure more sustainable muscle foods. Special thanks go to the authors for their valuable contributions to this Special Issue and to our colleagues who devoted their time to review the papers. We sincerely hope that readers will find this Special Issue on meat OMICS-based approaches motivating and informative.

Funding: This research received no external funding.

Acknowledgments: The author acknowledges the support of the Marie Sklodowska-Curie grant agreement No. 713654 under the project number MF20180029.

Conflicts of Interest: The author declares no conflict of interest.

References

1. Munekata, P.E.S.; Pateiro, M.; López-Pedrouso, M.; Gagaoua, M.; Lorenzo, J.M. Foodomics in meat quality. *Curr. Opin. Food Sci.* **2021**, *38*, 79–85. [CrossRef]
2. Purslow, P.P.; Gagaoua, M.; Warner, R.D. Insights on meat quality from combining traditional studies and proteomics. *Meat Sci.* **2021**, *174*, 108423. [CrossRef] [PubMed]
3. Valdés, A.; Álvarez-Rivera, G.; Socas-Rodríguez, B.; Herrero, M.; Ibáñez, E.; Cifuentes, A. Foodomics: Analytical Opportunities and Challenges. *Anal. Chem.* **2022**, *94*, 366–381. [CrossRef] [PubMed]
4. Gagaoua, M.; Terlouw, E.M.C.; Mullen, A.M.; Franco, D.; Warner, R.D.; Lorenzo, J.M.; Purslow, P.P.; Gerrard, D.; Hopkins, D.L.; Troy, D.; et al. Molecular signatures of beef tenderness: Underlying mechanisms based on integromics of protein biomarkers from multi-platform proteomics studies. *Meat Sci.* **2021**, *172*, 108311. [CrossRef] [PubMed]
5. Gagaoua, M.; Hughes, J.; Terlouw, E.M.C.; Warner, R.D.; Purslow, P.P.; Lorenzo, J.M.; Picard, B. Proteomic biomarkers of beef colour. *Trends Food Sci. Technol.* **2020**, *101*, 234–252. [CrossRef]
6. Gagaoua, M.; Monteils, V.; Picard, B. Data from the farmgate-to-meat continuum including omics-based biomarkers to better understand the variability of beef tenderness: An integromics approach. *J. Agric. Food Chem.* **2018**, *66*, 13552–13563. [CrossRef] [PubMed]
7. González-Blanco, L.; Diñeiro, Y.; Díaz-Luis, A.; Coto-Montes, A.; Oliván, M.; Sierra, V. Impact of Extraction Method on the Detection of Quality Biomarkers in Normal vs. DFD Meat. *Foods* **2021**, *10*, 1097. [CrossRef] [PubMed]
8. Gagaoua, M.; Warner, R.D.; Purslow, P.; Ramanathan, R.; Mullen, A.M.; López-Pedrouso, M.; Franco, D.; Lorenzo, J.M.; Tomasevic, I.; Picard, B.; et al. Dark-cutting beef: A brief review and an integromics meta-analysis at the proteome level to decipher the underlying pathways. *Meat Sci.* **2021**, *181*, 108611. [CrossRef] [PubMed]
9. Sierra, V.; González-Blanco, L.; Diñeiro, Y.; Díaz, F.; García-Espina, M.J.; Coto-Montes, A.; Gagaoua, M.; Oliván, M. New Insights on the Impact of Cattle Handling on Post-Mortem Myofibrillar Muscle Proteome and Meat Tenderization. *Foods* **2021**, *10*, 3115. [CrossRef] [PubMed]

10. Picard, B.; Gagaoua, M. Meta-proteomics for the discovery of protein biomarkers of beef tenderness: An overview of integrated studies. *Food Res. Int.* **2020**, *127*, 108739. [CrossRef] [PubMed]

11. Zhu, Y.; Gagaoua, M.; Mullen, A.M.; Kelly, A.L.; Sweeney, T.; Cafferky, J.; Viala, D.; Hamill, R.M. A Proteomic Study for the Discovery of Beef Tenderness Biomarkers and Prediction of Warner–Bratzler Shear Force Measured on Longissimus thoracis Muscles of Young Limousin-Sired Bulls. *Foods* **2021**, *10*, 952. [CrossRef] [PubMed]

12. Gagaoua, M.; Bonnet, M.; Picard, B. Protein Array-Based Approach to Evaluate Biomarkers of Beef Tenderness and Marbling in Cows: Understanding of the Underlying Mechanisms and Prediction. *Foods* **2020**, *9*, 1180. [CrossRef] [PubMed]

13. Gao, X.; Zhao, D.; Wang, L.; Cui, Y.; Wang, S.; Lv, M.; Zang, F.; Dai, R. Proteomic Changes in Sarcoplasmic and Myofibrillar Proteins Associated with Color Stability of Ovine Muscle during Post-Mortem Storage. *Foods* **2021**, *10*, 2989. [CrossRef] [PubMed]

14. Yuan, Z.; Ge, L.; Zhang, W.; Lv, X.; Wang, S.; Cao, X.; Sun, W. Preliminary Results about Lamb Meat Tenderness Based on the Study of Novel Isoforms and Alternative Splicing Regulation Pathways Using Iso-seq, RNA-seq and CTCF ChIP-seq Data. *Foods* **2022**, *11*, 1068. [CrossRef] [PubMed]

15. Cai, Z.; Zhou, S.; Liu, Q.; Ma, H.; Yuan, X.; Gao, J.; Cao, J.; Pan, D. A Simple and Reliable Single Tube Septuple PCR Assay for Simultaneous Identification of Seven Meat Species. *Foods* **2021**, *10*, 1083. [CrossRef]

16. Dobrovolny, S.; Uhlig, S.; Frost, K.; Schlierf, A.; Nichani, K.; Simon, K.; Cichna-Markl, M.; Hochegger, R. Interlaboratory Validation of a DNA Metabarcoding Assay for Mammalian and Poultry Species to Detect Food Adulteration. *Foods* **2022**, *11*, 1108. [CrossRef] [PubMed]

17. Chen, X.; Ji, Y.; Li, K.; Wang, X.; Peng, C.; Xu, X.; Pei, X.; Xu, J.; Li, L. Development of a Duck Genomic Reference Material by Digital PCR Platforms for the Detection of Meat Adulteration. *Foods* **2021**, *10*, 1890. [CrossRef] [PubMed]

18. Mallouchos, A.; Mikrou, T.; Gardeli, C. Gas Chromatography–Mass Spectrometry-Based Metabolite Profiling for the Assessment of Freshness in Gilthead Sea Bream (*Sparus aurata*). *Foods* **2020**, *9*, 464. [CrossRef] [PubMed]

19. Bottje, W.G.; Lassiter, K.R.; Kuttappan, V.A.; Hudson, N.J.; Owens, C.M.; Abasht, B.; Dridi, S.; Kong, B.C. Upstream Regulator Analysis of Wooden Breast Myopathy Proteomics in Commercial Broilers and Comparison to Feed Efficiency Proteomics in Pedigree Male Broilers. *Foods* **2021**, *10*, 104. [CrossRef] [PubMed]

Article

Impact of Extraction Method on the Detection of Quality Biomarkers in Normal vs. DFD Meat

Laura González-Blanco [1,2], Yolanda Diñeiro [1,2], Andrea Díaz-Luis [3], Ana Coto-Montes [2,4], Mamen Oliván [1,2] and Verónica Sierra [1,2,*]

[1] Área de Sistemas de Producción Animal, Servicio Regional de Investigación y Desarrollo Agroalimentario (SERIDA), Ctra. AS-267, PK 19, 33300 Villaviciosa, Asturias, Spain; lgblanco@serida.org (L.G.-B.); ydineiro@serida.org (Y.D.); mcolivan@serida.org (M.O.)

[2] Instituto de Investigación Sanitaria del Principado de Asturias (ISPA), Av. del Hospital Universitario, s/n, 33011 Oviedo, Asturias, Spain; acoto@uniovi.es

[3] Department of Molecular Biology, Faculty of Medicine, University of Cantabria, Av. Herrera Oria, s/n, 39011 Santander, Cantabria, Spain; ANDYLUIS@hotmail.com

[4] Department of Morphology and Cell Biology, Faculty of Medicine, University of Oviedo, Av. Julián Clavería, 6, 33006 Oviedo, Asturias, Spain

[*] Correspondence: veroniss@serida.org; Tel.: +34-985-890-066

Citation: González-Blanco, L.; Diñeiro, Y.; Díaz-Luis, A.; Coto-Montes, A.; Oliván, M.; Sierra, V. Impact of Extraction Method on the Detection of Quality Biomarkers in Normal vs. DFD Meat. *Foods* **2021**, *10*, 1097. https://doi.org/10.3390/foods10051097

Academic Editor: Hanne Christine Bertram

Received: 6 April 2021
Accepted: 13 May 2021
Published: 15 May 2021

Abstract: The objective of this work was to demonstrate how the extraction method affects the reliability of biomarker detection and how this detection depends on the biomarker location within the cell compartment. Different extraction methods were used to study the sarcoplasmic and myofibrillar fractions of the *Longissimus thoracis et lumborum* muscle of young bulls of the Asturiana de los Valles breed in two quality grades, standard (Control) or dark, firm, and dry (DFD) meat. Protein extractability and the expression of some of the main meat quality biomarkers—oxidative status (lipoperoxidation (LPO) and catalase activity (CAT)), proteome (SDS-PAGE electrophoretic pattern), and cell stress protein (Hsp70)—were analyzed. In the sarcoplasmic fraction, buffers containing Triton X-100 showed significantly higher protein extractability, LPO, and higher intensity of high-molecular-weight protein bands, whereas the TES buffer was more sensitive to distinguishing differences in the protein pattern between the Control and DFD meat. In the myofibrillar fraction, samples extracted with the lysis buffer showed significantly higher protein extractability, whereas samples extracted with the non-denaturing buffer showed higher results for LPO, CAT, and Hsp70, and higher-intensity bands in the electrophoretic pattern. These findings highlight the need for the careful selection of the extraction method used to analyze the different biomarkers considering their cellular location to adapt the extractive process.

Keywords: cellular compartments; protein extractability; sarcoplasmic proteins; myofibrillar proteins; meat quality biomarkers; DFD meat; oxidative stress; proteomics

1. Introduction

Variations in meat quality depend on the specific changes that occur at the muscle cellular structure and metabolism levels, which rely on metabolic pathways triggered during the *post-mortem* conversion of muscle into meat. Changes in the protein profile of the muscle tissue can be key to understanding these processes; as such, proteomics has become a useful tool in this field [1].

Muscle is more complex than other tissues, as the subcellular architecture of skeletal muscle is different from that of mononucleated cells [2]. Therefore, the extraction of the meat proteome is influenced by the interaction of multiple factors such as the extraction method, the protein solubility, the protein location, and the *post-mortem* changes that occur during the transformation of muscle into meat [3]. To address this complexity meat scientists commonly divide the whole proteome in two fractions, sarcoplasmic and myofibrillar, which require different extraction methods due to their different extractabilities and water solubilities.

Sarcoplasmic proteins represent the 30%–35% of the total protein content of skeletal muscle and are mainly composed of metabolic proteins located in the sarcoplasm of the muscle fibers that are soluble in water or in low-ionic-strength solutions (<0.15 M). The myofibrillar proteins account for about 50% of total proteins and are mainly composed of contractile proteins that, because of their high molecular masses, structure, and being highly interconnected [4], require the use of denaturing solutions containing urea, thiourea, reducing agents (dithiothreitol (DTT) and beta-mercaptoethanol), detergents (sodium dodecyl sulfate (SDS)), and salts for their extraction and solubilization [5–7]. However, Chen et al. [5] reported the use of water or low-ionic-strength media for the extraction and solubilization of myofibrillar proteins from skeletal muscle.

Considering the above, we hypothesized that the analysis of biomarkers of the conversion of muscle into meat and the ultimate meat quality may be significantly affected by the muscle extraction method. Extraction conditions, such as buffer pH, ionic strength, type of salt, extraction volume, and homogenization, influence muscle protein extractability [8–10]. Furthermore, some of the extraction factors (reagents, pH, and ionic strength) may not be compatible with some of the analytical procedures used to determine the presence and/or abundance of the most common biomarkers. The structure of the muscle cells results in some portions of the sarcoplasm remaining between the myofibrils, complicating their protein extraction and, therefore, the analysis of some of these biomarkers. Finally, the same extractive method may perform differently depending on whether the muscle shows a compact or deteriorated structure.

To the best of our knowledge, no previous study has determined the effect of the extraction method on the reliability of the determination of the main meat quality biomarkers in different muscle cell fractions. Therefore, the objective of this work was to identify the optimal methodology to be used for the extraction and detection of the main families of meat quality biomarkers such as those related to oxidative status, metabolic and structural proteins, and cell stress. We aimed to compare the reliability of protein extraction for a meat of standard-quality grade (Control) with that for a type of defective meat (dark, firm, and dry (DFD)), which exhibits alterations in the *post-mortem* muscle metabolism that produce a dark color and poor processing characteristics, such as higher water-holding capacity, unstructured texture, and higher spoilage [11–13].

2. Materials and Methods

2.1. Animals

A total of 80 young bulls from the autochthonous beef breed Asturiana de los Valles (AV) were included in this work. This breed is the second-most important in the Spanish market of protected geographical indication (PGI) fresh meat, both in production and economic value. Animals were slaughtered at 14–18 months of age, according to the commercial local market and PGI requirements, in two different slaughter batches (of 42 and 38 animals, respectively) with a one-week interval. Carcasses were transferred to a cold room at 3 °C within 2 h after slaughter. At 24 h *post-mortem*, the pH (pH24) was measured at the 13th, 10th, and 6th rib level of the *Longissimus thoracis et lumborum* (LTL) muscle of the left-half carcass using a penetration electrode coupled with a temperature probe (InLab Solids Go-ISM, Mettler-Toledo S.A.E., Barcelona, Spain). The average of the triplicate measurements was used to categorize the carcasses into two groups: Control (pH24 $\leq$ 6.2) and DFD (pH24 > 6.2). The pH24 threshold was set to 6.2 to ensure that the samples considered DFD were unambiguous [14]. DFD samples accounted for 9% of the total carcasses sampled. For each DFD carcass detected (n = 7), a carcass from the same farm, diet, transport, and weight but with a normal pH24 (5.4 to 5.5) was selected for the Control (n = 7) group.

2.2. Muscle Sample Collection

Muscle samples (20 g) were taken from the LTL at the 13th rib level at 24 h *post-mortem* for analysis of protein extractability and different biomarkers (oxidative status, sarcoplasmic and myofibrillar proteins, and stress protein). These samples were immediately snap-frozen in liquid nitrogen and stored at −80 °C until analysis.

At 24 h *post-mortem*, the LTL muscle was removed between the 6th and the 13th ribs and transported to the laboratory where it was divided into 2.5-cm steaks for the determination of the meat quality traits. The first steak was used for instrumental color and drip loss determination. The second steak was cut under sterile conditions and divided into three portions for subsequent microbiological analysis (mesophilic and *Enterobacteriaceae* total viable counts) at 3, 7, and 14 days *post-mortem*. The following three steaks were used for meat toughness measurement using the Warner–Bratzler shear force test at 3, 7, and 14 days *post-mortem*. Finally, the last steak was divided into three portions for subsequent proteomic analysis at 3, 7, and 14 days. The steaks intended for aging were vacuum-packed in 20 μm polyamide/70 μm polyethylene bags and aged in darkness under refrigerated conditions (4 ± 1 °C). After the corresponding aging period, the steaks were frozen and stored at −20 °C (−80 °C in the case of proteomics) for subsequent analysis.

2.3. Meat Quality Trait Measurements

Meat color was recorded on three 10 mm diameter spots on the exposed cut surface of the LTL muscle at the 7th rib level at 24 h *post-mortem*. Indicators of lightness (L*), redness (a*), and yellowness (b*) were taken after 60 min of blooming using a Minolta CM-2300d spectrophotometer, with a D65 illuminant, and a 10° standard observer in the CIE space (Konica Minolta Inc., Madrid, Spain), and the average value of the three determinations was used [15].

Meat drip loss (percent exudates) was determined by duplicates on 50 g of fresh samples taken 24 h *post-mortem* and placed in a container (Meat juice collector, Sarstedt, Germany) at 4 °C, according to the method of Honikel [16].

Meat toughness was measured on cooked meat using the Warner−Bratzler (WB) shear test as described by Diaz et al. [17]. Results are expressed as the mean WB shear force maximum load (kg) for each steak.

For microbiological analyses, meat samples were processed according to ISO 7218 (International Organization for Standardization, 2007). Firstly, each vacuum-packed portion of meat was opened (after 3, 7, and 14 days aging), a portion of 10 g was aseptically taken, and 90 mL of sterile buffered peptone water (PW, Oxoid, Basingstoke, UK) was added. The mixture was homogenized in a stomacher (IUL instruments, Barcelona, Spain) for 2 min. For microbial counts, the appropriate decimal dilutions of the samples were prepared and placed onto the corresponding medium Petri dishes. Total mesophilic aerobic microorganism counts were determined on plate count agar (PCA; Oxoid, Basingstoke, U.K.), incubated at 30 °C for 72 h; *Enterobacteriaceae* were determined on violet red bile glucose (VRBG; Oxoid, Basingstoke, UK), incubated at 37 °C for 24 h. After incubation, microbial counts were performed as described in ISO 7218:2007.

2.4. Muscle Extraction Methods

Figure S1 shows the flowchart of the extraction procedure for both sarcoplasmic and myofibrillar protein fractions from muscle.

2.4.1. Sarcoplasmic Protein Extraction

For each sample, eight different sarcoplasmic extraction methods, resulting from different combinations of four extraction buffers and two different centrifugation steps, were tested.

The four extraction buffers used were:

1. TES buffer (TES): 10 mM Tris (pH 7.6), 1 mM EDTA (pH 8.0), 0.25 M sucrose, and 0.6% protease inhibitor cocktail (P8340, Sigma-Aldrich Co., St. Louis, MO, USA) [18].

2. Sodium buffer (Na): 50 mM sodium phosphate buffer (pH 7.5) and 0.6% protease inhibitor cocktail (P8340, Sigma-Aldrich Co., St. Louis, MO, USA) [19].
3. Sodium with Triton buffer (Na + T): 50 mM sodium phosphate buffer (pH 7.5), 0.1% Triton X-100, and 0.6% protease inhibitor cocktail (P8340, Sigma-Aldrich Co., St. Louis, MO, USA) [20].
4. Potassium with Triton buffer (K + T): 10 mM potassium phosphate buffer (pH 7.4), 50 mM NaCl, 0.1% Triton X-100, and 0.6% protease inhibitor cocktail (P8340, Sigma-Aldrich Co., St. Louis, MO, USA) [21].

The homogenization of extracts was followed by two different speed centrifugation methods:

(a) $1000 \times g$, 6 min at 4 °C;
(b) $20,000 \times g$, 20 min at 4 °C.

For each meat sample and extraction method, 0.5 g of muscle was homogenized in 4 mL of the corresponding extraction buffer using a Polytron PT1200 E (Kinematica Inc., Luzern, Switzerland) two times for 15 s at maximum speed. The supernatants of the seven individuals of each sample group (Control and DFD) were collected and pooled (one pool for Control and one for DFD), aliquoted, and stored at −80 °C.

2.4.2. Myofibrillar Protein Extraction

For each sample, two different myofibrillar extraction methods were tested, using denaturing or non-denaturing solutions.

1. The denaturing extraction was performed on the sample residue after the extraction of sarcoplasmic proteins with the TES buffer and 20 min centrifugation at $20,000 \times g$ and 4 °C, as proposed by Bjarnadottir et al. [22]. The resulting pellet was homogenized into 4 mL of lysis buffer (10 mM Tris-HCl (pH 7.6), 7 M urea, 2 M thiourea, 2% CHAPS, and 10 mM DTT) with the polytron 2×15 s at 20,000 rpm. Subsequently, this solution was stirred for 1 h in a Multi Reax stirrer (Heidolph Instruments, Schwabach, Germany) and was centrifuged at 20,000 rpm for 20 min at 4 °C. The supernatant containing the myofibrillar proteins was collected and filtered through a nylon filter (5 mm), aliquoted, and stored at −80 °C.
2. The non-denaturing myofibrillar extraction was based on the method reported by Hashimoto et al. [23], with the following modifications: 0.5 g of muscle samples were homogenized in 4 mL of non-denaturing extraction buffer (30 mM of sodium phosphate buffer (pH 7)) and 0.6% protease inhibitor cocktail (Sigma-Aldrich Co., St. Louis, MO, USA) using a Polytron PT1200 E (Kinematica Inc., Luzern, Switzerland) two times for 15 s at maximum speed. The homogenates obtained were centrifuged at $8000 \times g$ for 20 min at 4 °C. The recovered pellet was resuspended in 4 mL of KCl phosphate buffer ((pH 7.5); 0.45 M KCl, 15.6 mM Na_2PO_4, and 3.5 mM KH_2PO_4) and vortexed. Subsequently, this solution was stirred for 30 min in a Multi Reax stirrer (Heidolph Instruments, Schwabach, Germany). The mixture was centrifuged twice at $5000 \times g$ for 15 min at 4 °C. After the centrifugation, the supernatant containing the myofibrillar proteins was recovered, aliquoted, and stored at −80 °C.

From now on, the eight different sarcoplasmic extracts are referred to as: TES 1000, TES 20,000, Na 1000, Na 20,000, Na + T 1000, Na + T 20,000, K + T 1000, and K + T 20,000, and the two myofibrillar extracts are referred to as "lysis" for the denaturing extraction and "ND" for the non-denaturing extraction.

2.5. Protein Extractability

The solubility of muscle proteins is the amount of protein remaining in a solution of defined characteristics after the application of a specific centrifugal force for a determined duration. The terms solubility and extractability are frequently interchanged, assuming that once the protein is solubilized, it can be readily extracted from muscle fibers or myofibrils into a solution [24]. The protein content of the different extracts was measured by the Bradford method [25].

2.6. Oxidative Stress

The oxidative status of the muscle tissue was assessed by the measurement of lipid oxidative damage (lipoperoxidation (LPO)) and catalase activity (CAT). LPO was analyzed by measuring the reactive aldehyde malondialdehyde (MDA) and 4-hydroxy-2-(E)-nonenal (4-HNE) using the LPO assay kit from Calbiochem (No.437634, San Diego, CA, USA) [26], which measures lipid hydroperoxides directly using redox reactions with ferrous ions, and the results are expressed as nmol MDA + 4-HNE/g protein.

Catalase activity (CAT; EC 1.11.1.6) was analyzed according to the method developed by Lubinsky and Bewley [27] using hydrogen peroxidase (H_2O_2) as the substrate. The results are expressed as μmol H_2O_2/min mg protein.

2.7. Sarcoplasmic and Myofibrillar Subproteome Analysis

The separation of proteins obtained with the different extraction buffers was performed using SDS-PAGE gels as described by Díaz et al. [17], with minor modifications. Sarcoplasmic (15 μg of protein) and myofibrillar (30 μg of protein) muscle extracts were denatured with sample buffer (65.8 mM Tris/HCl (pH 6.8), 2% SDS, 21% glycerol, 5% beta-mercaptoethanol, and 0.026% bromophenol blue) and boiled at 100 °C for 5 min. Samples were loaded into 1-mm dual vertical slab gels (Mini-PROTEAN® Tetra Cell, Bio-Rad Laboratories Inc., Hercules, CA, USA) and run for 2.50 h (sarcoplasmic extracts) or 2.20 h (myofibrillar extracts) at 150 V for one-dimensional electrophoresis (1D-SDS-PAGE). The resolving gel contained 12% and the stacking gel 4% of acrylamide/bis (30% acrylamide), 10% (*w/v*) SDS, 1.5 M Tris/HCl (pH 8.8), 0.5 M Tris/HCl (pH 6.8), 10% (*w/v*) ammonium persulphate, and 0.1% TEMED. Prestained molecular weight standards (Precision Plus Protein™ All Blue Standards, Bio-Rad Laboratories Inc., Hercules, CA, USA) were run on each gel to determine the protein band molecular weights. Gels were stained (50% methanol, 10% acetic acid, and QC Colloidal Coomassie from Bio-Rad) and afterward de-stained with distilled water. Three gels per sample were performed.

Stained-gel images were captured using the UMAX ImageScanner (Amersham Biosciences, Buckinghamshire, UK). SDS-PAGE densitometry analysis and band quantification were performed as described by Díaz et al. [17].

2.8. Stress Protein: Hsp70

Stressors, such as high temperature, hypoxia, ischemia, and oxidation, can induce the synthesis of stress proteins like the heat shock proteins (Hsps) to protect cellular proteins against denaturation [27]. Among the best-known and most-investigated Hsps is the Hsp70 family. Hsp70 is abundantly induced in the response to cellular stress in muscles [28] and it was proposed to be a key biomarker of the process of conversion of muscle into meat and, therefore, of the ultimate meat quality, as it can simultaneously indicate the tenderness, color, and WHC of meat [29], which are some of the quality traits that are more affected in DFD meat. Therefore, the expression of Hsp70-1A/B was measured by Western blotting. The homogenized tissue (90 μg protein per sample) was mixed with Laemmli sample buffer (65.8 mM Tris/HCl (pH 6.8)), 2% SDS, 21% glycerol, 5% beta-mercaptoethanol, and 0.026% bromophenol blue) and denatured by boiling at 100 °C for 5 min. The extracts were fractionated using SDS-PAGE at 200 V, and then proteins were transferred to polyvinylidene fluoride membranes (Immobilon TM-P; Millipore Corp., Burlington, MA, USA) at 350 mA. Once the membranes were blocked at 4 °C overnight with 10% (*w/v*) bovine seroalbumin (BSA) dissolved in Tris-buffered saline (TBS) (50 mM Tris-HCl and 150 mM NaCl, (pH 7.5)), they were incubated at 4 °C overnight with the primary antibody anti-Hsp70 (A5A) (ab2787, Abcam, Cambridge, U.K.), which detects Hsp70-1A/B (UniProtKB: P0DMV8/P0DMV9). The antibody was pre-diluted in TBS buffer containing 5% (*w/v*) BSA. After three washes in TBS-T (50 mM Tris-HCl (pH 7.5), 150 mM NaCl, and 0.05% Tween-20), the membranes were incubated with the corresponding horseradish-peroxidase-conjugated secondary antibody (Cell Signaling, Danvers, MA, USA) and diluted in TBS buffer with BSA 2% (*w/v*) for 1 h at room temperature. After three washes in TBS-T, the immunoconjugates were detected using a chemiluminescent horseradish peroxidase substrate (WBKLS0500, Millipore Corp., Darmstadt, Germany) according to the manufacturer's instructions. The Image Studio Lite 5.2.5 program (LI-COR Biosciences, Lincoln, NE, USA) allowed us to quantify the optical density of the bands. The densitometry results are expressed as semi-quantitative optical density (in arbitrary units) of blot bands, normalized to Ponceau bands as the loading control. Three replicates per sample were performed.

2.9. Statistical Analysis

The normality of variables was tested using the Kolmogorov–Smirnoff test. The effect of sample type (Control vs. DFD) on the different quality traits was analyzed by a *t*-test of independent samples. For variables measured at different *post-mortem* times (WBSF, microbiological loads), the effect of aging time (with animal as the random factor) was tested. For the rest of the variables included in the study, the effect of extraction method E (eight different extraction methods for sarcoplasmic extracts and two for myofibrillar extracts) and the effect of sample type T (Control vs. DFD), and their interaction (E × T) were analyzed by ANOVA using the general linear model procedure in SPSS v. 22.0 (SPSS Inc., Chicago, IL, USA). Once the interaction between E and T was established, the effects of the extraction method and the type of sample were tested independently. When significant, differences between extraction methods were analyzed by means of Tukey's post hoc test, and the Games–Howell test when variances were not homogeneous.

3. Results and Discussion

3.1. Meat Quality Traits

As expected, DFD meat had a higher pH24 ($p < 0.001$), darker color (L *, $p < 0.001$), was less red (a *, $p < 0.01$) and less yellow (b *; $p < 0.001$), and had a higher growth of mesophilic ($p < 0.001$) and *Enterobacteriaceae* ($p < 0.005$) microorganisms at 14 days of aging (Table 1).

Table 1. Effect of quality grade (Control vs. DFD) on meat quality traits (mean ± standard deviation).

Variable	Time *post-mortem*	Control ($n = 7$)	DFD ($n = 7$)	Sig.
pH	24 h	5.48 ± 0.05	6.49 ± 0.27	***
Drip loss (%)	48 h	1.19 ± 0.64	1.06 ± 0.31	NS
L*	48 h	34.35 ± 2.53	27.71 ± 2.34	***
a*	48 h	9.84 ± 2.82	5.83 ± 0.97	**
b*	48 h	11.87 ± 2.45	6.14 ± 2.38	***
Meat toughness (WBSF, kg)	3 days	7.15 ± 1.74b	6.63 ± 2.50	NS
	7 days	6.02 ± 1.31ab	5.56 ± 1.85	NS
	14 days	4.97 ± 1.01a	5.33 ± 1.35	NS
Mesophilic (log UFC/kg)	3 days	3.73 ± 0.37a	3.72 ± 1.06a	NS
	7 days	4.31 ± 0.84a	4.42 ± 1.71a	NS
	14 days	6.05 ± 0.37b	7.22 ± 0.64b	***
Enterobacteriaceae (log UFC/kg)	3 days	1.26 ± 1.23	1.53 ± 1.32a	NS
	7 days	1.45 ± 1.43	2.13 ± 1.64a	NS
	14 days	3.08 ± 1.59	4.91 ± 0.83b	*

For variables measured at different times *post-mortem* (meat toughness, mesophilic, and *Enterobacteriaceae*), means in the same column followed by different letters differ statistically. DFD: dark, firm and dry; Sig.: Significance; NS: not significant; * $p < 0.05$; ** $p < 0.01$; *** $p < 0.001$.

Previous studies comparing meat from three different Spanish autochthonous breeds reported similar results for color traits with significantly higher values of L*, a*, and b* in high-pH (>6) meat from Asturiana de los Valles and Rubia Gallega [30]. Poleti et al. [31] reported lower values of the color parameters in high-pH meat when comparing beef from Nellore cattle classified into two different pH groups: high (≥6.0) and normal (<5.8). It is known that due to the high pH, DFD meat is more prone to microbial spoilage than normal-pH meat [13]; accordingly, we found a faster reduction in the shelf life of the DFD meat at 14 days *post-mortem*. In agreement with our results, García-Torres et al. [30] found similar results with higher mesophlic loads in Rubia Gallega and Asturiana de los Valles breeds at 7 and 14 days of aging. No significant differences were found for WBSF between the Control and DFD samples in this study; however, an anomalous tenderization process was observed in DFD meat as meat toughness did not significantly decrease with aging.

Samples with a high pH24 (>6.2) in the present study were darker and their microbial spoilage was higher, so they were therefore of defective quality compared with those with a lower pH. These differences in quality traits may reflect differences at the muscle cell level (structure and metabolism), which have to be considered to understand the results obtained in this study.

3.2. Protein Extractability

The protein contents of the sarcoplasmic and myofibrillar fractions obtained by the different extraction methods tested are shown in Figure 1.

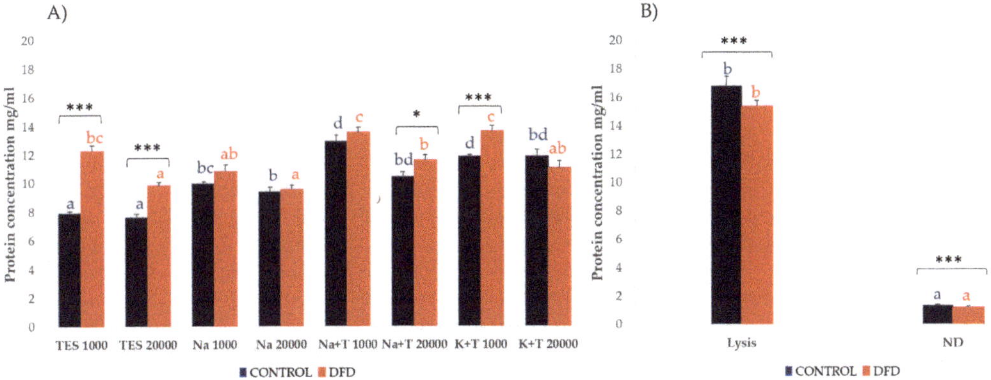

Figure 1. Protein content (mean ± SEM) of the (**A**) sarcoplasmic and (**B**) myofibrillar fractions from Control (blue) and DFD (red) meat samples and the different extraction methods. Charts with different letters (blue for Control and red for DFD) were significantly different between extraction methods at $p < 0.05$. Asterisks indicate significant differences between Control and DFD samples within the same extraction method. * $p < 0.05$; *** $p < 0.001$. TES 1000: TES buffer and 1000× g, 6 min; TES 20,000: TES buffer and 20,000× g, 20 min; Na 1000: sodium phosphate buffer and 1000× g, 6 min; Na 20,000: sodium phosphate buffer and 20,000× g, 20 min; Na + T 1000: sodium phosphate buffer with Triton X-100 and 1000× g, 6 min; Na + T 20,000: sodium phosphate buffer with Triton-X100 and 20,000× g, 20 min; K + T 1000: potassium phosphate buffer with Triton X-100 and 1000× g, 6 min; K + T 20,000: potassium phosphate buffer with Triton X-100 and 20,000× g, 20 min; Lysis: denaturing extraction with lysis buffer; ND: non-denaturing extraction.

In the sarcoplasmic fraction, higher extractability was obtained with buffers containing Triton X-100 (Na + T and K + T), which is a type of non-ionic detergent used for cell lysis, that is, for the disruption of cell membranes and the consequent release of intracellular materials that breaks protein–lipid and lipid–lipid associations, and generally does not denature proteins. The higher protein content in these extracts could be explained by Triton X-100 helping to solubilize most membrane proteins in their native and active form, retaining their protein interactors. In the sarcoplasmic fraction, the centrifugation speed only affected the protein solubility of some DFD extracts (TES, Na + T, and K + T), being significantly higher ($p < 0.05$) at 1000× g, whereas no significant differences were found for the Control samples. This could be related to a higher disintegration of the muscle structure in DFD meat, which resulted in the higher extraction capability of sarcoplasmic proteins retained within the sarcoplasm portions embedded between the myofibrils.

In the myofibrillar fraction, the lysis buffer showed significantly higher protein extractability than ND for both Control and DFD samples. Lysis buffer contains some agents such as urea, thiourea, CHAPS, and DTT, which may have been responsible for these extractability differences [32]. Urea is a chaotropic agent that denatures proteins by disrupting noncovalent and ionic links between amino acids [33], whereas thiourea improves the solubilization of hydrophobic membrane proteins [34]; therefore, their combination is used to extract proteins that are otherwise insoluble. CHAPS prevents hydrophobic interaction and DTT aids in the solubilization of complex mixtures by reduction of disulfide bonds, avoiding protein aggregation or precipitation [35]. The combination of these components increases solubilization and proteins extractability [36,37].

3.3. Oxidative Stress

Figures 2 and 3 show the results for LPO and CAT in both the sarcoplasmic and myofibrillar fractions. The TES buffer was incompatible with some reagents present in the LPO assay kit (probably EDTA) and produced unstable results, so the results of the TES extracts were not considered for this assay.

DFD samples showed higher LPO values (Figure 2) in all extracts and in both cellular fractions. This could be related to a higher pre-slaughter stress situation, which increases the oxidative damage of lipids in the cells of the animals that finally produced DFD carcasses. Within the sarcoplasmic fraction, the K + T 1000 extraction method showed higher LPO values ($p < 0.001$) in both the Control and DFD samples, whereas in the myofibrillar fraction, the ND buffer showed higher LPO values ($p < 0.001$).

CAT activity was higher in the Control samples in both the sarcoplasmic and myofibrillar extracts, which seems to be related to a higher antioxidant defense in the muscle of standard-quality meat obtained in the absence of pre-slaughter stress. In the sarcoplasmic extracts, higher CAT activity was found in the samples extracted with Na buffer independent of the centrifugation speed. However, its determination in extracts containing Triton X-100 was difficult due to the non-ionic detergents interfering with ultra-violet (UV) spectrophotometry, thus producing unstable results.

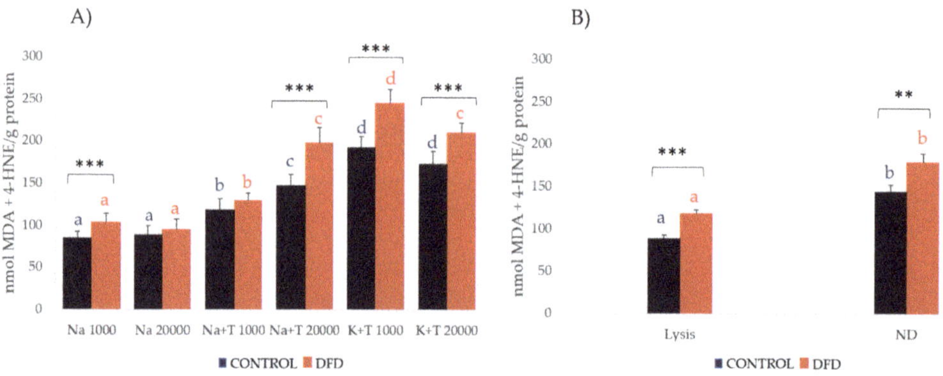

Figure 2. Lipoperoxidation (mean ± SEM) of (**A**) sarcoplasmic and (**B**) myofibrillar fractions from Control (blue) and DFD (red) meat samples. Charts with different letters (blue for and red for DFD) were significantly different between extraction methods at $p < 0.05$. Asterisks indicate significant differences between the Control and DFD samples within the same extraction procedure. ** $p < 0.01$; *** $p < 0.001$. Na 1000: sodium phosphate buffer and 1000× g, 6 min; Na 20,000: sodium phosphate buffer and 20,000× g, 20 min; Na + T 1000: sodium phosphate buffer with Triton X-100 and 1000× g, 6 min; Na + T 20,000: sodium phosphate buffer with Triton-X100 and 20,000× g, 20 min; K + T 1000: potassium phosphate buffer with Triton X-100 and 1000× g, 6 min; K + T 20,000: potassium phosphate buffer with Triton X-100 and 20,000× g, 20 min; Lysis: denaturing extraction with lysis buffer; ND: non-denaturing extraction.

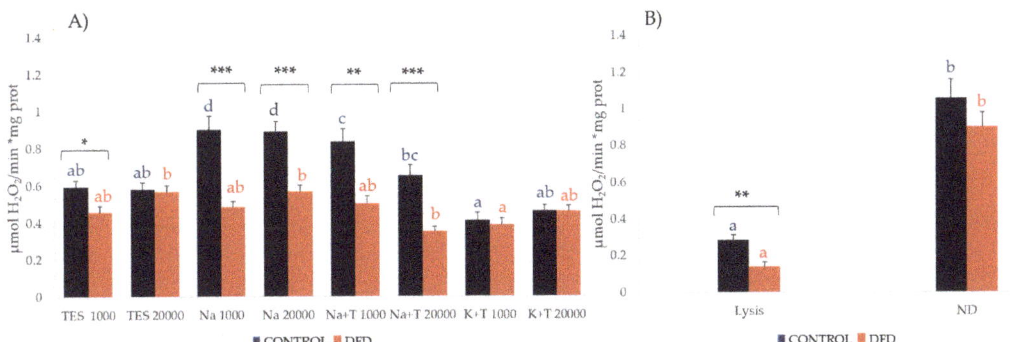

Figure 3. Catalase activity (mean $\pm$ SEM) of the (**A**) sarcoplasmic and (**B**) myofibrillar fractions from Control (blue) and DFD (red) meat samples. Charts with different letters (blue for Control and red for DFD) were significantly different between extraction methods at $p < 0.05$. Asterisks indicate significant differences between the Control and DFD samples within the same extraction procedure. * $p < 0.05$; ** $p < 0.01$; *** $p < 0.001$. TES 1000: TES buffer and $1000\times g$, 6 min; TES 20,000: TES buffer and $20,000\times g$, 20 min; Na 1000: sodium phosphate buffer and $1000\times g$; 6 min; Na 20,000: sodium phosphate buffer and $20,000\times g$, 20 min; Na + T 1000: sodium phosphate buffer with Triton X-100 and $1000\times g$, 6 min; Na + T 20,000: sodium phosphate buffer with Triton-X100 and $20,000\times g$, 20 min; K + T 1000: potassium phosphate buffer with Triton X-100 and $1000\times g$, 6 min; K + T 20,000: potassium phosphate buffer with Triton X-100 and $20,000\times g$, 20 min; Lysis: denaturing extraction with lysis buffer; ND: non-denaturing extraction.

In the myofibrillar fraction, CAT activity was significantly higher in the extracts obtained by the non-denaturing method ($p < 0.05$).

3.4. Sarcoplasmic and Myofibrillar Subproteome

SDS-PAGE gels allowed for the separation of 26 protein bands (ranging from 15 to 200 kDa) from the muscle sarcoplasmic subproteome, as shown in Figure 4, which shows the protein pattern obtained with the different extraction methods at the maximum centrifugation speed tested ($20,000\times g$) for both types of meat samples (Control and DFD).

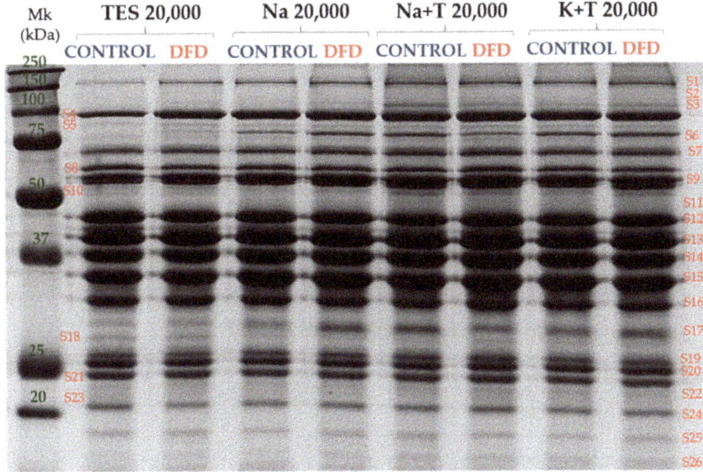

Figure 4. The 1D-SDS-PAGE gel image of sarcoplasmic subproteome from the Control and DFD meat samples extracted with different buffers (TES, Na, Na + T, and K + T) at $20,000\times g$. Mk: prestained molecular weight marker (All Blue prestained, Biorad). Band names are denoted by S (sarcoplasmic protein) followed by a number.

The complete details of the results (means $\pm$ SEM) for the significant sarcoplasmic bands obtained with the different extraction methods and type of samples are provided in Table S1. The analysis of the main factors studied (extraction method, type of sample, and their interaction) showed a significant interaction for four bands (S1, S4, S18, and S21). Once these bands were discarded, the differences between extraction methods were analyzed including all the samples regardless of being Control or DFD. Table 2 shows a total of 13 sarcoplasmic bands with significant differences in band intensity between extraction methods.

Table 2. Effect of the extraction method on the sarcoplasmic subproteome bands intensity (optical density in arbitrary units).

Sarcoplasmic Bands (MWe [1])	TES 1000	TES 20,000	Na 1000	Na 20,000	Na + T 1000	Na + T 20,000	K + T 1000	K + T 20,000	SEM	Sig.
S2 (137.9 kDa)	0.188	0.211	0.246	0.262	0.431	0.446	0.42	0.316	0.061	**
S3 (115.8 kDa)	0.305a	0.249a	0.406ab	0.344a	0.753bc	0.911c	0.602abc	0.421ab	0.084	***
S6 (81.31 kDa)	0.55a	0.572a	1.556b	1.256b	1.442b	1.559b	1.705b	1.482b	0.146	***
S10 (53.60 kDa)	0.895a	0.992ab	0.998ab	1.031ab	1.336b	1.325b	1.211ab	1.197ab	0.087	**
S11 (50.70 kDa)	1.264abc	1.107a	1.244abc	1.223ab	1.438abcd	1.68d	1.603bcd	1.611cd	0.086	***
S12 (45.55 kDa)	8.244ab	8.814b	8.101ab	7.837ab	7.536ab	7.049a	7.474ab	6.874a	0.351	**
S13 (40.72 kDa)	10.805b	10.448ab	10.159ab	10.154ab	9.258a	9.34ab	8.513a	8.98a	0.427	**
S14 (37.6 kDa)	8.775ab	8.707ab	9.287b	9.16ab	8.524ab	8.413ab	8.18a	8.345ab	0.242	*
S15 (34.74 kDa)	10.859bcd	10.457abc	11.59d	11.376cd	10.74abcd	9.77a	10.341abc	9.857ab	0.241	***
S16 (32.14 kDa)	8.128b	8.079b	6.672a	6.878a	6.53a	6.136a	6.446a	6.38a	0.211	***
S17 (29.74 kDa)	1.576a	1.826a	2.399c	2.835cd	2.649cd	2.677cd	3.28d	3.179d	0.109	***
S19 (26.68 kDa)	2.889b	2.89b	2.258a	2.546ab	2.419ab	2.585ab	2.22a	2.521ab	0.128	***
S20 (25.76 kDa)	4.162b	4.103b	3.635ab	3.661ab	3.425a	3.329a	3.482a	3.483a	0.133	***

Means within a row followed by different letters were significantly different at * $p < 0.05$; ** $p < 0.01$; *** $p < 0.001$. [1] MWe is the experimental molecular weight (kDa); SEM: standard error of the mean; Sig.: significance.

Bands of higher molecular weight (over 50 kDa) showed higher intensities when using the Na + T 20,000 method, whereas protein bands under 50 kDa showed higher intensities with the TES buffer. The majority of the sarcoplasmic bands separated by 1D SDS-PAGE fell into the <50 kDa molecular weight range; the TES buffer seems to be a better option for studying the sarcoplasmic subproteome.

When studying the effect of sample type, the extractions made with the TES buffer showed more protein bands (S1, S3, S4, S5, S9, S10, S15, S16, S18, S19, S20, and S21) with significant differences ($p < 0.05$) between the Control and DFD samples (Table S2), which reinforces the conclusion that this method of extraction is the most suitable for the electrophoretic analysis of the sarcoplasmic fraction of the muscle tissue. When skeletal muscle is homogenized in a sucrose medium, as in the case of TES buffer, it forms a gelatinous consistency that inhibits the disruption of the myofibrils; therefore, the differences found between the Control and DFD extracts reflect the differences in the *post-mortem* evolution of the myofibril disruption (faster in the defective and unstructured DFD meat), thus reporting an essential information of the differences in biomarker patterns between both meat types.

In the myofibrillar fraction, despite the large differences ($p < 0.001$) in protein extractability, both extracts provided a similar and adequate separation of 34 well-defined protein bands in the range of molecular weights from 15 to 250 kDa (Figure 5).

The complete details of the results (means $\pm$ SEM) for significant myofibrillar bands are provided in Table S3. The analysis of the main factors (extraction method, type of sample, and their interaction) showed a significant interaction of extraction method and sample type for three bands (M16, M17, and M26). Once these bands were discarded, differences between extraction methods were compared including all the samples regardless of being Control or DFD. Table 3 shows a total of 14 myofibrillar bands with significant differences in band intensity between the extraction methods.

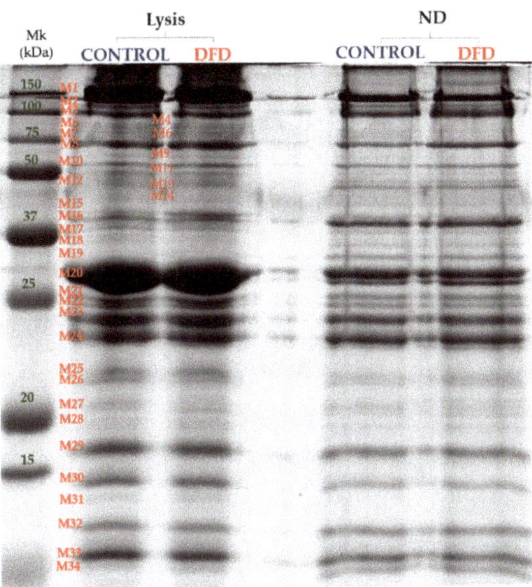

Figure 5. The 1D-SDS-PAGE gel image of myofibrillar subproteome from Control and DFD meat samples extracted with different buffers: lysis and non-denaturing buffer (ND). Mk: prestained molecular weight marker (All Blue prestained, Biorad). Band names are denoted by M (myofibrillar protein) followed by a number.

Table 3. Effect of extraction method on the myofibrillar subproteome band intensity (optical density in arbitrary units).

Myofibrillar Bands (MWe [1])	Lysis	ND	SEM	Sig.
M2 (170.8 kDa)	1.667	2.464	0.142	**
M3 (143.58 kDa)	3.139	5.893	0.417	***
M6 (110.53 kDa)	0.719	1.066	0.096	***
M11 (74.77 kDa)	0.896	0.500	0.063	**
M18 (49.7 kDa)	0.698	1.245	0.113	***
M19 (47.58 kDa)	0.899	1.717	0.104	**
M20 (41.07 kDa)	14.276	8.959	1.033	**
M23 (34.80 kDa)	5.503	4.660	0.242	*
M24 (32.76 kDa)	4.874	7.404	0.303	***
M27 (26.31 kDa)	1.466	2.128	0.085	***
M30 (19.46 kDa)	3.033	2.388	0.128	***
M31 (18.40 kDa)	0.693	0.406	0.051	***
M32 (17.09 kDa)	2.254	3.100	0.117	***
M34 (14.94 kDa)	0.817	2.314	0.123	***

Means within a row were significantly different at * $p < 0.05$; ** $p < 0.01$; *** $p < 0.001$. [1]: Mwe: the experimental molecular weight (kDa); Sig.: significance; Lysis: denaturing extraction with lysis buffer; ND: non-denaturing extraction.

Overall, 10 bands (M2, M3, M6, M18, M19, M24, M27, M32, and M34) showed higher intensity in the non-denaturing extracts, which indicated that the ND method, even despite its lower protein extractability, adequately separated well-defined myofibrillar protein bands by SDS-PAGE. Lysis buffer resulted in major band intensity for M11, M20, M23, M30, and M31 bands.

The effect of sample type was also analyzed for each buffer independently. In the lysis extracts, seven bands (M10, M16, M17, M26, M31, M32, and M34) showed significant differences between the Control and DFD meat, whereas only two bands (M26 and M31) showed significant differences in the ND extracts (Table S4). Therefore, it seems that the lysis buffer was more sensitive to changes in the muscle structure, probably due to the denaturing conditions increasing the extraction of proteins of a low molecular weight [32], leading to differences between samples of different muscle structure and compactness.

Considering the results for the sarcoplasmic and myofibrillar subproteome, it seems that some proteins were easily extracted with most of the buffers, whereas others remained linked to cellular organelles and membranes or in the sarcoplasm portions embedded within the myofibrils, which complicated their extraction. The intensity of these effects depends on the evolution of the muscle dismantlement in the process of conversion of muscle into meat and the resulting meat quality grade.

3.5. Stress Protein: Hsp70

Proteomic studies have reported the differential expression of Hsp70 in meat with variable quality traits [22,38]. Several studies have correlated meat quality with Hsp70 under stress situations [39–41].

Our results (Figure 6) showed clear differences in Hsp70 expression between the sarcoplasmic and myofibrillar fractions. For the sarcoplasmic fraction, higher Hsp70 expression was found in DFD samples, with these differences being significant for the TES 1000, TES 20,000, and Na 1000 extraction methods. However, the sarcoplasmic fraction of the Control samples showed higher Hsp70 expression when extracted with sodium buffers and higher centrifugal speed (Na 20,000 and Na + T 20,000).

For the myofibrillar fraction, higher Hsp70 expression was found for ND extracts, but contrary to what was observed for the sarcoplasmic fraction, the Control samples showed higher Hsp70 expression independent of the extractive method used (lysis and ND). This higher expression in Control samples from the myofibrillar extracts may have been due to the protective role that Hsp70 played against muscle dismantlement in the early *post-mortem* stages. Previous studies showed that the majority of Hsp70 is readily diffusible within the cytoplasm in non-stressed muscle fibers; after stress, Hsp70 primarily binds to and stabilizes the structure and function of cell membranes [42]. Furthermore, under stress situations such as during muscle-damaging exercise, Hsp70 translocates and accumulates to the cytoskeletal and myofibrillar proteins [43]. Xing et al. [44] found that Hsp70 was present in the cytoplasm and on the surface membranes of cells from the *Pectoralis major* muscle in normal-quality chicken meat following stress. Hsp70 was present on the surface membranes and extracellular matrix but was barely visible in the cytoplasm of the PSE-like samples, that is, low-quality meat due to stress. This diffusion capacity of Hsp70 may explain the differences found in this work between extracts. Under the stressful situation produced from slaughter, Hsp70 may translocate to the myofibrillar fraction in an attempt of stabilize the muscle structure, which would explain its increased expression in the myofibrillar fraction of the Control samples. However, in the DFD meat, more Hsp70 is easily removed and extracted from the sarcoplasmic fraction, showing the movement of Hsp70 from the inner myofibrillar compartment to the sarcoplasm due to the faster dismantlement of the muscle.

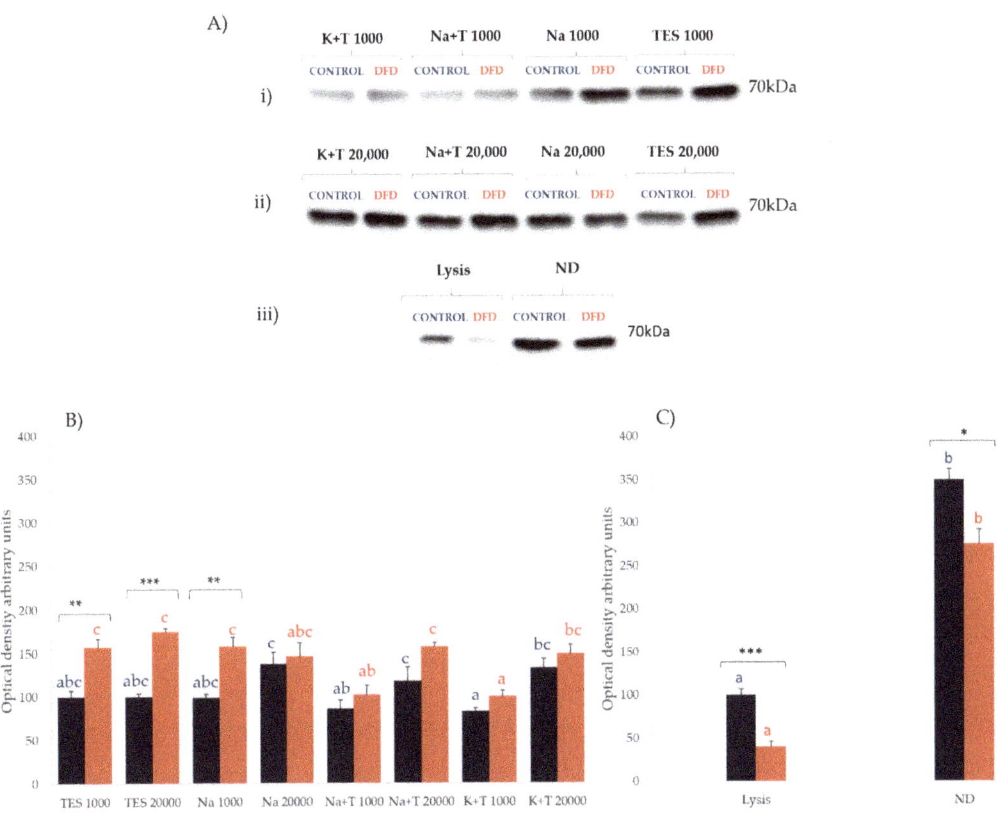

Figure 6. Hsp70 Western blotting results. (**A**) Representative immunoblot analyses of Hsp70 protein expression of sarcoplasmic extracts at (i) 1000× *g* and (ii) 20,000× *g*, and (iii) Hsp70 expression of myofibrillar extracts. Ponceau staining was used as a loading control. (**B**) Expression of Hsp70 (mean ± SEM from three independent experiments) of sarcoplasmic extracts (left) and (**C**) myofibrillar extracts (right) from Control (blue) and DFD (red) meat samples. Charts with different letters (blue for Control and red for DFD) were significantly different between extraction methods at $p < 0.05$. Asterisks indicate significant differences between the Control and DFD samples within the same extraction procedure. * $p < 0.05$; ** $p < 0.01$; *** $p < 0.001$. TES 1000: TES buffer and 1000× *g*, 6 min; TES 20,000: TES buffer and 20,000× *g*, 20 min; Na 1000: sodium phosphate buffer and 1000× *g*, 6 min; Na 20,000: sodium phosphate buffer and 20,000× *g*, 20 min; Na + T 1000: sodium phosphate buffer with Triton X-100 and 1000× *g*, 6 min; Na + T 20,000: sodium phosphate buffer with Triton-X100 and 20,000× *g*, 20 min; K + T 1000: potassium phosphate buffer with Triton X-100 and 1000× *g*, 6 min; K + T 20,000: potassium phosphate buffer with Triton X-100 and 20,000× *g*, 20 min; Lysis: denaturing extraction with lysis buffer; ND: non-denaturing extraction.

4. Conclusions

Within the sarcoplasmic fraction, buffers containing Triton X-100 led to a higher protein extractability, LPO detection, and determination of proteins with high molecular weight. However, TES buffer was more sensitive for the detection of Hsp70 expression and the electrophoretic bands of lower molecular weight, showing increased ability to discriminate between the meat samples with different metabolisms and the degree of cell dismantlement (Control vs. DFD). In the myofibrillar fraction, the non-denaturing buffer reported higher LPO, CAT activity, and Hsp70 expression, and showed higher intensity bands in the electrophoretic pattern; however, the lysis buffer increased protein extractability and its electrophoretic pattern was more sensitive to differences between the Control and DFD samples.

These findings highlight the need to select the most appropriate extraction method for each biomarker family and muscle structure type, and the need to consider different cell fractions and the movements of proteins between cytoskeletal and myofibrillar structures, for an accurate and reliable study of the process of conversion of the muscle into meat.

Supplementary Materials: The following are available online at https://www.mdpi.com/article/10.3390/foods10051097/s1, Figure S1. Flowchart of the different extraction methods for sarcoplasmic and myofibrillar proteins from beef muscle.1: TES buffer; 2: soidum phosphate buffer; 3: Naa+T: sodium phosphate buffer with triton X-100; K+T: potassium phosphate buffer with Triton X-100; ND: non-denaturing extraction. Table S1: Effect of extraction method (TES 1000, TES 20,000, Na 1000, Na 20,000, Na+T 1000, Na+T 20,000, K+T 1000 and K+T 20,000), type of sample (CONTROL vs. DFD) and their interaction on sarcoplasmic subproteome bands' intensity (optical density in arbitrary units). Table S2: The *p*-values for the effect of sample type (CONTROL vs. DFD) on the sarcoplasmic subproteome bands intensitiy (optical density in arbitrary units) obtained with the different extration methods. Table S3: Effect of extraction method (Lysis and Non-denaturant), type of sample (CONTROL vs. DFD) and their interaction on myofibrillar subproteome bands' intensity (optical density in arbitrary units). Table S4: Effect of meat type (CONTROL vs. DFD) within each extraction method (Lysis vs. Non-denaturant) on myofibrillar subproteome bands' intensity (optical density in arbitray units).

Author Contributions: Conceptualization: M.O., V.S., and A.C.-M.; methodology: L.G.-B., M.O., and V.S.; investigation: L.G.-B., A.D.-L., and Y.D.; formal analysis: V.S. data curation. L.G.-B.; writing—original draft preparation: V.S.; writing—review and editing: M.O. and A.C.-M.; visualization. L.G.-B.; supervision. M.O.; funding acquisition, V.S. and M.O. All authors have read and agreed to the published version of the manuscript.

Funding: This research was funded by the Ministerio de Ciencia, Innovación y Universidades (MCIU) (Spain), Agencia Estatal de Investigación (Spain) and FEDER funds, under the project number RTI2018-096162-RC21. The APC was funded by RTI2018-096162-RC21. L.G-B acknowledges her contract to the MCIU (PRE2019-091053).

Data Availability Statement: Data are available upon request.

Acknowledgments: We thank the staff of the Area of Livestock Production Systems from SERIDA, Matadero Central de Asturias S. L., Alimerka S. L., and ASINCAR for their valuable help in this work. We also thank Vega-Naredo for his helpful suggestions on this work.

Conflicts of Interest: The authors declare no conflict of interest. The funders had no role in the design of the study; in the collection, analyses, or interpretation of data; in the writing of the manuscript; or in the decision to publish the results.

References

1. Picard, B.; Gagaoua, M. Meta-proteomics for the discovery of protein biomarkers of beef tenderness: An overview of integrated studies. *Food Res. Int.* **2020**, *127*, 108739. [CrossRef] [PubMed]
2. Lieber, R.L.; Friden, J. Functional and clinical significance of skeletal muscle architecture. *Muscle Nerve* **2000**, *23*, 1647–1666. [CrossRef]
3. Eady, M.; Samuel, D.; Bowker, B. Effect of pH and postmortem aging on protein extraction from broiler breast muscle. *Poult. Sci.* **2014**, *93*, 1825–1833. [CrossRef]
4. Au, Y. The muscle ultrastructure: A structural perspective of the sarcomere. *Cell. Mol. Life Sci.* **2004**, *61*, 3016–3033. [CrossRef] [PubMed]
5. Chen, X.; Zou, Y.; Han, M.; Pan, L.; Xing, T.; Xu, X.; Zhou, G. Solubilization of myosin in a solution of low ionic strength L-histidine: Significance of the imidazole ring. *Food Chem.* **2016**, *196*, 42–49. [CrossRef]
6. Anderson, M.J.; Lonergan, S.M.; Huff-Lonergan, E. Myosin light chain 1 release from myofibrillar fraction during postmortem aging is a potential indicator of proteolysis and tenderness of beef. *Meat Sci.* **2012**, *90*, 345–351. [CrossRef]
7. Bowker, B.C.; Fahrenholz, T.M.; Paroczay, E.W.; Eastridge, J.S.; Solomon, M.B. Effect of hydrodynamic pressure processing and ageing on the tenderness and myofibrillar proteins of beef strip loins. *J. Muscle Foods* **2008**, *19*, 74–97. [CrossRef]
8. Lan, Y.H.; Novakofski, J.; Carr, T.R.; McKeith, F.K. Assay and storage conditions affect yield of salt soluble protein from muscle. *J. Food Sci.* **1993**, *58*, 963–967. [CrossRef]

9.   Munasinghe, D.M.; Sakai, T. Sodium chloride as a preferred protein extractant for pork lean meat. *Meat Sci.* **2004**, *67*, 697–703. [CrossRef]
10. Thomas, R.; Gadekar, Y.P.; Kandeepan, G.; George, S.K.; Kataria, M. Effect of extraction conditions and postmortem ageing period on yield and salt soluble proteins from buffalo (Bubalus bubalis) lean meat. *Am. J. Food Technol.* **2007**, *2*, 313–317. [CrossRef]
11. Newton, K.G.; Gill, C.O. The microbiology of DFD fresh meats: A review. *Meat Sci.* **1981**, *5*, 223–232. [CrossRef]
12. Grayson, A.L.; Shackelford, S.D.; King, D.A.; McKeith, R.O.; Miller, R.K.; Wheeler, T.L. Effect of degree of dark cutting on tenderness and sensory attributes of beef. *J. Anim. Sci.* **2016**, *94*, 2583–2591. [CrossRef]
13. Ponnampalam, E.N.; Hopkins, D.L.; Bruce, H.; Li, D.; Baldi, G.; Bekhit, A.E. Causes and contributing factors to "dark cutting" meat: Current trends and future directions: A review. *Compr. Rev. Food Sci. Food Saf.* **2017**, *16*, 400–430. [CrossRef]
14. Adzitey, F.; Nurul, H. Pale soft exudative (PSE) and dark firm dry (DFD) meats: Causes and measures to reduce these incidences— A mini review. *Int. Food Res. J.* **2011**, *18*, 11–20.
15. AMSA. *Meat Color Measurement Guidelines*; American Meat Science Association: Champaign, IL, USA, 2012.
16. Honikel, K.O. Reference methods for the assessment of physical characteristics of meat. *Meat Sci.* **1998**, *49*, 447–457. [CrossRef]
17. Díaz, F.; Díaz-Luis, A.; Sierra, V.; Diñeiro, Y.; González, P.; García-Torres, S.; Tejerina, D.; Romero-Fernández, M.P.; de Vaca, M.C.; Coto-Montes, A.; et al. What functional proteomic and biochemical analysis tell us about animal stress in beef? *J. Proteom.* **2020**, *218*, 103722. [CrossRef] [PubMed]
18. Jia, X.; Veiseth-Kent, E.; Grove, H.; Kuziora, P.; Aass, L.; Hildrum, K.I.; Hollung, K. Peroxiredoxin-6—A potential protein marker for meat tenderness in bovine longissimus thoracis muscle. *J. Anim. Sci.* **2009**, *87*, 2391–2399. [CrossRef]
19. Coto-Montes, A.; Caballero, B.; Sierra, V.; Vega-Naredo, I.; Tomás-Zapico, C.; Hardeland, R.; Tolivia, D.; Ureña, F.; Rodríguez-Colunga, M.J. Actividad de los principales enzimas antioxidantes durante el periodo de Oreo de culones de la raza Asturiana de los Valles. *ITEA* **2004**, *100*, 43–45.
20. Potes, Y.; de Luxán-Delgado, B.; Rodriguez-González, S.; Guimarães, M.R.M.; Solano, J.J.; Fernández-Fernández, M.; Coto-Montes, A. Overweight in elderly people induces impaired autophagy in skeletal muscle. *Free Radic. Biol. Med.* **2017**, *110*, 31–41. [CrossRef] [PubMed]
21. Rubio-Gonzalez, A.; Potes, Y.; Illan-Rodriguez, D.; Vega-Naredo, I.; Sierra, V.; Caballero, B.; Fabrega, E.; Velarde, A.; Dalmau, A.; Olivan, M.; et al. Effect of animal mixing as a stressor on biomarkers of autophagy and oxidative stress during pig muscle maturation. *Animal* **2015**, *9*, 1188–1194. [CrossRef]
22. Bjarnadóttir, S.G.; Hollung, K.; Faergestad, E.M.; Veiseth-Kent, E. Proteome changes in bovine longissimus thoracis muscle during the first 48 h postmortem: Shifts in energy status and myofibrillar stability. *J. Agric. Food Chem.* **2010**, *58*, 7408–7414. [CrossRef]
23. Hashimoto, K.; Watabe, S.; Kono, M.; Shiro, K. Muscle protein composition of sardine and mackerel. *Bull. Jpn. Soc. Sci. Fish* **1979**, *45*, 1435–1441. [CrossRef]
24. Chen, X.; Tume, R.K.; Xu, X.; Zhou, G. Solubilization of myofibrillar proteins in water or low ionic strength media: Classic techniques, basic principles and novel functionalities. *Crit. Rev. Food Sci. Nutr.* **2015**, *57*, 3260–3280. [CrossRef]
25. Bradford, M.M. A rapid and sensitive method for the quantitation of microgram quantities of protein utilizing the principle of protein-dye binding. *Anal. Biochem.* **1976**, *72*, 248–254. [CrossRef]
26. Lubinsky, S.; Bewley, G.C. Genetics of catalase in Drosophila melanogaster: Rates of synthesis and degradation of the enzyme in flies aneuploid and euploid for the structural gene. *Genetics* **1979**, *91*, 723–742. [CrossRef] [PubMed]
27. Joseph, P.; Suman, S.P.; Rentfrow, G.; Li, S.; Beach, C.M. Proteomics of muscle-specific beef color stability. *J. Agric. Food Chem.* **2012**, *60*, 3196–3203. [CrossRef]
28. Picard, B.; Gagaoua, M. Chapter 11—proteomic investigations of beef tenderness. *Proteom. Food Sci.* **2017**, 177–197. [CrossRef]
29. Huang, C.; Hou, C.; Ijaz, M.; Yan, T.; Li, X.; Li, Y.; Zhang, D. Proteomics discovery of protein biomarkers linked to meat quality traits in post-mortem muscles: Current trends and future prospects: A review. *Trends Food Sci. Tech.* **2020**, *105*, 416–432. [CrossRef]
30. García-Torres, S.; de Vaca, M.C.; Tejerina, D.; Romero-Fernández, M.P.; Ortiz, A.; Díaz, F.; Sierra, V.; González, P.; Franco, D.; Zapata, C.; et al. ¿Es el pH un criterio suficiente para la clasificación de carne DFD en vacuno? In Proceedings of the I Congreso Iberoamericano de Marcas de Calidad de Carne y de Productos Cárnicos, Bragança, Portugal, 24–25 October 2019.
31. Poleti, M.; Moncau, C.; Silva-Vignato, B.; Fernandes-Rosa, A.; Lobo, A.; Cataldi, T.; Negrão, J.; Silva, S.; Eler, J.; Balieiro, J. Label-free quantitative proteomic analysis reveals muscle contraction and metabolism proteins linked to ultimate pH in bovine skeletal muscle. *Meat Sci.* **2018**, *145*, 209–219. [CrossRef]
32. Della Malva, A.; Albenzio, M.; Santillo, A.; Russo, D.; Figliola, L.; Caroprese, M. Methods for Extraction of Muscle Proteins from Meat and Fish Using Denaturing and Nondenaturing Solutions. *J. Food Qual.* **2018**, *2018*, 1–9. [CrossRef]
33. Saw, M.M.; Riederer, B.M. Sample preparation for two-dimensional gel electrophoresis. *Proteomics* **2003**, *3*, 1408–1417. [CrossRef]
34. Rabilloud, T. Use of thiourea to increase the solubility of membrane proteins in two-dimensional electrophoresis. *Electrophoresis* **1998**, *19*, 758–760. [CrossRef]
35. Malafaia, C.B.; Guerra, M.L.; Silva, T.D.; Paiva, P.M.; Souza, E.B.; Correia, M.T.; Silva, M.V. Selection of a protein solubilization method suitable for phytopathogenic bacteria: A proteomics approach. *Proteome Sci.* **2015**, *13*, 1–7. [CrossRef] [PubMed]
36. Chan, L.L.; Lo, S.C.; Hodgkiss, I.J. Proteomic study of a model causative agent of harmful red tide. Prorocentrum triestinum I: Optimization of sample preparation methodologies for analyzing with two-dimensional electrophoresis. *Proteomics* **2002**, *2*, 1168–1186. [CrossRef]

37. Molloy, M.P.; Herbert, B.R.; Walsh, B.J.; Tyler, M.I.; Traini, M.; Sanchez, J.C.; Hochstrasser, D.F.; Williams, K.L.; Gooley, A.A. Extraction of membrane proteins by differential solubilization for separation using two-dimensional gel electrophoresis. *Electrophoresis* **1998**, *19*, 837–844. [CrossRef]
38. Di Luca, A.; Mullen, A.M.; Elia, G.; Davey, G.; Hamill, R.M. Centrifugal drip is an accessible source for protein indicators of pork ageing and water-holding capacity. *Meat Sci.* **2011**, *88*, 261–270. [CrossRef]
39. Yu, J.; Tang, S.; Bao, E.; Zhang, M.; Hao, Q.; Yue, Z. The effect of transportation on the expression of heat shock proteins and meat quality of M. longissimus dorsi in pigs. *Meat Sci.* **2009**, *83*, 474–478. [CrossRef] [PubMed]
40. Xing, T.; Xu, X.L.; Zhou, G.H.; Wang, P.; Jiang, N.N. The effect of transportation of broilers during summer on the expression of heat shock protein 70, postmortem metabolism and meat quality. *J. Anim. Sci.* **2015**, *93*, 62–70. [CrossRef]
41. Díaz-Luis, A.; Díaz, F.; Diñeiro, Y.; González-Blanco, L.; Arias, E.; Coto-Montes, A.; Oliván, M.; Sierra, V. Nuevos indicadores de carnes (DFD): Estrés oxidativo, autofagia y apoptosis. *ITEA* **2020**, *117*, 3–18. [CrossRef]
42. Larkins, N.T.; Murphy, R.M.; Lamb, G.D. Influences of temperature, oxidative stress, and phosphorylation on binding of heat shock proteins in skeletal muscle fibers. *Am. J. Physiol. Cell Physiol.* **2012**, *303*, 654–665. [CrossRef] [PubMed]
43. Paulsen, G.; Vissing, K.; Kalhovde, J.M.; Ugelstad, I.; Bayer, M.L.; Kadi, F.; Schjerling, P.; Hallén, J.; Raastad, T. Maximal eccentric exercise induces a rapid accumulation of small heat shock proteins on myofibrils and a delayed HSP70 response in humans. *Am. J. Physiol. Regul. Integr. Comp. Physiol.* **2007**, *293*, 844–853. [CrossRef] [PubMed]
44. Xing, T.; Wang, M.F.; Han, M.Y.; Zhu, X.S.; Xu, X.L.; Zhou, G.H. Expression of heat shock protein 70 in transport-stressed broiler pectoralis major muscle and its relationship with meat quality. *Animal* **2017**, *11*, 1599–1607. [CrossRef] [PubMed]

Article

New Insights on the Impact of Cattle Handling on Post-Mortem Myofibrillar Muscle Proteome and Meat Tenderization

Verónica Sierra [1,2], Laura González-Blanco [1,2], Yolanda Diñeiro [1,2], Fernando Díaz [1], María Josefa García-Espina [1], Ana Coto-Montes [2,3], Mohammed Gagaoua [4,*] and Mamen Oliván [1,2,*]

[1] Área de Sistemas de Producción Animal, Servicio Regional de Investigación y Desarrollo Agroalimentario (SERIDA), Ctra. AS-267, PK 19, 33300 Villaviciosa, Spain; veroniss@serida.org (V.S.); lgblanco@serida.org (L.G.-B.); ydineiro@serida.org (Y.D.); ferdm89@gmail.com (F.D.); mjgarcia@serida.org (M.J.G.-E.)

[2] Instituto de Investigación Sanitaria del Principado de Asturias (ISPA), Av. del Hospital Universitario, s/n, 33011 Oviedo, Spain; acoto@uniovi.es

[3] Department of Morphology and Cell Biology, Faculty of Medicine, University of Oviedo, Av. Julián Clavería, 6, 33006 Oviedo, Spain

[4] Food Quality and Sensory Science Department, Teagasc Food Research Centre, Dublin 15, D15 KN3K Ashtown, Ireland

* Correspondence: Mohammed.Gagaoua@teagasc.ie or gmber2001@yahoo.fr (M.G.); mcolivan@serida.org (M.O.)

Citation: Sierra, V.; González-Blanco, L.; Diñeiro, Y.; Díaz, F.; García-Espina, M.J.; Coto-Montes, A.; Gagaoua, M.; Oliván, M. New Insights on the Impact of Cattle Handling on Post-Mortem Myofibrillar Muscle Proteome and Meat Tenderization. *Foods* **2021**, *10*, 3115. https://doi.org/10.3390/foods10123115

Academic Editor: Thierry Astruc

Received: 16 November 2021
Accepted: 13 December 2021
Published: 15 December 2021

Publisher's Note: MDPI stays neutral with regard to jurisdictional claims in published maps and institutional affiliations.

Abstract: This study investigated the effect of different cattle management strategies at farm (Intensive vs. Extensive) and during transport and lairage (mixing vs. non-mixing with unfamiliar animals) on the myofibrillar subproteome of *Longissimus thoracis* et *lumborum* (LTL) muscle of "Asturiana de los Valles" yearling bulls. It further aimed to study the relationships with beef quality traits including pH, color, and tenderness evaluated by Warner–Bratzler shear force (WBSF). Thus, comparative proteomics of the myofibrillar fraction along meat maturation (from 2 h to 14 days *post-mortem*) and different quality traits were analyzed. A total of 23 protein fragments corresponding to 21 unique proteins showed significant differences among the treatments ($p < 0.05$) due to any of the factors considered (Farm, Transport and Lairage, and *post-mortem* time ageing). The proteins belong to several biological pathways including three structural proteins (MYBPC2, TNNT3, and MYL1) and one metabolic enzyme (ALDOA) that were affected by both Farm and Transport/Lairage factors. ACTA1, LDB3, and FHL2 were affected by Farm factors, while TNNI2 and MYLPF (structural proteins), PKM (metabolic enzyme), and HSPB1 (small Heat shock protein) were affected by Transport/Lairage factors. Several correlations were found between the changing proteins (PKM, ALDOA, TNNI2, TNNT3, ACTA1, MYL1, and CRYAB) and color and tenderness beef quality traits, indicating their importance in the determination of meat quality and their possible use as putative biomarkers.

Keywords: intensive management; extensive management; mixing unfamiliar animals; myofibrillar proteins; pre-slaughter stress; protein biomarkers

1. Introduction

Improving beef production and meat quality to cope with meet consumer demands is a major concern of the livestock production sector. It is well known that cattle intrinsic factors, such as breed and genetics, have a decisive influence on beef production and on the ultimate meat quality, therefore, different breeding strategies and meat maturation procedures must be adapted to the genetic diversity of the animals [1,2]. In this sense, there is great interest in promoting the development of native cattle breeds, as they seem to be more adapted to regional production systems and for promoting proximity trade as a sustainability strategy [3]. Apart from the intrinsic factors, there are also extrinsic factors, overall, from farm-to-fork, related to the routine handling of animals and animal-human interactions that must be considered to ensure beef quality [4]. Among them, production system and feeding strategies play an important role, not only due to the effect that dietary components may exert on the animal's growth rate, and muscle/meat characteristics [2,5,6], but also due to their influence on the animal's physiology, social behavior, and reactivity to stress [7–10]. Moreover, psychological, and physiological status of the animals can affect final meat quality [10]. In fact, cattle are herd animals that establish social orders, so the regrouping of animals or mixing with unfamiliar animals during transport and lairage, despite being a common husbandry practice, can have a detrimental effect on animal welfare, increasing animal stress [11,12].

On the other hand, animal handling may affect animal's emotional state, hence inducing pre-slaughter stress (PSS), whose influence on the *post-mortem* process of muscle-to-meat conversion has been shown in pigs [13,14] and in cattle [3,9,10,15–18]. Those biochemical changes were evidenced using several high-throughput OMICs methods, including proteomics that revealed, for instance, that the meat tenderizing process involves myriad pathways, such as the degradation of structural proteins, energy metabolism pathways, response to stress, apoptosis, autophagy, and signaling pathways [19,20] as confirmed recently by the integromics meta-analysis study of Gagaoua et al. [21]. Comparative proteomics appeared to be a useful tool to study the biological pathways underpinning the effect of PSS on the ultimate meat quality. In this context, proteomic approaches point out the possible identification of putative biomarkers from the sarcoplasmic subproteome fraction of the *post-mortem* muscle [9,17,21,22]. Since tenderness and color are considered as important beef quality traits for consumers., the impact of PSS on these attributes is worthy of investigation. Indeed, PSS is proven to have a detrimental effect due to the changes that induces in the enzymatic processes that, for example, induce the breakdown of myofibrillar structure mainly composed of structural and contractile proteins [23–26].

Based on the above, this study aimed to apply a proteomics approach to investigate the effect of pre-slaughter factors such as mixing unfamiliar animals during the transport and lairage period on the myofibrillar subproteome of young "Asturiana de los Valles" bulls reared under two divergent rearing practices (intensive or extensive management systems). This trial further provides an opportunity to identify putative protein biomarkers [27] related to beef tenderization and PSS.

2. Materials and Methods

2.1. Animals and Experimental Design

This trial used 24 yearling bulls of "Asturiana de los Valles" (AV) breed that were slaughtered between 13 and 15 months of age. AV breed is a native breed from the north of Spain, with a high growth rate and low-fat content [28,29] and protected by the quality label "Ternera Asturiana", which is one of the most significant in terms of production and economic value [30]. Calves were managed with their mothers from birth to weaning, fed on concentrate and barley straw ad libitum during the winter, and assigned in spring to two different farm management systems:

(1) Intensive ("I") (*n* = 12), with animals managed indoors, in pens of 6 × 6 m (6 animals per pen) and finished for 100 days before slaughter with 8 kg/day of concentrate (84% barley meal, 10% soya meal, 3% fat, 3% minerals, vitamins and oligoelements) and 2 kg/day of barley straw

(2) Extensive ("E") (*n* = 12), with animals managed outdoors in two 1.5 ha plots (6 animals per plot) and finished for 100 days before slaughter grazing on ryegrass and clover pasture + 3.5 kg/day of supplementation with concentrate.

At an approximate slaughter weight of 500 kg, the animals were transported in groups of six to a commercial abattoir located at around 40 km from the farm where the animals were finished. Half of them from each rearing system (I and E) was mixed with unfamiliar animals from other pens/groups not belonging to the study (mixing treatment "M") and the other half (non-mixing treatment "NM") was maintained in their original group for transport and lairage. Thus, there were six animals assigned to each group (I-M, I-NM, E-M, E-NM).

The experimental procedures were in compliance with the RD 53/201, where no authorization is required for practices carried out for recognized zootechnical purposes (Art 2.5d) and those that do not cause more pain than the introduction of a needle (Art 2.5f).

The pre-slaughter management lasted 6 h from when the animals left the farm, including the process of loading, travelling, unloading, and lairage, and was in accordance with the Council Regulation (EU) Nr. 1/2005, which relates to protecting the welfare of animals during transport and related operations. Animals were stunned with a captive bolt, slaughtered by immediate exsanguination, and dressed according to the current EU regulations (Council Regulation (EC) No 1099/2009) in accredited abattoirs.

2.2. Muscle Sampling and Meat Quality Measurements

The carcasses were chilled at 3 °C within 2 h after slaughter. *Longissimus thoracis et lumborum* (LTL) muscle samples (20 g) were taken from the left-side carcass of each animal at the thirteenth rib level at 2 h, 8 h, and 24 h *post-mortem*. The muscle samples were immediately snap frozen in liquid nitrogen and stored at −80 °C until analysis.

At 24 h post-slaughter, the LTL muscle was removed from the left half carcass between the sixth and the tenth ribs, and transported to the laboratory. The LTL temperature and pH were recorded (pH24) at the sixth rib using a digital portable pH meter equipped with a penetration electrode coupled with a temperature probe (InLab Solids Go-ISM, Mettler-Toledo S.A.E., Barcelona, Spain).

Meat color was recorded at 24 h *post-mortem* on three 10 mm diameter spots on the exposed cut surface of the LTL muscle at the seventh rib level after 60 min blooming. The coordinates lightness (L^*), redness (a^*), and yellowness (b^*) were obtained using a Minolta CM-2300d portable Spectrophotometer, with an illuminant C and D65 illuminant, 10° standard observer angle geometry and 8 mm aperture size in the CIE space (Konica Minolta Inc., Osaka, Japan), and the average value of the three spots was calculated. Further, both Chroma (C^*) and Hue angle (h^*) were calculated according to the next equations: $C^* = \sqrt{(a^{*2} + b^{*2})}$ and $h^* = \tan^{-1} b^*/a^*$ [31].

The rest of the LTL striploin was sliced into 3.5 cm steaks that were vacuum packed in polyamide 20 μm/polyethylene 70 μm bags and aged in darkness under refrigerated conditions (4 °C ± 1 °C) at different *post-mortem* ageing times (3, 7, and 14 days). After the corresponding ageing period, steaks for comparative proteomics were frozen at −80 °C, while the steaks for meat toughness analysis were frozen and stored at −20 °C for subsequent analysis. Meat toughness was measured by the Warner–Bratzler (WB) shear test on meat cooked at 75 °C from 30 min by immersion in a water bath. After cooling, eight cores (1 cm^2 in cross-section) from each steak were subjected to a perpendicular cut by the WB blade set HDP/WBV with a "V" slot using the TA.XT Plus instrument (Stable Micro Systems, London, UK). The maximum load (kg) required for total split was recorded, the results were subjected to detection of outliers by box plot and the extreme values were deleted. Results were expressed as the mean WB shear force maximum load for each steak. Tenderization rate (TR, %) was calculated as the percentage of decrease in WB shear force in a given period of time (3 to 7 days, 7 to 14 days, 3 to 14 days).

2.3. Myofibrillar Protein Extraction

Proteomic analysis was performed on the muscle samples of the 24 animals. From each animal, muscle myofibrillar extracts were obtained at 2 h, 8 h, 24 h, 3 days, 7 days, and 14 days *post-mortem*, following the method described by Bjarnadottir et al. [32]. Briefly, 0.5 g muscle samples were homogenized in 4 mL of Tris-EDTA-Sucrose (TES) buffer containing 10 mM Tris [pH 7.6], 1 mM EDTA [pH 8.0], 0.25 M sucrose, and 0.6% protease inhibitor cocktail [P8340, Sigma-Aldrich Co., St. Louis, MO, USA], using a Polytron PT1200 E (Kinematica Inc., Luzern, Switzerland) two times for 15 s at maximum speed. The homogenate was centrifuged (20 min at 20,000× *g*) at 4° C. The resulting pellet was homogenized into 4 mL of lysis buffer containing 10 mM Tris-HCl pH [7.6], 7 M urea, 2 M thiourea, 2% CHAPS, and 10 mM DTT with the polytron 2 × 15 s at maximum speed.

Subsequently, the solution was stirred at room temperature for 1 h in a Multi Reax stirrer (Heidolph Instruments, Schwabach, Germany) and was centrifuged at $20,000 \times g$ for 20 min at 4 °C. The supernatant containing the myofibrillar proteins was collected and filtered through a nylon filter (5 μm), aliquoted, and stored at −80 °C. The protein content of the extract was measured by the Bradford method [33].

2.4. Myofibrillar Subproteome Analysis (1D SDS-PAGE) and Protein Identification

The myofibrillar muscle extracts (30 μg) were prepared for SDS-PAGE as follows. First, they were denatured using a solution containing 65.8 mM Tris/HCl pH 6.8, 21% glycerol, 5% beta-mercaptoethanol, 2% SDS, 0.026% of bromophenol blue that were subsequently) heated at 100 °C for 5 min. Second, the denatured samples were loaded into 1 mm dual vertical slab gels (Mini-protean, Bio-Rad Laboratories Inc., Hercules, CA, USA) for separation using a 12% resolving gel and 4% stacking gel. Pre-stained molecular weight standards (Precision Plus Protein™ All Blue Standards, Bio-Rad Laboratories Inc., Hercules, CA, USA) were added on each gel.

Overall, three gels per sample were performed. The stained gel images were captured using the UMAX ImageScanner (Amersham Biosciences, Buckinghamshire, UK). The densitometry analysis and band quantification were carried out using Image Studio Lite 5.2.5 program (LI-COR Biosciences, Lincoln, NE, USA). To account for slight variations in protein loading, the optical density of protein bands was expressed as relative abundance (normalized volume) and expressed in arbitrary units.

Bands of interest (with significant differences among the groups) were manually excised from the gels and prepared for identification by MALDI-TOF/TOF mass spectrometry. The details of this procedure were previously described by Díaz et al. [9].

2.5. Statistical and Bioinformatics Analyses

Raw data were scrutinized for data entry errors and outliers by boxplot. Normality of variables was tested by a Kolmogorov–Smirnoff test. For meat quality traits (pH, meat color and WBSF), the effect of animal management at Farm "F" (I vs. E), during the Transport and Lairage "TL" (M vs. NM) and the interaction (F × TL) were analyzed using a General Lineal model procedure (SPSSv22.0 SPSS Inc., Chicago, IL, USA). Further a repeated measure ANOVA was used to investigate the effect of Treatment (I-M, I-NM, E-M, E-NM) on WBSF measured at 3, 7, and 14 days *post-mortem*.

For myofibrillar bands intensities measured at different *post-mortem* times (2 h, 8 h, 24 h, 3 days, 7 days, and 14 days), the ANOVA model included Farm (F), Transport and Lairage (TL), ageing time (t), and their interactions as fixed factors and animal as covariate.

Significant differences among *post-mortem* times were studied by the Tukey's post-hoc test (Games–Howell when the variances were not homogeneous) at a significant level of $p \leq 0.05$.

The relationships between meat quality traits and the myofibrillar subproteome at different *post-mortem* times were calculated by bivariate Pearson's correlations. Moreover, principal component analysis (PCA) was performed to study the relationships among the meat quality traits and the differential proteins along meat tenderization.

The significantly changing protein bands (differential proteins) were investigated using Metascape open-source tool to identify the main enriched Gene Ontology (GO) terms among the proteins following the procedures described by Gagaoua et al. [22,34]. The STRING database (Search Tool for Retrieval of Interacting Genes, ver. 11.0 at https://string-db.org/, accessed on 10 October 2021) was further used to construct the Protein–Protein Interactions (PPI) relating the differential proteins according to the pathways to which they belong. Moreover, the list of the identified proteins that differ among the groups were compared to the repertoire of Gagaoua et al. [21] to identify the extent of overlap with the previously identified beef tenderness biomarkers in LTL muscle.

3. Results

3.1. Meat Quality Attributes

The handling conditions (F and TL) had no significant effect on pH24 that showed normal values within the range 5.43–5.52, whatever the treatments. However, animals' farm management (F) affected meat color parameters (Table 1), being L^* ($p < 0.05$) b^*, C^*, and h^* ($p < 0.001,$) significantly lower in meat from the extensive system, that was brownish and darker, which agrees with previous literature that describes that meat from grass-fed animals is darker than meat from grain-fed animals, and attributes these differences to diet, physical activity, or a combination of both [35–37].

Table 1. Effect of Farm (F) and Transport and Lairage (TL) and their interaction on meat color parameters.

Quality Traits	Farm Management (F)			Transport and Lairage (TL)			Significance		
	I	E	SEM	NM	M	SEM	F	TL	F × TL
pH	5.6	5.46	0.05	5.49	5.58	0.05	NS	NS	NS
L^*	41.86	38.20	0.99	38.73	41.33	0.99	*	NS	NS
a^*	12.03	10.82	0.523	10.4	12.45	0.523	NS	*	*
b^*	15.49	10.28	0.522	12.19	13.58	0.522	***	NS	NS
C^*	19.64	14.97	0.681	16.13	18.49	0.681	***	*	*
h^*	52.49	43.61	1.04	49.07	47.03	1.04	***	NS	NS

F: Management at farm; TL: Management during transport and lairage; NM: Non-mixing; M: Mixing; NS: not significant; * $p < 0.05$; *** $p < 0.001$; SEM: standard error of the mean.

The effect of animal mixing during transport and lairage (TL) and the F × TL interaction were significant for a^* and C^* ($p < 0.05$). In both parameters, meat from mixed animals had higher values compared to the meat from non-mixed animals, being a^* (12.45 vs. 10.4) and C^* (18.49 vs. 16.13) for M and NM, respectively.

Farm rearing system were found to significantly affect meat toughness ($p < 0.001$) at 14 days *post-mortem* when meat from the intensive treatment showed lower values. These findings are in agreement to those of previous studies that found a negative effect of extensive treatment on tenderness [38,39].

When comparing the evolution of meat toughness in the different handling treatments (I-M, I-NM, E-M and E-NM) (Figure 1A), no significant differences were found at 3 days *post-mortem*; however lower values of WBSF were found at 7 ($p < 0.05$) and 14 days ($p < 0.001$) in meat from the I-NM animals. When looking to the tenderization rate (Figure 1B), a decrease in meat toughness along ageing, it can be seen that the meat from animals of the intensive treatment had a higher tenderization rate in the global studied period (3 to 14 days) being higher for I-NM animals (22%) in the earliest period from 3 to 7 days, and for I-M animals in the last period of the *post-mortem* ageing (22%) from 7 to 14 days.

Overall, meat from animals of the I-M group showed redder meat and a lower tenderization rate, which could together be related to higher PSS. In fact, previous studies of the serum biomarkers of stress (cortisol, lactate, glucose, amyloid A, and haptoglobin) in this group of animals [9,18] evidenced that I-M animals were the most sensitive to stress reactivity, as indicated by its high serum haptoglobin levels.

3.2. Separation and Identification of Myofibrillar Subproteome

1D SDS-PAGE of the myofibrillar proteins allowed the visualization of 36 protein bands (ranging from 15 to 200 kDa) from the muscle myofibrillar subproteome, as shown in Figure 2.

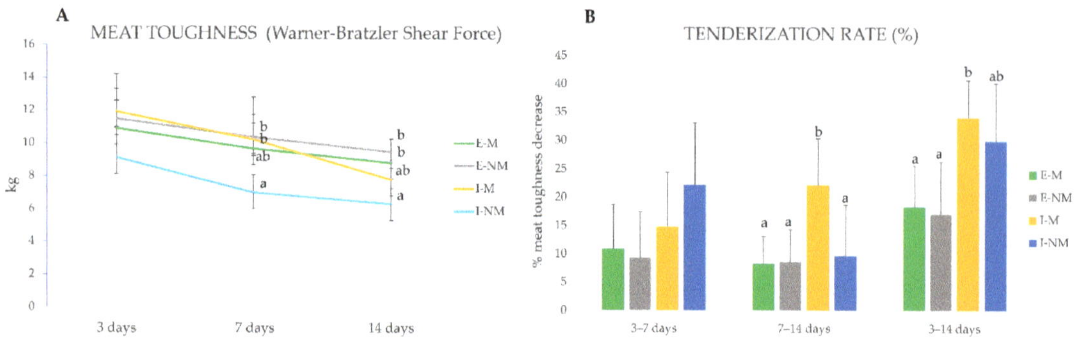

Figure 1. *Post-mortem* evolution of instrumental toughness for meat from the different handling treatments. (**A**) Warner–Braztler Shear force at 3, 7, and 14 days *post-mortem*. (**B**) Tenderization rate (%) calculated as the percentage of decrease in toughness in a given period of time (3 to 7 days, 7 to 14 days, 3 to 14 days). Different letters indicate significant differences ($p < 0.05$) among treatments (I-M, I-NM, E-M and E-NM) at 3, 7 or 14 days.

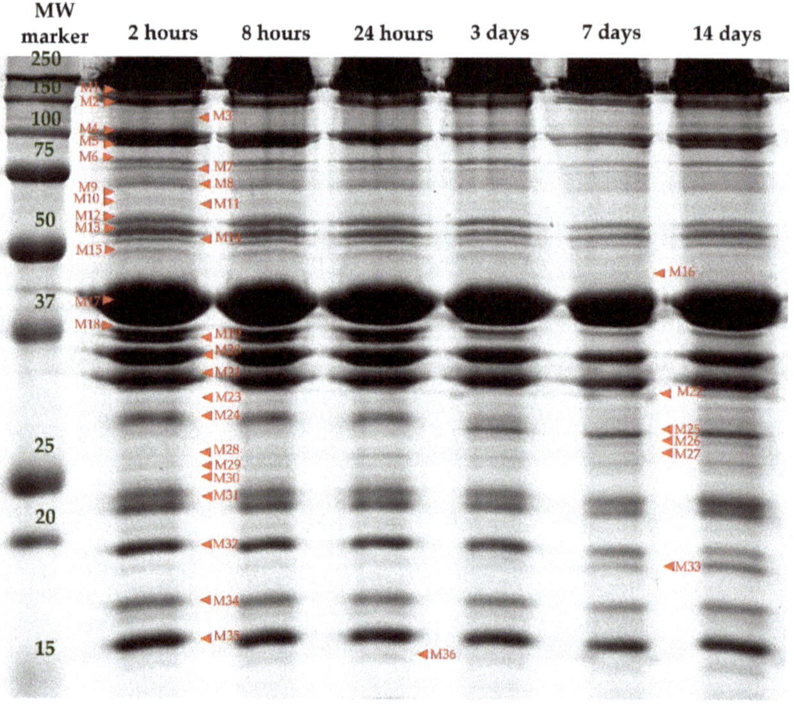

Figure 2. A representative 1D SDS-PAGE electrophoretic pattern of the myofibrillar subproteome profile of the LTL muscle of a yearling bull from "Asturiana de los Valles" at different *post-mortem* times (2 h, 8 h, 24 h, 3 days, 7 days, 14 days). MW marker: pre-stained molecular weight marker (All Blue pre-stained, Biorad). Band names are denoted by M (Myofibrillar proteins) followed by a number.

Among the 36 bands analyzed in the myofibrillar subproteome, 23 bands were significantly affected by at least one of the factors analyzed in this study (F, TL, and t), as shown in Table 2.

Table 2. Effect of handling factors (animal management at Farm (F) and during Transport and Lairage (TL)) and the *post-mortem* time (t) and their interactions (F × TL, F × t, TL × t and F × TL × t) on the myofibrillar subproteome (arbitrary units).

Band [MWe]	Farm Management (F)			Transport and Lairage (TL)			Post-Mortem Time (t)							Significance						
	I	E	SEM	NM	M	SEM	2 h	8 h	24 h	3 d	7 d	14 d	SEM	F	TL	t	F × TL	F × t	TL × t	F × TL× t
M1 (155.03)	4.66	3.67	0.25	4.76	3.87	0.15	4.08	4.39	4.22	4.4	4.36	3.57	0.24	*	**	NS	NS	NS	NS	NS
M4 (99.16)	5.94	4.97	0.29	5.48	5.43	0.17	6.04 b	5.99 b	5.71 ab	5.03ab	5.03ab	4.93 a	0.28	NS	NS	**	NS	NS	NS	NS
M6 (83.56)	2.18	2.0	0.09	2.102	2.08	0.06	2.05 ab	2.24 b	2.28 b	2.13 b	2.07 ab	1.78 a	0.09	NS	NS	**	NS	NS	NS	NS
M9 (71.02)	1.23	1.36	0.06	1.27	1.32	0.04	1.37 b	1.43 b	1.33 b	1.37 b	1.21 a	1.05 a	0.06	NS	NS	***	***	NS	NS	NS
M12 (59.93)	2.63	2.22	0.15	2.63	2.22	0.09	2.35	2.53	2.61	2.60	2.35	2.12	0.15	NS	**	NS	***	NS	NS	NS
M13 (56.24)	2.42	2.49	0.09	2.47	2.44	0.05	2.47 ab	2.70 b	2.59 b	2.37 ab	2.37 ab	2.25 a	0.09	NS	NS	**	NS	NS	NS	NS
M15 (50.74)	1.5	1.57	0.06	1.49	1.58	0.04	1.47 ab	1.62 b	1.62 b	1.64 b	1.52 ab	1.34 a	0.06	NS	NS	**	*	NS	NS	NS
M17 (40.53)	16.14	19.3	0.79	17.64	17.8	0.47	19.19 ab	16.23 a	17.08 a	16.93 a	16.91 a	19.98 b	0.76	*	NS	**	NS	NS	NS	NS
M18 (37.64)	2.96	2.98	0.12	2.96	2.98	0.07	3.27 bc	3.49 c	3.38 c	2.82 ab	2.52 a	2.34 a	0.12	NS	NS	***	NS	NS	NS	NS
M19 (36.88)	2.34	1.94	0.07	2.24	2.05	0.04	2.61 c	2.73 c	2.67 c	1.99 b	1.49 a	1.38 a	0.07	**	**	***	NS	**	NS	NS
M20 (35.03)	7.47	7.04	0.22	7.28	7.23	0.13	7.09 abc	7.58 bc	8.02 c	7.68 b	6.84 ab	6.33 a	0.21	NS	NS	***	NS	NS	NS	NS
M21 (32.95)	6.1	5.9	0.17	6.01	5.99	0.1	5.26 a	5.7 ab	5.74 ab	6.56 c	6.62 c	6.1 c	0.16	*	NS	***	NS	NS	NS	NS
M23 (30.89)	1.03	1.25	0.05	1.15	1.13	0.03	1.05 a	1.04 a	1.04 a	1.06 a	1.21 ab	1.44 b	0.05	*	NS	***	NS	NS	NS	NS
M24 (29.75)	2.4	2.6	0.1	2.49	2.51	0.06	3.04 d	2.89 d	2.74 cd	2.32 bc	2.17 ab	1.84 a	0.10	NS	NS	***	***	NS	NS	NS
M25 (28.67)	1.58	1.71	0.08	1.61	1.68	0.05	0.89 a	0.92 a	0.97 a	1.68 b	2.51 c	2.89 d	0.08	NS	NS	***	NS	*	NS	NS
M26 (27.82)	0.65	0.8	0.03	0.71	0.73	0.02	0.54 a	0.53 a	0.51 a	0.73 b	0.95 c	1.08 c	0.03	*	NS	***	*	*	NS	NS
M27 (27.25)	0.56	0.69	0.02	0.6	0.65	0.01	0.47 a	0.45 a	0.45 a	0.62 b	0.83 c	0.93 d	0.02	**	*	***	NS	NS	NS	NS
M30 (25.10)	1.22	1.33	0.06	1.35	1.2	0.04	1.33 bc	1.37 bc	1.42 c	1.24 abc	1.16 ab	1.11 a	0.06	NS	**	**	***	NS	NS	NS
M31 (23.20)	5.53	6.04	0.13	5.61	5.96	0.08	5.81	5.77	5.59	5.69	5.86	5.97	0.12	*	**	NS	NS	NS	NS	NS
M32 (19.89)	3.83	3.7	0.13	3.66	3.86	0.07	4.11 bc	4.09 bc	3.98 bc	4.23 c	3.64 b	2.52 a	0.12	NS	**	***	NS	***	NS	NS
M33 (19.07)	1.17	1.22	0.08	1.2	1.18	0.05	0.79 a	0.77 a	0.79 a	1.05 a	1.54 b	2.21 c	0.08	NS	NS	***	NS	***	NS	NS
M34 (17.54)	3.84	3.92	0.11	3.96	3.8	0.07	3.82	3.81	3.78	3.86	3.94	4.06	0.13	NS	NS	NS	*	NS	NS	NS
M35 (16.05)	5.76	5.37	0.14	5.43	5.7	0.08	5.45	5.49	5.34	5.62	5.73	5.76	0.05	NS	*	NS	NS	NS	NS	NS

Mwe: the experimental molecular weight (kDa), SEM: standard error of the mean; F: Management at farm; TL: Management during the transport and lairage; t: *Post-mortem* time; I: Intensive, E: Extensive; NM: Non-mixing with unfamiliar animals, M: Mixing with unfamiliar animals; Significance: NS: not significant; *: $p < 0.001$; **: $p < 0.05$; ***: $p < 0.01$. The variables not followed by the same letter in the same row (a, b, c, and d) are statistically different ($p < 0.05$).

It is important to note that these band intensity differences are due to the effect that handling factors may have in either the synthesis of a determined protein and/or to variations in the muscle *post-mortem* metabolism, which may result in increased proteolysis of that protein decreasing its relative abundance or causing its disappearance and the consequent increases of smaller protein fragments/peptides. Therefore, the relative abundance of a given (intact) protein is a balance between synthesis and degradation [22].

Table 3 shows the identification of protein bands with significant intensity differences among treatments. These proteins belong to three major biological pathways (Figure 3A):

➢ Muscle contraction, structure and associated proteins: M1 (Myosin-binding protein C, fast-type isoform X2 "MYBPC2"), M4 (Alpha-actinin-3 "ACTN3"), M13 (Desmin, partial "DES"), M17 (Actin, alpha skeletal muscle "ACTA1"), M21 (Tropomyosin alpha-1 chain "TPM1"), M23 (LIM domain-binding protein 3 isoform X5 "LDB3"), M24 (Four and a half LIM domains protein 1 isoform 1 "FHL1"), M25 and M27 (Troponin T, fast skeletal muscle isoform X31 "TNNT3"), M26 (Four and a half LIM domains protein 1 isoform 2 "FHL2"), M31 (Myosin light chain 1/3 skeletal muscle isoform "MYL1"), M32 (Troponin I, fast skeletal muscle "TNNI2"), M34 (Troponin C, skeletal muscle "TNNC1" and M35 (Myosin regulatory light chain 2, skeletal muscle isoform "MYLPF");

➢ Energy metabolism and associated pathways: M6 (ATP-dependent 6-phosphofructokinase, muscle type "PFKM"), M12 (Pyruvate kinase PKM isoform X1 "PKM"), M15 (ATP synthase subunit beta, mitochondrial precursor "ATP5F1B"), M18 y M19 (Fructose-biphosphate aldolase A "ALDOA") and M20 (Glyceraldehide-3-phosphate dehydrogenase, "GAPDH");

➢ Heat shock proteins: M9 (Heat Shock 70 kDa protein 1A "HSPA1A"), M30 (Heat Shock protein family B member 1 variant 1 "HSPB1") and M33 (Alpha-crystallin B chain "CRYAB").

Table 3. Protein identification of myofibrillar bands separated by 1D-SDS-PAGE that showed significant differences with treatments (Farm, Transport and Lairage, and/or *post-mortem* time).

Band: Gene Name	Identification	Accession Number	MOWSE Scores	Sequence Coverage (%)	Matched Queries	MWt
M1: MYBPC2	Myosin-binding protein C, fast-type isoform X2	E1BNV1	295	24	25	128.5
M4: ACTN3	Alpha-actinin-3	Q0III9	506	38	35	103.7
M6: PFKM	ATP-dependent 6-phosphofructokinase, muscle type	Q0IIG5	352	40	42	86.1
M9: HSPA1A	Heat Shock 70 kDa protein 1A	Q27975	288	38	24	70.5
M12: PKM	Pyruvate kinase PKM, isoform X1	A5D984	822	66	46	58.5
M13: DES	Desmin, partial	O62654	246	54	21	52.6
M15: ATP5F1B	ATP synthase subunit beta, mitochondrial precursor	P00829	445	48	25	56.2
M17: ACTA1	Actin, alpha skeletal muscle	P68138	522	52	24	42.4
M18: ALDOA	Fructose-biphosphate aldolase A	A6QLL8	430	62	24	39.9
M19: ALDOA	Fructose-biphosphate aldolase A	A6QLL8	286	60	21	39.9
M20: GAPDH	Glyceraldehide-3-phosphate dehydrogenase	P10096	394	45	20	36.1
M21: TPM1	Tropomyosin alpha-1 chain	Q5KR49	199	41	15	32.7
M23: LDB3	LIM domain-binding protein 3 isoform X5	G3N3C9	180	55	19	30.9
M24: FHL1	Four and a half LIM domains protein 1 isoform 1	G3MZ95	671	87	34	35.5
M25: TNNT3	Troponin T, fast skeletal muscle isoform X31	Q8MKI3	206	45	18	28.9
M26: FHL2	Four and a half LIM domains protein 1 isoform 2	Q2KI95	95	37	11	33.8
M27: TNNT3	Troponin T, fast skeletal muscle isoform X31	Q8MKI3	178	43	15	28.9
M30: HSPB1	Heat Shock protein family B member 1 variant 1	Q3T149	368	73	14	22.4
M31: MYL1	Myosin light chain 1/3 skeletal muscle isoform	A0JNJ5	425	77	18	21.1
M32: TNNI2	Troponin I, fast skeletal muscle	F6QIC1	194	70	23	21.6
M33: CRYAB	Alpha-crystallin B chain	P02510	163	69	13	20.1
M34: TNNC1	Troponin C, skeletal muscle	P63315	333	56	15	18.3
M35: MYLPF	Myosin regulatory light chain 2, skeletal muscle isoform	Q0P571	517	66	21	19.11

The MOWSE score: numeric descriptor of the likelihood that the identification is correct. Protein scores greater than 94 are significant ($p < 0.05$); Mwt: theoretical molecular weight (kDa).

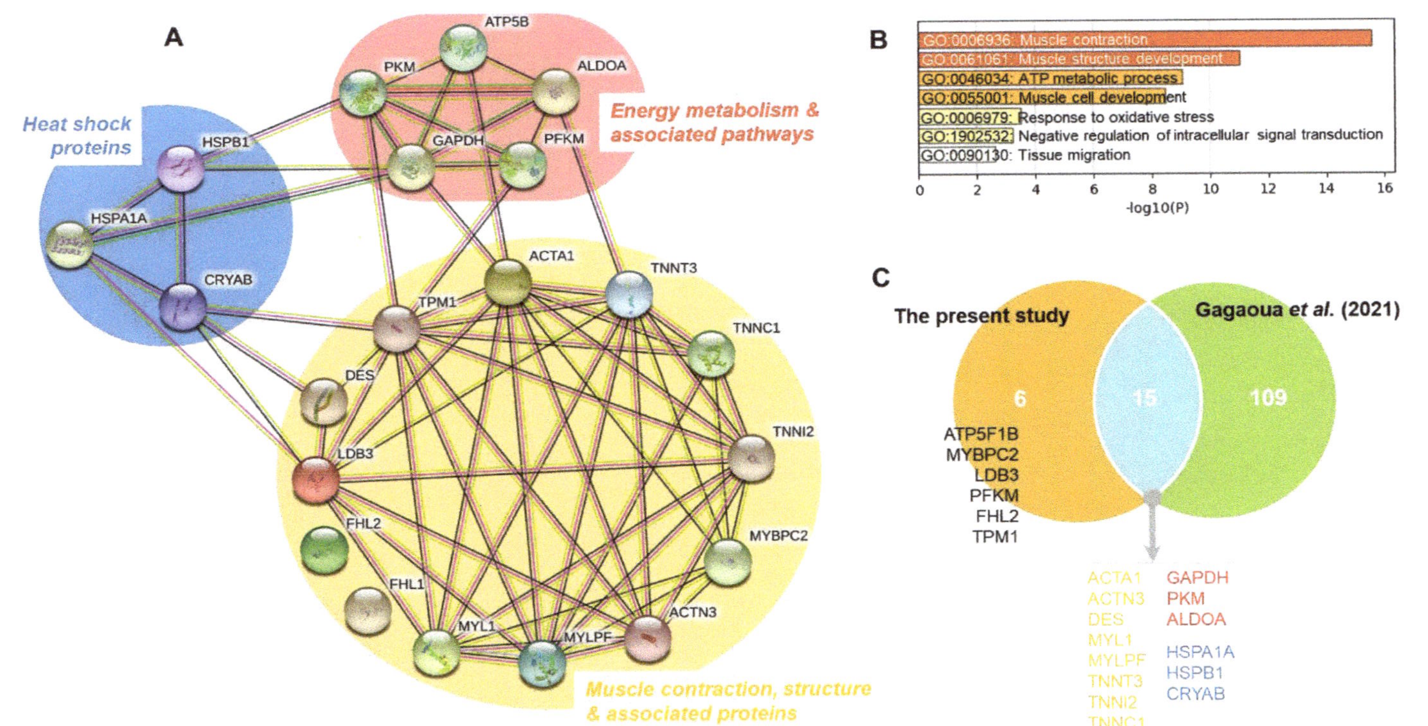

Figure 3. Bioinformatics analyses of the differential proteins affected with the treatments (Farm, Transport and Lairage and/or *post-mortem* time). (**A**) Protein-Protein Interactions of the 21 proteins using String database, highlighting 3 major pathways these being Muscle contraction, structure & associated proteins, Energy metabolism & associated pathways and Heat Shock proteins. (**B**) Significant enriched Gene Ontology (GO) terms obtained using Metascape tool. (**C**) Overlap among the 21 proteins with the repertoire of beef tenderness biomarkers reported in the database of Gagaoua et al. [18]. The 15 common proteins were highlighted by the corresponding molecular pathway color as in (**A**).

It is important to note that some proteins appeared in more than one band, as was the case of ALDOA (M18 and M19) and TNNT3 (M25 and M27), probably due to differences in the molecular weight or to conformational changes of the different proteolytic fragments originated from the same protein. The analyses of these differential proteins, which finally constitute 21 unique proteins (Table 3 and Figure 3A), allowed for the construction of an interconnected network (Figure 3A), highlighting the importance of muscle contraction and structure pathways. The Gene Ontology (GO) analysis allowed for the identification of seven enriched GO terms (Figure 3B), from which the top two enriched terms being GO:0006936: Muscle contraction and GO:0061061: Muscle structure development. These were followed by GO:0046034: ATP metabolic process, GO:0055001: Muscle cell development, GO:0006979: Response to oxidative stress, GO:1902532: Negative regulation of intracellular signal transduction and GO:0090130: Tissue migration.

It must be noted that most protein bands from the myofibrillar subproteome (61%) correspond to structural insoluble proteins, but soluble proteins such as glycolytic enzymes (26%) or HSPs (13%) were also found in the myofibrillar fraction (Figure 3A). This can be due to a decrease of solubility, maybe partly as a consequence of the early pH drop while the muscle temperature is still high, and also to the relationships that exist among the proteins [34], as confirmed in the network of Figure 3A. In fact, such conditions may cause denaturation of proteins which became insoluble, and their aggregation and precipitation onto myofibrils [24,40]. In support of this, the recent study by Gagaoua et al. showed that the maturation process involves interconnected molecular pathways in a pH-dependent manner leading, for instance, to the concomitant appearance of two major proteolytic fragments at 110 and 30 kDa, based on 1DE electrophoresis [34]. These two protein bands appearing during ageing, also observed in this study, increase in their intensity as a function of *post-mortem* time in a pH decline-dependent manner. LC-MS/MS analysis yielded 22 unique proteins for the 110 kDa fragment and 13 for the 30 kDa, with four common proteins related to both actin and fibrinogen complex. The Gene Ontology analysis revealed that a myriad of biological pathways are influential with many of them, as confirmed in the present study (Figure 3), and were related to proteins involved primarily in muscle contraction and structure. Other pathways were apoptotic mitochondrial changes, calcium and ion transport, energy metabolism, etc. Interestingly, most of the proteins composing these two fragments among others that appear or disappear during the tenderization process and in line to the results of this study have been so far identified as biomarkers of beef tenderness (Gagaoua et al., 2021) [21]. In addition, HSPs can translocate and accumulate in the cytoskeleton and myofibrillar proteins during early *post-mortem* stages, as they exert a protective role against muscle degradation [40]. These facts reinforce the need to consider different cell fractions and the movements of proteins between cytoskeletal and myofibrillar structures, for an accurate and reliable study of the process of conversion of the muscle into meat, as has been highlighted by previous studies [21,41].

3.2.1. Handling Effects on the Muscle Contraction, Structure and Associated Proteins

Among the protein bands with structural and contraction functions, only M27 (27.25 kDa), identified as Troponin T, fast skeletal muscle isoform X31 (TNNT3), was affected by the three factors analyzed in this study (F, TL, and t). Higher intensities of M27 were found in the muscle of the animals from the extensive treatment ($p < 0.01$) and mixed group animals during transport and lairage ($p < 0.05$). Further, a significant increase ($p < 0.001$) of TNNT3 band intensity with *post-mortem* time was found. Band M25 with a molecular weight of 28.67 kDa, was also identified as TNNT3, but it did not show significant differences with F or TL, but a significant increase with *post-mortem* time ($p < 0.001$). Troponin T is the most frequently identified differential biomarker of ongoing proteolysis and tenderization due to the appearance of degradation fragments of 30 kDa and 28 kDa correlated to meat tenderness [25,34,42–44]. TNNT3 is one of the proteins of the Troponin complex, composed of three regulatory proteins (Troponin T, C and I) that are integral to muscle contraction. It is well known that tenderization acts on all the proteins of the complex and it was identified as a robust biomarker of beef tenderness in the study of Gagaoua et al. [21], as evidenced in the Venn diagram of Figure 3C. Accordingly, our results showed changes in all the proteins of the Troponin complex. Bands M32 and M34 were identified as TNNI2 and TNNC1 respectively. TNNI2 was affected by TL, with higher intensity ($p < 0.01$) in meat from the mixed animals, and also a significant effect of *post-mortem* time was observed with a significant decrease ($p < 0.001$) after 3 days *post-mortem* in meat from E-M, E-NM and I-NM treatments, but delayed for I-M, which seems to indicate slower tenderization rate.

Other bands with significant changes correspond to myosin related proteins M1 (Myosin-binding protein C, fast-type isoform X2 "MYBPC2") and M31 (Myosin light chain 1/3 skeletal muscle isoform "MYL1") were affected by both F and TL but not by ageing time. MYBPC2 showed higher intensity ($p < 0.05$) in meat from the indoor reared animals and in meat from the non-mixed group ($p < 0.01$), which could be related to differences in the level of physical exercise, as found for MYOM2, a major component of the myofibrillar subproteome that appeared in the sarcoplasmic subproteome of these animals, in agreement with the recent studies of Diaz et al., [9] and Gagaoua et al. [34]. MYBPC2 belongs to the Myosin Binding Proteins family formed by sarcomeric proteins located in the A-band in close association with the thick filaments that are known as regulators of the myofilament contractility [45]. MYBPC exists in three main isoforms: skeletal slow (MYBPC1), skeletal fast (MYBPC2), and cardiac (MYBPC3). MYPBC1 and MYBPC2 were recently identified to be the major components of the 110 kDa fragment appearing during the tenderization of beef in a pH dependent manner [34]. It is also important to note that a closely member, the myosin binding protein H (MYBPH) has been previously identified as a negative biomarker of color and beef tenderness, whatever the gender, due to its significant effect on length, thickness, and lateral alignment of myosin filaments [21,46,47]. In this work, MYBPC2 showed higher intensity levels in the I-NM meat, which was the tenderer one, hence confirming the findings by Gagaoua et al. [21,34] proposing this protein as a good marker of meat tenderization.

MYBPC has a theoretical molecular weight of approximately 130 kDa; however, the band identified as MYBPC2 in our study shows a higher experimental molecular weight (155 kDa) what could be indicative of aggregation of this protein or to its interaction with nebulin and other proteins as observed in earlier studies [34,48]. These modifications may explain the loss of its function in the alignment of myosin filaments, and therefore the positive role it might play on tenderness. On the other hand, MYL1 is a member of the Myosin light chains that are crucial for muscle function in terms of contraction velocity and power. In this work, MYL1 band showed higher intensities in the meat reared outdoors ($p < 0.05$) and in meat from mixed animals ($p < 0.01$). Apart from MYL1, another band (M35) from the same family was identified as Myosin regulatory light chain 2, skeletal muscle isoform "MYLPF" that was not different within farm management, but was significantly affected by TL with higher values at the M treatment ($p < 0.05$) in line with the findings for MYL1. *Post-mortem* disruption of myosin light and heavy chains and actin in the actomyosin complex plays a central role in the muscle to meat conversion and may have a direct effect on tenderness [49]. The *post-mortem* concentration of these proteins has been previously associated with pork and beef tenderness [50,51], and belong to the robust list of biomarkers of beef tenderness shortlisted by Gagaoua et al. [21]. Increased proteolysis of MYL1 has also been described in dark-cutting beef [22,26]. It is worthy to note that myosin lights chains were also identified as biomarkers of several beef color traits [47].

Apart from the aforementioned, management at farms affected other three structural proteins: M17 (Actin, alpha skeletal muscle "ACTA1"), M23 (LIM domain-binding protein 3 isoform X5 "LDB3) and M26 (Four and a half LIM domains protein 1 isoform 2 "FHL2"). All of them were more intense ($p < 0.05$) in the meat from animals reared outdoors. Actin is the second most abundant myofibrillar protein, after myosin, and was described as the top biomarker of beef tenderness (Figure 3C) [21]. The breakdown of transverse cytoskeletal actin filaments can cause detachment of the sarcolemma from the basal lamina and the extracellular matrix network, causing muscle cells degradation and hence increasing tenderness; therefore, actin has been found to be a good biomarker of tenderness [21,40,49,52]. The other two proteins affected by farm management were the LIM-domain containing proteins, LDB3 and FHL2. The LIM domain is a cysteine-histidine rich, zinc-coordinating domain, consisting of two tandemly repeated zinc fingers. The LIM domain-containing proteins are known to play critical roles in vertebrate development and cellular differentiation. The LDB3 protein, located in the sarcomere, is essential for maintaining Z line structure and muscle integrity [53]. FHL2 is a member of the four and a half LIM domain protein family (FHL), with an important role in muscle development [54]. To the best of our knowledge, this protein has never been related to meat quality before; however another protein from the FHL family, the Four and a half LIM domains protein 1 isoform 1 (FHL1,) also known as Cypher protein, has been related to the release of α-actinin and the weakening of the Z-disc during meat tenderization [55,56]. In the present study, band M24 was identified as FHL1 and, contrary to what was found for FHL2, only a significant decrease with *post-mortem* ageing and a significant interaction F × TL (E-NM > E-M = I-M > I-NM) was observed ($p < 0.001$). Previous studies found increased intensity of FHL1 in DFD meat [16,22,57].

The other structural protein bands found in the myofibrillar extract were affected only by *post-mortem* time, with a significant decrease ($p < 0.01$) of M4 (Alpha-actinin-3 "ACTN3") and M13 (Desmin, partial "DES") and a significant increase ($p < 0.001$) of M21 (Tropomyosin alpha-1 chain "TPM1"). It is important to note that the *post-mortem* ageing time is the factor that causes the greatest differences in the intensity of the muscle contraction and structural proteins, with 10 out of the 14 structural bands showing significant differences in agreement to the very recent findings of Gagaoua et al. [34]. Most of these band's intensities remain similar at the earliest *post-mortem* period (from 0 h to 24 h), but then increase or decrease drastically as a consequence of proteolysis.

3.2.2. Handling Effects on the Energy Metabolism and Associated Pathways Proteins

Most of the bands related to energy metabolism found in the myofibrillar fraction were glycolytic enzymes (Figure 3), usually found in the sarcoplasmic fraction. Among them, only the band M19 (36.88 kDa), corresponding to Fructose-biphosphate aldolase A "ALDOA", was affected by the three factors studied: F, TL, and ageing time, showing higher intensity in meat from the Intensive rearing system ($p < 0.01$) and from the NM animals ($p < 0.01$), and a significant decrease along *post-mortem* ageing ($p < 0.001$) that starts at 8 h *post-mortem* in the meat from the Intensive treatment and at 24 h from the Extensive system (significant interaction ($p < 0.01$) F × t). Band M18 (37.64 kDa), also identified as ALDOA, showed a significant *post-mortem* decrease. In contrast with these results in the myofibrillar fraction, previous studies in the same animals showed higher intensity of ALDOA (37.1 kDa) in the sarcoplasmic subproteome of meat from the extensive reared animals [9]. ALDOA catalyzes the conversion of fructose 1, 6-diphosphate to glyceraldehyde 3-phosphate during glycolysis, therefore its lower intensity in meat from Intensive reared animals in the sarcoplasmic fraction can be associated with a faster glycogenolysis exhaustion or degradation of the enzyme in these animals that were found to be more susceptible to pre-slaughter stress [9]. However, it is also known that ALDOA in association with other metabolic enzymes, assists in the creation of cross-links between adjacent actin filaments or in binding troponin to the thin filaments, to enhance energy provision, where it is actively needed during contraction, hence affecting the distance between myofibrils, and therefore light scattering and tenderness [57]. This could at least partially explain the differences found between sarcoplasmic and myofibrillar ALDOA contents, as, due to the higher glycolytic metabolism of Intensive reared animals at slaughter, more energy provision may be needed for contraction and more ALDOA can be retained within the interstitial spaces of the myofibrils during the extraction. It is worthy to note that ALDOA was identified as a robust biomarker of beef tenderness (Figure 3C) [18] and of color variation [47].

Apart from ALDOA, Farm management did not affect significantly any of the other proteins from the energy metabolic pathway, however TL affected significantly to the band M12 (Pyruvate kinase PKM isoform X1 "PKM") with increased intensity ($p < 0.01$) in the meat from the NM group under a significant F $\times$ TL interaction ($p < 0.001$). This band was the only one corresponding to metabolic enzymes that did not show significant differences with *post-mortem* time. PKM is a glycolytic enzyme implicated in the last phases of the glycolysis that catalyzes the dephosphorylation of phosphoenolpyruvate to pyruvate, yielding one molecule of pyruvic acid and one of ATP. Previous studies have shown that animals mixed during transport and lairage are more affected by pre-slaughter stress and they have also been related to higher *post-mortem* glycolytic metabolism [9,58]. Moreover, lower abundance of PKM was previously found in myofibrillar subproteome of high pH24 meat [26]. Thus, the lower levels of PKM found in the myofibrillar fraction of mixed animals in the present study can be explained by the glycogen depletion occurring due to PSS caused by the mixing procedure before slaughter. This can lead to a decrease in the glycolysis rate after slaughter due to the early depletion of glycogen and a downregulation of proteins involved in the glycolytic pathway in conjunction with oxidative pathways driven by mitochondria (for review see Gagaoua et al. [22]).

The other metabolic enzymes, PFKM, ATP5F1B, and GAPDH, showed significant differences ($p < 0.01$) only with *post-mortem* time, with increasing band intensities during the early *post-mortem* until 24 h or 3 days *post-mortem* and decreasing afterwards. This reflects the cell metabolism behavior, with high levels of this proteins at early *post-mortem* due to the trigger of the glycolytic metabolism but decreasing later due to the impact of pH that might desaturate them or significantly reduce their activity.

3.2.3. Handling Effects on the Heat Shock Proteins

Overall, three bands from the myofibrillar subproteome were identified as members of the Heat Shock proteins (HSPs) family: M9 was identified as the large Heat shock 70 kDa protein 1A "HSPA1A", and the other two (M30 and M33) as members of the small HSPs subfamily, identified as Heat Shock protein family B member 1 variant 1 "HSPB1" (also known as HSP27) and Alpha-crystallin B chain "CRYAB". Functional proteomic studies have confirmed differential expression of HSPs in meat from different breeds, handling systems and quality traits [16,21,32,47,59,60]. In the present study, all the HSPs bands showed significant changes along the *post-mortem* meat maturation, with a significant increase ($p < 0.01$) up to 24 h *post-mortem* followed by a progressive decrease afterwards in the case of HSPB1, a significant increase after three days ($p < 0.001$) for HSPA1A and a significant decrease ($p < 0.001$) after three days for CRYAB. Surprisingly, none of these bands were affected by management at farm, and the mixing treatment affected significantly only HSPB1 band that was more intense ($p < 0.001$) in meat from the non-Mixed group. Moreover, a significant interaction F $\times$ TL ($p < 0.001$) was found for this band.

Among the small HSPs, HSPB1 is one of the most frequently related protein to beef tenderness but in different directions (positive or negative manner) depending on the studied factors (PSS, breed, gender, post-translational modifications (isoform), rearing factor, etc.), as explained by Gagaoua et al. [21]. HSPs exert protective functions as chaperone proteins from proteases, therefore they could reduce degradation of myofibrillar proteins [61]. Several studies reported positive relationship between degradation of HSPB1 and meat tenderness improvements, as degraded HSPs may no longer prevent irreversible damage to myofibrillar proteins [1,60,62,63]. Accordingly, our results showed higher intensity of HSPB1 band in the myofibrillar subproteome of meat from the E-NM treatment, which was the treatment that showed higher WBSF, and thus a lower tenderization rate. Moreover, the evolution of HSPB1 along meat ageing is in agreement with previous studies in bovine muscles [50,64,65] that have demonstrated that muscle HSPB1 increases in abundance shortly after slaughter but decreases during meat storage. Similarly, our results showed a significant decrease ($p < 0.001$) of the large HSPA1A with *post-mortem* ageing. Members of the HSP70 family were previously found in the sarcoplasmic subproteome of these animals with a significant effect of Farm management, showing higher intensity in meat from the extensive reared animals [9]. Under stress situations, HSPA1A, which is an inducible protein that translocates and accumulates in the cytoskeletal and myofibrillar proteins in an attempt of stabilizing the muscle structure [40]. This HSP diffusion capacity between cellular fractions may explain the differences found between rearing treatments. As aforementioned, animals from intensive rearing system are suggested to be more susceptible to handling and pre-slaughter stress [9,18] so that more HSP may translocate to myofibrils while in the extensively reared group more HSPA-A is probably easily removed and extracted from the sarcoplasmic fraction.

CRYAB was affected by *post-mortem* time showing a significant ($p < 0.001$) increase in intensity after seven days of storage. In contrast, previous studies demonstrate a decrease of intact CRYAB with ageing in total extracts from different muscles (Longissimus lumborum, Semimembranosus and Psoas major) of Angus x Simmental beef cattle [60]. Our results showing late *post-mortem* increases of this band could be explained by solubility changes due to protein modifications such as fragmentation, oxidation, precipitation, or aggregation thereby going from soluble to an insoluble state. This was previously pointed out by Bjarnadóttir et al. [32], who discovered that some of the small HSPs proteins increased their abundance in the insoluble protein fraction, possibly as a result of aggregation onto myofibrillar proteins, thereby following them during extraction. It is also important to note that CRYAB behaves also as a structural protein and therefore it can be susceptible of degradation along ageing.

3.3. Relationship between Meat Quality Traits and the Significantly Changing Myofibrillar Proteins

The correlations between the differential protein bands from the myofibrillar subproteome and meat quality traits, measured at different *post-mortem* times were analyzed, and significant correlations with color traits and meat toughness are shown in Tables 4 and 5 respectively.

In the present study beef color traits were correlated (Table 4) with two metabolic enzymes (PKM and ALDOA), two Heat Shock proteins (HSPA1A and HSPB1) and three structural proteins (DES, TNNI2 and MYLPF). This confirms the knowledge about the importance of cell glycolytic rate and its consequences on *post-mortem* modifications of proteins such as myosin, actin, troponin, and other metabolic proteins, particularly glycolytic enzymes in the sarcoplasm, and therefore influences the ultimate meat color.

A recent integromics study evidenced that both glycolytic enzymes and HSPs pathways have important roles in beef color determination where several relevant biomarkers were shortlisted [21]. Accordingly, PKM showed significant ($p < 0.05$) positive correlations with L^*, b^* and h^* and negative with a^* and C^*. The highest correlation coefficients were found at 8 h *post-mortem* with a^* (-0.850, $p < 0.01$) and h^* (0.800, $p < 0.01$). ALDOA showed the highest correlations with a^* and h^* at 7 days *post-mortem* (0.821, $p < 0.01$ and -0.695, $p < 0.05$, respectively). Other correlations were found between color traits and GAPDH (0.701, $p < 0.05$ with a^* and -0.705, $p < 0.01$ with h^*) at 7 days *post-mortem*. PKM and ALDOA may exert their effect on muscle color due to their involvement in the glycolytic pathway providing energy to the muscle contraction. They can also assist in the creation of crosslinks between actin filaments or in binding troponin to the thin filaments, hence affecting the distance between myofibrils and therefore light scattering [21,22,66–69]. In the case of HSPs, we found positive correlations between HSPA1A and L^* and h^* ($+0.731$, $p < 0.05$ at 8 h *post-mortem*) and negative between HSPB1 and L^* and b^* (-0.73, $p < 0.01$ at 7 days *post-mortem*). Many studies have related HSPs with color [70–72] probably due to their protective action against stress-induced denaturation of muscle proteins, that would affect reflectance, light scattering, and myoglobin, hence influencing color parameters [57,73–77].

Table 4. Significant Pearson correlations coefficients between myofibrillar subproteome bands and meat color traits.

		PFKM	HSPA1A	PKM	DES	ALDOA M18	ALDOA M19	GAPDH	TPM1	LDB3	FHL1	TNNT3 M25	TNNT3 M27	HSPB1	TNNI2	CRYAB	MYLPF
*L**(60)	2 h		0.691 *							−0.654 *				−0.651 *	0.624 *		0.702 *
	8 h		**0.731** **	0.613 *	0.608 *									−0.581 *	0.604 *		
	24 h				0.615 *										0.697 *		
	3 d													−0.687 *	**0.768** **	−0.630*	0.610 *
	7 d							−0.652 *			−0.588 *	0.583 *		−**0.777** **			.578 *
	14 d				0.599 *		.633*					0.615 *					
*a**(60)	2 h			−0.695 *	−0.683 *												−0.593 *
	8 h	−0.588 *		−**0.850** **													
	24 h	−0.694 *		−0.675 *													
	3 d			−0.726 *		0.685 *									−0.581 *		−0.707 *
	7 d					**0.821** **		0.701 *				−0.592 *	−0.686*				−0.611 *
	14 d			−0.585 *								−0.632 *					
*b**(60)	2 h		0.645 *							−0.633 *					0.627 *		0.608 *
	8 h		0.612 *						0.606 *						0.683 *		
	24 h				0.588 *												
	3 d									−0.664 *				−0.624 *	0.634 *	−0.597 *	
	7 d			0.581*							−0.650 *			−**0.738** **			
	14 d																
*C**(60)	2 h			−0.605 *	−0.584 *		−0.592 *										
	8 h			−0.613 *			−0.644 *										
	24 h	−0.714 *					−0.599 *										
	3 d			−0.637 *		0.651 *											
	7 d					0.666 *							−0.634*				
	14 d																
*h** (60)	2 h		0.592 *		0.588 *					−0.617 *				−0.599 *			**0.741** **
	8 h		0.700 *	**0.800** **													
	24 h			0.670 *											0.610 *		0.578 *
	3 d			0.676 *											**0.755** **		**0.721** **
	7 d					−0.695 *		−**0.751** **				0.626 *		−0.641 *			0.654 *
	14 d				0.689 *							0.618 *					

* *p* < 0.05; ** *p* < 0.01; PFKM: ATP-dependent 6-phosphofructokinase; HSPA1A: Heat shock 70 kDa protein 1A; PKM: Pyruvate kinase muscle; DES: Desmin; ALDOA: Fructose-biphosphate aldolase A; GAPDH: Glyceraldehide3-phosphate dehydrogenase; TPM1: Tropomyosin alpha 1; LDB3: LIM domain-binding protein 3; FHL1: Four and a half LIM domains protein 1 isoform 1; TNNT3: Troponin T fast skeletal muscle isoform x31; HSPB1: Heat shock protein family B member 1 variant 1; TNNI2: Troponin I fast skeletal muscle; CRYAB: Alpha-crystallin B chain; MYLPF: Myosin regulatory light chain 2. The higher and significant correlation coefficients are in bold.

Table 5. Significant Pearson correlations coefficients between myofibrillar subproteome bands and tenderness evaluated by Warner–Bratzler shear force and tenderization rate (%).

		ACTN3	HSPA1A	PKM	DES	ATP5F1B	ACTA1	ALDOA M18	ALDOA M19	GAPDH	FHL1	TNNT3 M25	FHL2	TNNT3 M27	MYL1	TNNI2	CRYAB	TNNC1
WB3 d	2 h			−0.482 *	−0.427 *				−0.437 *			0.539 **			0.415 *	0.497 *	0.423 *	
	8 h								−0.493 *						0.439 *			
	24 h		0.415 *			0.472 *			−0.545 **								0.428 *	
	3 d	0.495 *																
	7 d											−0.450 *						
	14 d						−0.521 **					−0.417 *						
WB7 d	2 h											0.556 **			0.503 *			
	8 h						−0.411 *					0.529 **			0.516 **			
	24 h								−0.502 *						0.453 *			
	3 d														0.472 *			
	7 d											−0.451 *						
	14 d	0.447 *					−0.599 **				0.413 *						0.514 *	
WB14 d	2 h								−0.441 *			0.514 *			0.483 *			0.417 *
	8 h								−0.524 **			0.625 **			0.503 *	0.405 *		0.441 *
	24 h								−0.421 *					0.445 *	0.456 *			
	3 d														0.542 **			
	7 d																0.521 *	
	14 d						−0.567 **				0.429 *						0.614 **	
% TR. 3–14 d	2 h																0.592 **	
	8 h																	
	24 h													−0.505 *				
	3 d							0.442 *					−0.491 *	−0.564 **				
	7 d												−0.546 **	−0.538 **				
	14 d									0.453 *			−0.557 **	−0.469 *				

* $p < 0.05$; ** $p < 0.01$. WB3 d: Warner–Braztler shear force 3 days; WB7 d: Warner–Braztler shear force 7 days; WB14 d: Warner–Braztler shear force 14 days; % TR. 3–14 d: Tenderization rate from 3 to 14 days *post-mortem*. ACTN3: Alpha A-acinin-3; HSPA1A: Heat Shock 70 kDa protein 1A; PKM: Pyruvate kinase muscle; DES: Desmin; ATP5F1B: ATP synthase subunit beta, mitochondrial precursor; ACTA1: Actin, alpha skeletal muscle; ALDOA: Fructose-biphosphate aldolase A; GAPDH: Glyceraldehide3-phosphate dehydrogenase; FHL1: Four and a half LIM domains protein 1 isoform 1; TNNT3: Troponin T fast skeletal muscle isoform x31; FHL-2: Four and a half LIM domains protein 1 isoform 2; MYL1: Myosin light chain 1/3 skeletal muscle isoform; TNNI2: Troponin I fast skeletal muscle; CRYAB: Alpha-crystallin B chain; TNNC1: Troponin C. The higher and significant correlation coefficients are in bold.

Finally, strong positive correlations were found between some structural proteins and color, such as between MYLPF with L^* and h^* at 2 h *post-mortem* (0.7, $p < 0.01$) and between TNNI2 and L^* and h^* at 3 days post-mortem (0.75, $p < 0.01$), while negative correlations were found between a^* and DES at 2 h post-mortem (-0.683, $p < 0.05$) and MYLPF at 3 days post-mortem (-0.707, $p < 0.01$). Hughes et al. [76] found that meat color was not determined only by chromatic heme pigments, but also by the physical structure and achromatic light scattering properties of the muscle. Therefore, the effect of structural proteins in meat color is related to their denaturation and degradation during the *post-mortem* process that affect the protein density along the sarcomere, and therefore light scattering from the structural elements as evidenced in the recent integromics proteomics meta-analyses of beef color and dark-cutting beef by Gagaoua et al. [22,47]. The current insights as revealed by both proteomics and conventional biochemical studies were further recently discussed by Purslow et al. [77]. The authors stated that it is increasingly likely that omics techniques, including proteomics, will be used to discover more of the complex interactions between pathways behind the qualities of meat and their determination.

Regarding tenderness, a total of 18 bands showed significant correlations with WBSF and/or with meat tenderization rate (% of WBSF decrease from 3 days to 14 days (Table 5). ACTA1, TNNT3, MYL1, ALDOA, and CRYAB were correlated with tenderness. These proteins showed the highest correlations with WBSF during the first 24 h *post-mortem*, being negative in the case of ALDOA, mainly with WBSF 3 days and ALDOA measured at 24 h *post-mortem* (-0.549, $p < 0.05$), and positive in the case of TNNT3 and WBSF at 14 days *post-mortem* (0.625, $p < 0.01$). Thus, these findings support the previous knowledge of the importance of these proteins (Figure 3C) as early biomarkers of meat tenderization. In agreement with these results, ALDOA was positively associated with tenderness [18]. Finally, and with respect to tenderness, CRYAB, showed a positive correlation at 2 h *post-mortem* with the tenderization rate from 3 to 14 days *post-mortem* (0.592, $p < 0.01$). Increased CRYAB levels were associated with delayed myofibril degradation in beef with ultimate pH < 5.7 [61].

With the aim to summarize the complex relationships between meat quality traits and the myofibrillar subproteome, a PCA was performed including only the variables with higher correlation loadings (over 50% of explained variance). Figure 4 shows the biplot obtained by PCA between variables (loadings), individual meat samples (scores), and treatments (centroids). The first PC1 and PC2 explained 62% of the variability. The PC1 separated in the positive side the meat samples from the Extensive treatment, with higher meat WBSF values and overexpression of MYL1 at 2, 8, and 24 h, MYLPF at 24 h, LDB3 at 3 days, FHL2 at 7 and 14 days, TNNT3 (29 kDa) at 8 h, TNNT3 (27 kDa) at 3 days, TNNI2 at 24 h and CRYAB at 2 h. The meat samples from the Intensive treatment were located in the negative side, showing higher b^* values, h^* and C^* and ACTA1 at 8 h *post-mortem*.

Multivariate analysis showed that the handling factors (F and TL) had a clear effect in the different variables analyzed. Thus, animal's farm management produced a clear separation between meat samples from animals reared under Intensive or Extensive systems. On the other hand, the effect of Mixing animals during Transport and Lairage was significant in the intensively reared animals with a clear separation in two groups: (1) meat of the I-NM animals with lower WBSF values and higher intensity of the ACTA1 band at 8 h *post-mortem*; and (2) meat of the I-M animals with higher b^*, h^*, and C^*. Our previous studies pointed out that these I-M animals suffered from higher PSS as they showed higher serum haptoglobin and glucose levels at slaughter and lower muscle ATP levels, thus resulting in the blockage of the muscle antioxidant defense and slower *post-mortem* autophagic rate [9,18]. In fact, response to oxidative stress was found as a significant enriched term in this study (Figure 3B), which need further studies in the future to better elucidate the underlying mechanisms about its role in relation to the factors we investigated in this study.

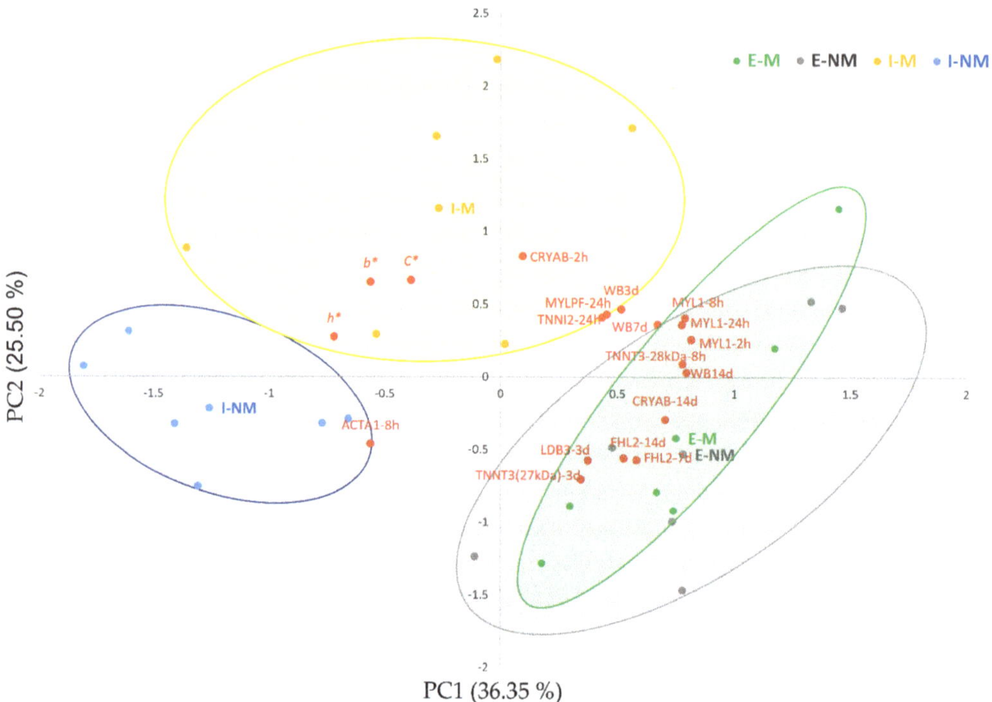

PC2 (25.50 %)

PC1 (36.35 %)

Figure 4. Biplot of variables and individuals (meat samples). The centroids of the animal treatments are shown in squares denoted with codes: I-M (Intensive-Mixed), I-NM (Intensive-Non-Mixed), E-M (Extensive Mixed), E-NM (Extensive-Non-Mixed). Individual samples are shown in yellow bullets (I-M), blue bullets (I-NM), green bullets (E-M) and grey bullets (E-NM). WB3 d: Warner–Bratzler Shear Force at 3 days, WB7 d: Warner–Bratzler Shear Force at 7 days, WB14 d: Warner–Bratzler Shear Force at 14 days; b^*: yellowness; C^*: Chroma; h^*: hue angle; CRYAB: Alpha-crystallin B chain; MYL1: Myosin light chain 1/3 skeletal muscle isoform; MYLPF: Myosin regulatory light chain 2, skeletal muscle isoform; TNNT3: Troponin T, fast skeletal muscle isoform X31; TNNI2: Troponin I, fast skeletal muscle; LDB3: LIM domain-binding protein 3 isoform X5; FHL2: Four and a half LIM domains protein 1 isoform 2; ACTA1: Actin, alpha skeletal muscle. 2 h: 2 h *post-mortem*, 8 h: 8 h post-mortem; 24 h: 24 h *post-mortem*; 3 d: 3 days *post-mortem*; 7 d: 7 days *post-mortem*.

Within the extensively reared animals, the effect of social mixing was not very clear as mixed and non-mixed animals were overlapped in the positive side of the PC1, indicating a lower effect of mixing unfamiliar animals, as previously described [9]. It is difficult to determine if these differences between treatments are due to differences in diet, physical activity or higher PSS derived from the different animal's handling treatments, but there are clear differences and the changes they produce in the myofibrillar subproteome at early *post-mortem* could provide putative biomarkers of the final meat quality.

4. Conclusions

The findings of this study confirmed that meat quality of young "Asturiana de los Valles" bulls is affected by handling practices at Farm and during Transport and Lairage before slaughter and during the tenderization of the meat. At the farm level, the production system (Intensive vs. Extensive) significantly affected meat color parameters (L^*, b^*, C^*, and h^*), highlighting that the meat from the Extensive treatment was brownish and darker. The Transport and Lairage factor (Mixed vs. Non-Mixed) affected also color traits mainly redness (a^*) and Chroma (C^*) leading to lower values in non-mixed animals. The tenderization rate of the meats of the investigated groups was higher but delayed in the meat samples from I-M animals. The comparative proteomics extended our knowledge and revealed that Farm, Transport/Lairage and *post-mortem* ageing has huge but different effects on the *post-mortem* muscle myofibrillar subproteome. The major pathway that was impacted by this factor was related to muscle structure. In fact, farm management affected six structural proteins (MYBPC2, TNNT3, MYL1, ACTA1, LDB3, and FHL2) and one metabolic enzyme (ALDOA) while Transport and Lairage prior to slaughter induced changes in five structural protein bands (MYBPC2, TNNT3, TNNI2, MYL1, and MYLPF), two metabolic enzymes (PKM and ALDOA), and one Heat shock protein (HSPB1).

Post-mortem ageing was the most important factor affecting the myofibrillar proteome, among the different handling practices confirming the importance of monitoring subproteome changes along meat ageing for an accurate understanding of the effects. Several correlations were found between the protein changing in this trial at early *post-mortem* times with meat color and tenderness parameters (PKM, ALDOA, HSPA1A, HSPB1, CRYAB, DES, TNNT3, TNNI2, and MYLPF), confirming that they could be used as meat quality biomarkers in comparison to the largest and recent beef tenderness biomarkers database recently published by Gagaoua et al. [21].

Author Contributions: Conceptualization: M.O., V.S., M.G. and A.C.-M.; methodology: L.G.-B., F.D., M.O. and V.S.; investigation: L.G.-B., F.D., M.J.G.-E. and Y.D.; formal analysis: V.S. and M.G.; data curation: L.G.-B.; writing—original draft preparation: V.S.; writing—review and editing: M.O., M.G. and A.C.-M. visualization. L.G.-B. and M.G.; supervision. M.O.; funding acquisition V.S. and M.O. All authors have read and agreed to the published version of the manuscript.

Funding: This research was funded by Instituto Nacional de Investigación y Tecnología Agraria y Alimentaria (INIA) and FEDER funds under the project RTA2014-00034-C04-01, and grant RTI2018-096162-RC21 funded by MCIN/AEI/10.13039/501100011033 and by "ERDF a way making Europe". The APC was funded by RTI2018-096162-RC21. F.D. acknowledges the contract PEJ-2014-P-01094 awarded by MINECO (Spain). L.G.-B acknowledges her grant PRE2019-091053 funded by MCIN/AEI/10.13039/501100011033 and by ESF "investing in your future".

Institutional Review Board Statement: The consideration of the ethical and welfare aspects by the Animal Care & Ethics Committee (ACEC) was not required for the development of the current study, since the animals were subjected to standard production practices during growing and finishing phases in compliance with the Spanish Law 32/2007 for the care of animals during their exploitation, transport, experimentation, and sacrifice. No additional measures were required.

Informed Consent Statement: Not applicable.

Data Availability Statement: Data available on request.

Acknowledgments: We thank the staff of the Area of Livestock Production Systems from SERIDA and the staff from ASEAVA for skilled management of Animals. Mohammed Gagaoua acknowledges the funding support of the Marie Sklodowska-Curie grant agreement No. 713654 under the project number MF20180029.

Conflicts of Interest: The authors declare no conflict of interest. The funders had no role in the design of the study; in the collection, analyses, or interpretation of data; in the writing of the manuscript or in the decision to publish the results.

References

1. Picard, B.; Gagaoua, M.; Micol, D.; Cassar-Malek, I.; Hocquette, J.F.; Terlouw, C.E. Inverse relationships between biomarkers and beef tenderness according to contractile and metabolic properties of the muscle. *J. Agric. Food Chem.* **2014**, *62*, 9808–9818. [CrossRef] [PubMed]
2. Gagaoua, M.; Monteils, V.; Picard, B. Data from the farmgate-to-meat continuum including omics-based biomarkers to better understand the variability of beef tenderness: An integromics approach. *J. Agric. Food Chem.* **2018**, *66*, 13552–13563. [CrossRef]
3. López-Pedrouso, M.; Rodríguez-Vázquez, R.; Purriños, L.; Oliván, M.; García-Torres, S.; Sentandreu, M.Á.; Lorenzo, J.M.; Zapata, C.; Franco, D. Sensory and physicochemical analysis of meat from bovine breeds in different livestock production systems, pre-slaughter handling conditions, and ageing time. *Foods* **2020**, *9*, 176. [CrossRef]
4. Gagaoua, M.; Picard, B.; Soulat, J.; Monteils, V. Clustering of sensory eating qualities of beef: Consistencies and differences within carcass, muscle, animal characteristics and rearing factors. *Livest. Sci.* **2018**, *214*, 245–258. [CrossRef]

5. Priolo, A.; Micol, D.; Agabriel, J. Effects of grass feeding systems on ruminant meat colour and flavour. A review. *Anim. Res.* **2001**, *50*, 185–200. [CrossRef]

6. Andersen, H.J.; Oksbjerg, N.; Young, J.F.; Therkildsen, M. Feeding and meat quality—A future approach. *Meat Sci.* **2005**, *70*, 543–554. [CrossRef]

7. Bouissou, M.F.; Boissy, A.; le Neindre, P.; Veissier, I. The social behaviour of cattle. *Soc. Behav. Farm Anim.* **2001**, 113–145. [CrossRef]

8. Marco-Ramell, A.; Arroyo, L.; Saco, Y.; García-Heredia, A.; Camps, J.; Fina, M.; Piedrafita, J.; Bassols, A. Proteomic analysis reveals oxidative stress response as the main adaptive physiological mechanism in cows under different production systems. *J. Proteom.* **2012**, *75*, 4399–4411. [CrossRef]

9. Díaz, F.; Díaz-Luis, A.; Sierra, V.; Diñeiro, Y.; González, P.; García-Torres, S.; Oliván, M. What functional proteomic and biochemical analysis tell us about animal stress in beef? *J. Proteom.* **2020**, *218*, 103722. [CrossRef] [PubMed]

10. Terlouw, E.M.C.; Picard, B.; Deiss, V.; Berri, C.; Hocquette, J.F.; Lebret, B.; Lefèvre, F.; Hamill, R.; Gagaoua, M. Understanding the determination of meat quality using biochemical characteristics of the muscle: Stress at slaughter and other missing keys. *Foods* **2021**, *10*, 84. [CrossRef] [PubMed]

11. Bøe, K.E.; Færevik, G. Grouping and social preferences in calves, heifers and cows. *Appl. Anim. Behav. Sci.* **2003**, *80*, 175–190. [CrossRef]

12. Disanto, C.; Celano, G.; Varvara, M.; Fusiello, N.; Fransvea, A.; Bozzo, G.; Celano, G.V. Stress factors during cattle slaughter. *Ital. J. Food Saf.* **2014**, *3*, 143–144. [CrossRef]

13. Rubio-González, A.; Potes, Y.; Illán-Rodríguez, D.; Vega-Naredo, I.; Sierra, V.; Caballero, B.; Fàbrega, E.; Velarde, A.; Dalmau, A.; Oliván, M.; et al. Effect of animal mixing as a stressor on biomarkers of autophagy and oxidative stress during pig muscle maturation. *Animal* **2015**, *9*, 1188–1194. [CrossRef] [PubMed]

14. Potes, Y.; Oliván, M.; Rubio-González, A.; de Luxán-Delgado, B.; Díaz, F.; Sierra, V.; Arroyo, L.; Peña, R.; Bassols, A.; González, J.; et al. Pig cognitive bias affects the conversion of muscle into meat by antioxidant and autophagy mechanisms. *Animal* **2017**, *11*, 2027–2035. [CrossRef] [PubMed]

15. Franco, D.; Mato, A.; Salgado, F.J.; López-Pedrouso, M.; Carrera, M.; Bravo, S.; Parrado, M.; Gallardo, J.M.; Zapata, C. Tackling proteome changes in the Longissimus thoracis bovine muscle in response to pre-slaughter stress. *J. Proteomics* **2015**, *122*, 73–85. [CrossRef] [PubMed]

16. Díaz-Luis, A.; Díaz, F.; Diñeiro, Y.; González-Blanco, L.; Arias, E.; Coto-Montes, A.; Oliván, M.; Sierra, V. Nuevos indicadores de carnes (DFD): Estrés oxidativo, autofagia y apoptosis. *ITEA* **2020**, *117*, 3–18. [CrossRef]

17. Fuente-García, C.; Aldai, N.; Sentandreu, E.; Oliván, M.; Franco, D.; García-Torres, S.; Barron, L.J.R.; Sentandreu, M.Á. Caspase activity in post-mortem muscle and its relation to cattle handling practices. *J. Sci. Food Agric.* **2021**, *101*, 6258–6264. [CrossRef]

18. García-Torres, S.; Cabeza de Vaca, M.; Tejerina, D.; Romero-Fernández, M.P.; Ortiz, A.; Franco, D.; Sentandreu, M.Á.; Oliván, M. Assessment of stress by serum biomarkers in calves and their relationship to ultimate pH as an indicator of meat quality. *Animals* **2021**, *11*, 2291. [CrossRef] [PubMed]

19. Ouali, A.; Gagaoua, M.; Boudida, Y.; Becila, S.; Boudjellal, A.; Herrera-Mendez, C.; Sentandreu, M.Á. Biomarkers of meat tenderness: Present knowledge and perspectives in regard to our current understanding of the mechanisms involved. *Meat Sci.* **2013**, *95*, 854–870. [CrossRef]

20. Picard, B.; Gagaoua, M. Chapter 11-proteomic investigations of beef tenderness. In *Proteomics in Food Science: From Farm to Fork*; Colgrave, M.L., Ed.; Academic Press: Cambridge, MA, USA, 2017; pp. 177–197. [CrossRef]

21. Gagaoua, M.; Terlouw, E.M.C.; Mullen, A.M.; Franco, D.; Warner, R.D.; Lorenzo, J.M.; Purslow, P.P.; Gerrard, D.; Hopkins, D.L.; Troy, D.; et al. Molecular signatures of beef tenderness: Underlying mechanisms based on integromics of protein biomarkers from multi-platform proteomics studies. *Meat Sci.* **2021**, *172*, 108311. [CrossRef] [PubMed]

22. Gagaoua, M.; Warner, R.B.; Purslow, P.; Ramanathan, R.; Mullen, A.M.; López-Pedrouso, M.; Franco, D.; Lorenzo, J.M.; Tomasevic, I.; Picard, B.; et al. Dark-cutting beef: A brief review and an integromics meta-analysis at the proteome level to decipher the underlying pathways. *Meat Sci.* **2021**, *181*, 108611. [CrossRef]

23. Warner, R.; Wheeler, T.L.; Ha, M.; Li, X.; Bekhit, A.E.D.; Morton, J.; Vaskoska, R.; Dunshea, F.; Liu, R.; Purslow, P.; et al. Meat tenderness: Advances in biology, biochemistry, molecular mechanisms and new technologies. *Meat Sci.* **2021**, 108657. [CrossRef]

24. Zapata, I.; Zerby, H.N.; Wick, M. Functional proteomic analysis predicts beef tenderness and the tenderness differential. *J. Agric. Food Chem.* **2009**, *57*, 4956–4963. [CrossRef] [PubMed]

25. Sierra, V.; Fernández-Suárez, V.; Castro, P.; Osoro, K.; Vega-Naredo, I.; García-Macía, M.; Rodríguez Colunga, P.; Coto-Montes, A.; Oliván, M. Identification of biomarkers of meat tenderisation and its use for early classification of Asturian beef into fast and late tenderising meat. *J. Sci. Food Agric.* **2012**, *92*, 2727–2740. [CrossRef]

26. Fuente-Garcia, C.; Sentandreu, E.; Aldai, N.; Oliván, M.; Sentandreu, M.Á. Characterization of the myofibrillar proteome as a way to better understand differences in bovine meats having different ultimate pH values. *J. Proteomics* **2020**, *20*, 2000012. [CrossRef]

27. Gagaoua, M. The path from protein profiling to biomarkers: The potential of proteomics and data integration in beef quality research. *IOP Conf. Ser. Earth Environ. Sci.* **2021**, *854*, 012029. [CrossRef]

28. Piedrafita, J.; Quintanilla, R.; Sañudo, C.; Olleta, J.L.; Campo, M.M.; Panea, B.; Renand, G.; Turin, F.; Jabet, S.; Osoro, K.; et al. Carcass quality of ten beef cattle breeds of the south-west of europe. *Livest. Prod. Sci.* **2003**, *82*, 1–13. [CrossRef]

29. Sierra, V.; Guerrero, L.; Fernández-Suárez, V.; Martínez, A.; Castro, P.; Osoro, K.; Rodríguez-Colunga, M.J.; Coto-Montes, A.; Oliván, M. Eating quality of beef from biotypes included in the PGI "Ternera Asturiana" showing distinct physicochemical characteristics and tenderization pattern. *Meat Sci.* **2010**, *86*, 343–351. [CrossRef]

30. Mapama, I.G.P. Ternera Asturiana. Available online: https://www.mapa.gob.es/es/alimentacion/temas/calidad-diferenciada/dop-igp/carnes/IGP_Ternera_Asturiana.aspx (accessed on 7 August 2021).

31. AMSA (American Meat Science Association). Meat Color Measurement Guidelines, Champaign, IL, USA, 2017. Available online: http://www.meatscience.org (accessed on 10 October 2021).

32. Bjarnadóttir, S.G.; Hollung, K.; Faergestad, E.M.; Veiseth-Kent, E. Proteome changes in bovine *Longissimus thoracis* muscle during the first 48 h postmortem: Shifts in energy status and myofibrillar stability. *J. Agric. Food Chem.* **2010**, *58*, 7408–7414. [CrossRef]

33. Bradford, M.M. A rapid and sensitive method for the quantitation of microgram quantities of protein utilizing the principle of protein-dye binding. *Anal. Biochem.* **1976**, *72*, 248–254. [CrossRef]

34. Gagaoua, M.; Troy, D.; Mullen, A.M. The extent and rate of the appearance of the major 110 and 30 kDa proteolytic fragments during post-mortem aging of beef depend on the glycolysing rate of the muscle and aging time: An LC–MS/MS approach to decipher their proteome and associated pathways. *J. Agric. Food Chem.* **2021**, *69*, 602–614. [CrossRef]

35. Vestergaard, M.; Oksbjerg, N.; Henckel, P. Influence of feeding intensity, grazing and finishing feeding on muscle fibre characteristics and meat colour of semitendinosus, longissimus dorsi and supraspinatus muscles of young bulls. *Meat Sci.* **2000**, *54*, 177–185. [CrossRef]

36. Realini, C.E.; Duckett, S.K.; Brito, G.W.; dalla Rizza, M.; de Mattos, D. Effect of pasture vs. concentrate feeding with or without antioxidants on carcass characteristics, fatty acid composition, and quality of Uruguayan beef. *Meat Sci.* **2004**, *66*, 567–577. [CrossRef]

37. Gagaoua, M.; Monteils, V.; Couvreur, S.; Picard, B. Identification of biomarkers associated with the rearing practices, carcass characteristics, and beef quality: An integrative approach. *J. Agric. Food Chem.* **2017**, *65*, 8264–8278. [CrossRef]

38. Vestergaard, M.; Therkildsen, M.; Henckel, P.; Jensen, L.R. Influence of feeding intensity, grazing and finishing feeding on meat and eating quality of young bulls and the relationship between muscle fibre characteristics, fibre fragmentation and meat tenderness. *Meat Sci.* **2000**, *54*, 187–195. [CrossRef]

39. Nuernberg, K.; Dannenberger, D.; Nuernberg, G.; Ender, K.; Voigt, J.; Scollan, N.D.; Wood, J.D.; Nute, G.R.; Richardson, R.I. Effect of a grass-based and a concentrate feeding system on meat quality characteristics and fatty acid composition of *Longissimus* muscle in different cattle breeds. *Livest. Prod. Sci.* **2005**, *94*, 137–147. [CrossRef]

40. Laville, E.; Sayd, T.; Morzel, M.; Blinet, S.; Chambon, C.; Lepetit, J.; Renand, G.; Hocquette, J.F. Proteome changes during meat aging in tough and tender beef suggest the importance of apoptosis and protein solubility for beef aging and tenderization. *J. Agric. Food Chem.* **2009**, *57*, 10755–10764. [CrossRef]

41. Paulsen, G.; Vissing, K.; Kalhovde, J.M.; Ugelstad, I.; Bayer, M.L.; Kadi, F.; Schjerling, P.; Hallén, J.; Raastad, T. Maximal eccentric exercise induces a rapid accumulation of small heat shock proteins on myofibrils and a delayed HSP70 response in humans. *Am. J. Physiol. Regul. Integr. Comp. Physiol.* **2007**, *293*, 844–853. [CrossRef] [PubMed]

42. González-Blanco, L.; Diñeiro, Y.; Díaz-Luis, A.; Coto-Montes, A.; Oliván, M.; Sierra, V. Impact of extraction method on the detection of quality biomarkers in normal vs. DFD meat. *Foods* **2021**, *10*, 1097. [CrossRef]

43. Ho, C.Y.; Stromer, M.H.; Robson, R.M. Identification of the 30 kDa polypeptide in post mortem skeletal muscle as a degradation product of troponin-T. *Biochimie* **1994**, *76*, 369–375. [CrossRef]

44. Kemp, C.M.; Sensky, P.L.; Bardsley, R.G.; Buttery, P.J.; Parr, T. Tenderness—An enzymatic view. *Meat Sci.* **2010**, *84*, 248–256. [CrossRef]

45. Picard, B.; Gagaoua, M. Meta-proteomics for the discovery of protein biomarkers of beef tenderness: An overview of integrated studies. *Food Res. Int.* **2020**, *127*, 108739. [CrossRef] [PubMed]

46. Flashman, E.; Redwood, C.; Moolman-Smook, J.; Watkins, H. Cardiac myosin binding protein C: Its role in physiology and disease. *Circ. Res.* **2004**, *94*, 1279–1289. [CrossRef] [PubMed]

47. Guillemin, N.; Bonnet, M.; Jurie, C.; Picard, B. Functional analysis of beef tenderness. *J. Proteomics* **2011**, *75*, 352–365. [CrossRef]

48. Gagaoua, M.; Hughes, J.; Terlouw, E.M.C.; Warner, R.D.; Purslow, P.P.; Lorenzo, J.M.; Picard, B. Proteomic biomarkers of beef colour. *Trends Food Sci. Technol.* **2020**, *101*, 234–252. [CrossRef]

49. Wu, G.; Clerens, S.; Farouk, M.M. LC MS/MS identification of large structural proteins from bull muscle and their degradation products during post mortem storage. *Food Chem.* **2014**, *150*, 137–144. [CrossRef]

50. Huff-Lonergan, E.; Zhang, W.; Lonergan, S.M. Biochemistry of postmortem muscle—Lessons on mechanisms of meat tenderization. *Meat Sci.* **2010**, *86*, 184–195. [CrossRef] [PubMed]

51. Jia, X.; Hollung, K.; Therkildsen, M.; Hildrum, K.I.; Bendixen, E. Proteome analysis of early post-mortem changes in two bovine muscle types: *M. longissimus dorsi* and *M. semitendinosis*. *J. Proteom.* **2006**, *6*, 936–944. [CrossRef]

52. Sawdy, J.C.; Kaiser, S.A.; St-Pierre, N.R.; Wick, M.P. Myofibrillar 1-D fingerprints and myosin heavy chain MS analyses of beef loin at 36 h postmortem correlate with tenderness at 7 days. *Meat Sci.* **2004**, *67*, 421–426. [CrossRef]

53. Morzel, M.; Terlouw, C.; Chambon, C.; Micol, D.; Picard, B. Muscle proteome and meat eating qualities of *Longissimus thoracis* of "blonde d'Aquitaine" young bulls: A central role of HSP27 isoforms. *Meat Sci.* **2008**, *78*, 297–304. [CrossRef]

54. Zhou, Q.; Chu, P.H.; Huang, C.; Cheng, C.F.; Martone, M.E.; Knoll, G.; Shelton, G.D.; Evans, S.; Chen, J. Ablation of Cypher, a PDZ-LIM domain Z-line protein, causes a severe form of congenital myopathy. *J. Cell Biol.* **2001**, *155*, 605–612. [CrossRef]

55. Shathasivam, T.; Kislinger, T.; Gramolini, A.O. Genes, proteins and complexes: The multifaceted nature of FHL family proteins in diverse tissues. *J. Cell Mol. Med.* **2021**, *14*, 2702–2720. [CrossRef] [PubMed]

56. Morzel, M.; Chambon, C.; Hamelin, M.; Sante-Lhoutellier, V.; Sayd, T.; Monin, G. Proteome changes during pork meat ageing following use of two different pre-slaughter handling procedures. *Meat Sci.* **2004**, *67*, 689–696. [CrossRef]

57. Gagaoua, M.; Bonnet, M.; Ellies-Oury, M.P.; de Koning, L.; Picard, B. Reverse phase protein arrays for the identification/validation of biomarkers of beef texture and their use for early classification of carcasses. *Food Chem.* **2018**, *250*, 245–252. [CrossRef]

58. Hughes, J.; Clarke, F.; Li, Y.; Purslow, P.; Warner, R. Differences in light scattering between pale and dark beef *Longissimus thoracis* muscles are primarily caused by differences in the myofilament lattice, myofibril and muscle fibre transverse spacings. *Meat Sci.* **2019**, *149*, 96–106. [CrossRef] [PubMed]

59. Oliván, M.; Fernández-Suárez, V.; Díaz-Martínez, F.; Sierra, V.; Coto-Montes, A.; Luxán-Delgado, B.; Peña, R.; Bassols, A.; Fàbrega, E.; Dalmau, A.; et al. Identification of biomarkers of stress in meat of pigs managed under different mixing treatments. *Br. Biotechnol. J.* **2016**, *11*, 1–13. [CrossRef]

60. Yu, J.; Tang, S.; Bao, E.; Zhang, M.; Hao, Q.; Yue, Z. The effect of transportation on the expression of heat shock proteins and meat quality of *M. longissimus* dorsi in pigs. *Meat Sci.* **2009**, *83*, 474–478. [CrossRef]

61. Ma, D.; Kim, Y.H.B. Proteolytic changes of myofibrillar and small heat shock proteins in different bovine muscles during aging: Their relevance to tenderness and Water-holding capacity. *Meat Sci.* **2020**, *163*, 108090. [CrossRef]

62. Lomiwes, D.; Hurst, S.M.; Dobbie, P.; Frost, D.A.; Hurst, R.D.; Young, O.A.; Farouk, M.M. The protection of bovine skeletal myofibrils from proteolytic damage post-mortem by small heat shock proteins. *Meat Sci.* **2014**, *97*, 548–557. [CrossRef]

63. Balan, P.; Kim, Y.H.; Blijenburg, R. Small heat shock protein degradation could be an indicator of the extent of myofibrillar protein degradation. *Meat Sci.* **2014**, *97*, 220–222. [CrossRef] [PubMed]

64. Cramer, T.; Penick, M.L.; Waddell, J.N.; Bidwell, C.A.; Kim, Y.H.B. A new insight into meat toughness of callipyge lamb loins—The relevance of anti-apoptotic systems to decreased proteolysis. *Meat Sci.* **2018**, *140*, 66–71. [CrossRef] [PubMed]

65. Jia, X.; Hildrum, K.I.; Westad, F.; Kummen, E.; Aass, L.; Hollung, K. Changes in enzymes associated with energy metabolism during the early post-mortem period in *Longissimus thoracis* bovine muscle analyzed by proteomics. *J. Proteome Res.* **2006**, *5*, 1763–1769. [CrossRef]

66. Jia, X.H.; Ekman, M.; Grove, H.; Faergestad, E.M.; Aass, L.; Hildrum, K.I.; Hollung, K. Proteome changes in bovine *Longissimus thoracis* muscle during the early post-mortem storage period. *J. Proteome Res.* **2007**, *6*, 2720–2731. [CrossRef] [PubMed]

67. Canto, A.C.; Suman, S.P.; Nair, M.N.; Li, S.; Rentfrow, G.; Beach, C.M.; Silva, T.J.; Wheeler, T.L.; Shackelford, S.D.; Grayson, A.; et al. Differential abundance of sarcoplasmic proteome explains animal effect on beef *Longissimus lumborum* color stability. *Meat Sci.* **2015**, *102*, 90–98. [CrossRef]

68. Wu, W.; Gao, X.G.; Dai, Y.; Fu, Y.; Li, X.M.; Dai, R.T. Post-mortem changes in sarcoplasmic proteome and its relationship to meat color traits in *M. semitendinosus* of Chinese Luxi yellow cattle. *Food Res. Int.* **2015**, *72*, 98–105. [CrossRef]

69. Wu, W.; Yu, Q.Q.; Fu, Y.; Tian, X.J.; Jia, F.; Li, X.M.; Dai, R.T. Towards muscle specific meat color stability of Chinese Luxi yellow cattle: A proteomic insight into post-mortem storage. *J. Proteomics* **2016**, *147*, 108–118. [CrossRef]

70. Hughes, J.; Clarke, F.; Purslow, P.; Warner, R. High pH in beef longissimus thoracis reduces muscle fibre transverse shrinkage and light scattering which contributes to the dark colour. *Food Res. Int.* **2017**, *101*, 228–238. [CrossRef] [PubMed]

71. Yu, Q.; Wu, W.; Tian, X.; Jia, F.; Xu, L.; Dai, R.; Li, X. Comparative proteomics to reveal muscle-specific beef color stability of Holstein cattle during post-mortem storage. *Food Chem.* **2017**, *229*, 769–778. [CrossRef] [PubMed]

72. Zhang, Y.M.; Zhang, X.Z.; Wang, T.T.; Hopkins, D.L.; Mao, Y.W.; Liang, R.R.; Yang, G.F.; Luo, X.; Zhu, L.X. Implications of step-chilling on meat color investigated using proteome analysis of the sarcoplasmic protein fraction of beef *Longissimus lumborum* muscle. *J. Integr. Agric.* **2018**, *17*, 2118–2125. [CrossRef]

73. Mahmood, S.; Turchinsky, N.; Paradis, F.; Dixon, W.T.; Bruce, H.L. Proteomics of dark cutting *Longissimus thoracis* muscle from heifer and steer carcasses. *Meat Sci.* **2018**, *137*, 47–57. [CrossRef]

74. Gagaoua, M.; Terlouw, E.M.C.; Picard, B. The study of protein biomarkers to understand the biochemical processes underlying beef color development in young bulls. *Meat Sci.* **2017**, *134*, 18–27. [CrossRef] [PubMed]

75. Purslow, P.P.; Warner, R.D.; Clarke, F.M.; Hughes, J.M. Variations in meat colour due to factors other than myoglobin chemistry; a synthesis of recent findings (invited review). *Meat Sci.* **2020**, *159*, 107941. [CrossRef] [PubMed]

76. Hughes, J.; Oiseth, S.K.; Purslow, P.P.; Warner, R.D. A structural approach to understanding the interactions between colour, water-holding capacity and tenderness. *Meat Sci.* **2014**, *98*, 520–532. [CrossRef] [PubMed]

77. Purslow, P.P.; Gagaoua, M.; Warner, R.D. Insights on meat quality from combining traditional studies and proteomics. *Meat Science.* **2021**, *174*, 108423. [CrossRef] [PubMed]

Article

A Proteomic Study for the Discovery of Beef Tenderness Biomarkers and Prediction of Warner–Bratzler Shear Force Measured on *Longissimus thoracis* Muscles of Young Limousin-Sired Bulls

Yao Zhu [1,2,†], Mohammed Gagaoua [1,†], Anne Maria Mullen [1], Alan L. Kelly [2], Torres Sweeney [3], Jamie Cafferky [1], Didier Viala [4] and Ruth M. Hamill [1,*]

1   Food Quality and Sensory Science Department, Teagasc Food Research Centre, Ashtown, D15 KN3K Dublin 15, Ireland; yao.zhu@teagasc.ie (Y.Z.); gmber2001@yahoo.fr (M.G.); AnneMaria.mullen@teagasc.ie (A.M.M.); Jamie.Cafferky@teagasc.ie (J.C.)
2   School of Food and Nutritional Sciences, University College Cork, T12 K8AF Cork, Ireland; a.kelly@ucc.ie
3   School of Veterinary Sciences, University College Dublin, D04 V1W8 Dublin 4, Ireland; torres.sweeney@ucd.ie
4   Metabolomic and Proteomic Exploration Facility (PFEM), INRAE, F-63122 Saint-Genès-Champanelle, France; didier.viala@inrae.fr
*   Correspondence: ruth.hamill@teagasc.ie; Tel.: +353-(0)1-805-9933
†   Both authors contributed equally to this work.

Citation: Zhu, Y.; Gagaoua, M.; Mullen, A.M.; Kelly, A.L.; Sweeney, T.; Cafferky, J.; Viala, D.; Hamill, R.M. A Proteomic Study for the Discovery of Beef Tenderness Biomarkers and Prediction of Warner–Bratzler Shear Force Measured on *Longissimus thoracis* Muscles of Young Limousin-Sired Bulls. *Foods* 2021, 10, 952. https://doi.org/10.3390/foods10050952

Academic Editor: Hanne Christine Bertram

Received: 14 April 2021
Accepted: 23 April 2021
Published: 27 April 2021

Publisher's Note: MDPI stays neutral with regard to jurisdictional claims in published maps and institutional affiliations.

Abstract: Beef tenderness is of central importance in determining consumers' overall liking. To better understand the underlying mechanisms of tenderness and be able to predict it, this study aimed to apply a proteomics approach on the *Longissimus thoracis* (LT) muscle of young Limousin-sired bulls to identify candidate protein biomarkers. A total of 34 proteins showed differential abundance between the tender and tough groups. These proteins belong to biological pathways related to muscle structure, energy metabolism, heat shock proteins, response to oxidative stress, and apoptosis. Twenty-three putative protein biomarkers or their isoforms had previously been identified as beef tenderness biomarkers, while eleven were novel. Using regression analysis to predict shear force values, MYOZ3 (Myozenin 3), BIN1 (Bridging Integrator-1), and OGN (Mimecan) were the major proteins retained in the regression model, together explaining 79% of the variability. The results of this study confirmed the existing knowledge but also offered new insights enriching the previous biomarkers of tenderness proposed for *Longissimus* muscle.

Keywords: foodomics; beef tenderness; bovine biomarkers; muscle; proteome; liquid chromatography-tandem mass spectrometry (LC-MS/MS)

1. Introduction

Meat-eating quality consists of a complex set of sensory traits including tenderness, flavour, and juiciness, each of which plays an important role in defining the appeal of beef to consumers [1,2]. Amongst these quality attributes, however, tenderness is considered to be one of the most important factors in purchase decisions regarding beef, with negative experience on toughness contributing to a lower likelihood of repeat purchase [3]. To meet the expectations of consumers, beef producers must pursue the provision of consistent high-quality beef. The underlying mechanisms involved in dictating the final meat tenderness are intricate, with muscle biochemistry interacting with processing, influenced by several factors including breed [4,5], gender [6], age at slaughter [7], muscle type [8,9], cooking temperature [5], stress at slaughter [10], and post-slaughter management and many other factors from farm-to-fork [9,11].

There have been a number of studies using omics tools to, firstly, enhance our understanding of the pathways and processes contributing to beef tenderness variation [12,13] and secondly, to propose prediction equations to explain the observed variability in this important quality trait [14]. Thus, omics-related analytical technologies and bioinformatics tools have been applied in recent decades, resulting in a deeper understanding of gene expression, physiological responses, and other metabolic processes that are involved in meat quality determination, especially tenderness [2,12,15].

Foodomics is an emerging group of disciplines encompassing genomics, transcriptomics, proteomics, metabolomics, and lipidomics applied to food and parameters related to its quality and has been extensively used to study both fresh meat and meat products [16]. Among the many foodomics approaches, proteomics played an important role in the discovery of candidate biomarkers of several meat quality attributes [2,13,17]. A pipeline to search for proteomic biomarkers of beef tenderness was proposed [12,14]. Compared with traditional evaluation methods for beef tenderness using instrumental or sensory methods, an optimised protocol for quality monitoring using rapid methods to record the abundance of specific proteins of interest would offer an advantage to predict the meat quality before consumption. Moreover, these approaches have the potential to be developed further to allow advanced prediction of the future tenderness phenotype at a range of stages from farm-to-fork [11,15].

This study aimed to apply shotgun proteomics on muscle tissue of young Limousin-sired bulls to identify putative biomarkers of beef tenderness evaluated by Warner–Bratzler shear force (WBSF) [15]. We further aimed to propose regression models and identify the main biological interactions among the proteins underpinning WBSF variation to gain insights into the mechanisms of beef tenderness determination.

2. Materials and Methods

2.1. Meat Sample Collection

Eighteen young Limousin-sired bulls were obtained and finished at the Irish Cattle Breeders Federation Progeny Test Centre and slaughtered in an EU-licensed abattoir by electrical stunning (50 Hz) followed by exsanguination from the jugular vein. All 18 animals were finished to U- to E+ conformation score, 3- to 5= fat score and at an average age of 487 days ($\pm$24 days) and live weight of 678 kg ($\pm$58 kg) [6]. According to the muscle sampling method used by Zhu et al. [18], *Longissimus thoracis et lumborum* (LTL) samples from the 10th rib of each carcass were collected and finely macerated in 5 mL RNAlater® for 24 h. The RNAlater® was then removed, and the sample was subsequently transferred for storage at −80 °C until analysis. Loins were boned out at 48 h post-mortem, and steaks with a thickness of 2.54 cm were cut out from the right-side LTL of the carcass starting at the anterior end and packaged in vacuum bags. The steaks were then aged for 14 days and stored at −20 °C until Warner–Bratzler shear force (WBSF) analysis.

2.2. Warner–Bratzler Shear Force Measurement

Steaks were thawed at room temperature by immersion in a circulating water bath for 4 h. After that, external fat was trimmed from the steaks, and they were cooked in open bags in a circulating water bath (Grant Instruments Ltd., Cambridge, UK) set at 72 °C to reach an internal end-point cooking temperature of 71 °C. The cooked steaks were cooled down and stored in a refrigerator at 4 °C overnight. Shear force analysis was conducted following a modified version of the guideline of the American Meat Science Association (AMSA) [6]. For each steak, seven cores were taken with a 1.27 cm diameter parallel to the muscle fibre direction. The shear force was measured by an Instron 4464 Universal testing machine (Instron Ltd., Buckinghamshire, UK), and data analysed using Bluehill 2 Software (Instron Ltd., Buckinghamshire, UK). To reduce the standard deviation among the cores, the maximum and minimum shear values (Newton) were discarded, and the mean values of the remaining 5 cores were reported.

2.3. Muscle Protein Extraction

Frozen muscle tissue samples (80 mg) were first homogenised in 2 mL of 8.3 M urea, 2 M thiourea, 1% dithiothreitol, 2% 3-[(3-cholamidopropyl) dimethylammonio]-1-

propanesulfonate, 2% immobilised pH gradient (IPG) buffer pH 3–10 using a T 25 digital ULTRA-TURRAX® following the protocol of Bouley et al. [19]. To remove non-extracted cellular components, fat, insoluble proteins, the protein homogenates were incubated with shaking for 30 min at 4 °C followed by a 30 min centrifugation at 10,000× *g*. The supernatant was then transferred into Eppendorf tubes for protein quantification using the dye-binding protocol of Bradford [20].

2.4. Shotgun Proteomics

2.4.1. One Dimensional SDS-PAGE and Protein Bands Preparation

The protein extract was firstly mixed (1:1) with Laemmli sample buffer (Bio-Rad Laboratories, Deeside, UK), then concentrated on 1D stacking gel of sodium dodecyl sulphate-polyacrylamide gel electrophoresis (SDS-PAGE) using commercial Mini-PROTEAN® TGX™ precast gels of 8.6 × 6.7 × 0.1 cm and 12% polyacrylamide (Bio-Rad Laboratories, Deeside, UK). Twenty µg proteins were loaded in each gel lane, and the electrophoresis was run at 4 watts for about 15 min to concentrate the proteins in the stacking gel [21]. Subsequently, the gels were washed three times with Milli-Q water, stained with EZ Blue Gel staining reagent (Sigma, Saint Louis, MO, USA) with gentle shaking for 2 h, and then washed with Milli-Q water. The protein bands were excised from the washed gels using a sterile scalpel and immediately transferred into Eppendorf tubes containing 200 µL of 25 mM ammonium bicarbonate (Sigma, Saint Louis, MO, USA)-5% acetonitrile for 30 min. Then, bands were washed twice using 200 µL of 25 mM ammonium bicarbonate-50% acetonitrile for 30 min each. Finally, they were dehydrated with 100% acetonitrile for 10 min, and the liquid was discarded. Subsequently, the dried protein bands were stored at −80 °C until LC-MS/MS analysis. The immobilised proteins in the 1D gel bands were discoloured/reduced-alkylated, as described by Gagaoua et al. [22].

2.4.2. LC-MS/MS

The hydrolysis of the protein bands was carried out with 48 µL of a 25 mM ammonium bicarbonate buffer-12.5 ng/µL trypsin solution (Promega) per band for 5 h in an oven at 37 °C. Then, 30 µL buffer was added periodically during hydrolysis so that the bands were always covered with liquid. The extraction of the peptides was carried out under ultrasound (15 min) with acetonitrile and trifluoroacetic acid. Then, the supernatant was transferred into 500 µL Eppendorf tubes and dry concentrated using a Speedvac for 2 h. The volume was adjusted exactly to 20 µL with a solution of isotopologic peptides (50 pmol/µL) that was diluted 18 times in a 0.05% Trifluoroacetic acid (TFA) solution. After passing through the ultrasonic bath (10 min), the entire supernatant was transferred to the High performance liquid chromatography (HPLC) vial before LC-MS/MS analysis.

For the separation, the hydrolysate was injected into the nano-LC-MS/MS (Thermo Fisher Scientific) using an Ultimate 3000 system coupled to a QExactive HF-X mass spectrometer (MS) with a nanoelectrospray ion source. Briefly, 1 µL of hydrolysate was first preconcentrated and desalted at a flow rate of 30 µL/min on a C18 pre-column 5 cm length × 100 µm (Acclaim PepMap 100 C18, 5 µm, 100 Å nanoViper) equilibrated with trifluoroacetic acid 0.05% in water to remove contaminants that could potentially disrupt the efficiency of the mass spectrometry analysis. After 6 min, the concentration column was put in line with a nano debit analytical column operating at 400 nL/min. The peptides were then separated according to their hydrophobicity (column C18, length 25 cm, diameter 75 µm, SN 10711310), using a gradient of a solution of acetonitrile (ACN/FA-99.9/0.1) of 4 to 25% in 50 min.

2.4.3. LC-MS/MS Data Processing and Protein Identification

The raw files from the LC-MS/MS were aligned against the *Bos taurus* database (i.e., ref_bos_taurus, 23,970 sequences) with Mascot V.2.5.1 (http://www.matrixscience.com, accessed on 30 August 2020). The precursor and fragment mass tolerance were set up at 10 ppm and 0.02 Da, respectively. The variable modifications included carbamidomethylation (C), oxidation (M), and deamidation (NQ). Protein identification could be verified when at least two peptides derived from one protein showed statistically significant identity. The Mascot score was 33 with a False Discovery Rate of 1%, and the p-value was adjusted at a given threshold (0.0093).

2.5. Bioinformatics Analyses

2.5.1. Protein-Protein Interactions (PPI)

The protein-protein interactions between the putative protein biomarkers were analysed using the STRING web service database (https://string-db.org/, accessed on 28 November 2020). Default settings were used, i.e., medium confidence of 0.4 and 4 criteria for linkage: co-occurrence, experimental evidence, existing databases, and text mining. As the bovine Gene Ontology (GO) had limits, orthologous human Uniprot IDs, following the procedure by Gagaoua et al. [14], were used for this analysis to take advantage of the most complete annotations available.

2.5.2. Gene Ontology and Pathway and Process Enrichment Analyses

The pathway and Gene Ontology analyses were performed using two web-based tools. First, ProteINSIDE (http://www.proteinside.org/, accessed on 28 November 2020) was used to investigate GO terms for potential functions and molecular mechanisms [23]. For this analysis, the top 20 GO enrichment terms (p-value, Benjamini–Hochberg < 0.05) were considered and covered Biological Process (BP), Molecular Function (MF), and Cellular Component (CC) categories. The Metascape® (https://metascape.org/, accessed on 28 November 2020) web service tool was further used to investigate the pathway and process enrichment analyses using the list of 34 differential proteins. The statistically significant enriched ontology terms were displayed based on the hypergeometric test and Benjamini–Hochberg p-value correction algorithm [24].

2.6. Statistical Analyses

Statistical analyses of protein abundance were performed with XLSTAT 2018.2 (AddinSoft, Paris, France), as well as the online tools NormalyzerDE and MetaOmGraph, mainly for data standardisation. Raw data were scrutinised for data entry errors, any missing data, or outliers. Log2 transformation and mean normalisation were performed on protein abundance among replicate samples. For the comparison of protein abundance between the tender (low WBSF values) and tough meat samples (high WBSF values), a one-way analysis of variance was performed for each protein. Differences in protein abundance between the tender and tough groups were considered significant at $p < 0.05$, and significant proteins were considered as candidate protein biomarkers. Pearson correlations were computed between the individual WBSF values and protein abundances for those proteins significant following ANOVA. Correlations were considered significant at $p < 0.05$. To get an overview of the main proteins related to WBSF variability, Partial Least Squares (PLS) regressions on standardised data were conducted to generate explanatory models using the list of the candidate protein biomarkers and identify the most influential proteins based on the variable importance in projection (VIP) filter set at both VIP > 1.0 and >0.8, as described by Gagaoua et al. [9]. Moreover, a stepwise regression analysis was used to explain WBSF using the 34 differential proteins (as independent variables, x). The absence of collinearity was systematically tested [25], specifically, the variable was identified as collinear if it possessed a high condition index > 10. The regression model allowed the entry of no more than 3 explanatory variables based on the parsimony principle.

3. Results

3.1. Differential Proteins between Extreme Groups of High and Low WBSF Values

According to Huffman et al. [26], for beef cooked to 70 °C, meat with Warner–Bratzler shear force values of 4.1 kg (40.18 N) or less was correlated with high levels (98%) of consumer acceptability, while beef prepared under the same conditions with shear force values of 5.8 kg (56.84 N) or greater remained unacceptable. Two groups of beef samples with a large difference in shear force were selected from a panel of 107 beef animals collected and profiled under similar conditions. The mean shear force value in the lower shear force group was 33.21 N, while the mean shear force value for the other group was 63.96 N. These groups were classified as tender and tough, respectively. Putative protein biomarkers of beef tenderness that significantly differed in abundance in muscle samples were identified from these divergent groups (Table 1 and details in Table S1).

Table 1. Warner–Bratzler shear force (WBSF) values of the *Longissimus thoracis* muscles used in this trial.

Quality Traits	Min	Max	Mean	SD	CV (%)
WBSF (N) (n = 9)	27.70	38.85	33.21	3.24	9.75
WBSF (N) (n = 9)	59.25	71.40	63.96	3.98	6.22

N, Newtons; SD, Standard Deviation; CV, Coefficient of Variation.

A total of 34 proteins were different ($p < 0.05$) in their abundance between the tender and tough groups (Table 2). These 34 proteins belonged to five major biological pathways (Table 2), these being: (i) muscle contraction, structure, and associated proteins (n = 17; 50%); (ii) energy metabolism and associated pathways (n = 5; 15%); (iii) heat shock proteins (n = 4; 12%); (iv) oxidative stress (n = 2; 6%); and (v) other pathways including regulation of cellular processes, binding, apoptotic, and transport proteins (n = 6; 17%). The 34 proteins were then compared with a database of beef tenderness biomarkers by Gagaoua et al. [13], of which 23 overlapped with the database (Table 2).

Table 2. List of the 34 differential proteins organised by biological family, identified to significantly differ among the two WBSF (tenderness) groups.

Uniprot ID	Gene Name	Full Protein Name	Differences		Pearson Correlations [a]	Overlap with Gagaoua et al. Database [13]
			Fold Change (Log2)	*p*-Value	WBSF	
Muscle contraction, structure and associated proteins (n = 17)						
Q08DI7	MYOZ3 [b]	Myozenin 3	−0.53	0.002	0.741 ***	✓
E1BNG8	BIN1	Bridging Integrator-1	−0.25	0.003	0.616 **	
Q148F1	CFL2	Cofilin-2	−0.47	0.003	0.670 **	
E1BIN0	FHOD1	Formin homology 2 domain containing 1	−0.40	0.005	0.714 ***	
A6QLZ8	CORO6	Coronin	−0.57	0.007	0.607 **	
Q0P571	MYLPF	Myosin regulatory light chain 2	−0.71	0.009	0.613 **	✓
Q3SX40	PDLIM7 [b]	PDZ and LIM domain protein 7	−0.47	0.010	0.642 **	✓
Q0VC48	TMOD4	Tropomodulin-4	−0.42	0.010	0.640 **	TMOD1
Q2KJH4	WDR1	WD repeat-containing protein 1	−0.49	0.019	0.542 *	✓
P60712	ACTB	Actin, cytoplasmic 1	−0.98	0.023	0.524 *	✓
A0JNJ5	MYL1	Myosin light chain 1/3, skeletal muscle isoform	−0.47	0.024	0.554 *	✓
P02453	COL1A1	Collagen alpha-1(I) chain	0.89	0.025	−0.536 *	✓
Q3SYZ8	PDLIM3	PDZ and LIM domain protein 3	−0.51	0.029	0.512 *	PDLIM7/PDLIM1
Q0III9	ACTN3 [c]	Alpha-actinin-3	−0.34	0.034	0.529 *	✓
A4FV78	KLHL41	KBTBD10 protein	−0.38	0.037		✓
F1N789	VCL	Vinculin	−0.21	0.040		✓
Q32LP2	RDX	Radixin	−0.26	0.043	0.476 *	
Energy metabolism (n = 5)						
Q5E956	TPI1	Triosephosphate isomerase	−0.35	0.018	0.631 **	✓
Q3ZBY4	ALDOC	Fructose-bisphosphate aldolase	−0.33	0.018	0.621 **	✓
A5D984	PKM	Pyruvate kinase	−0.39	0.029	0.540 *	✓
A6QLL8	ALDOA	Fructose-bisphosphate aldolase	−0.29	0.029	0.554 *	✓
A3KN12	ADSL	Adenylosuccinate lyase	0.27	0.019	−0.629 **	

Table 2. *Cont.*

| Uniprot ID | Gene Name | Full Protein Name | Differences | | Pearson Correlations [a] | Overlap with Gagaoua et al. Database [13] |
			Fold Change (Log2)	*p*-Value	WBSF	
		Heat shock proteins (n = 4)				
P19120	HSPA8 [b]	Heat shock cognate 71 kDa protein	−0.41	0.006	0.590 **	✓
P31081	HSPD1	60 kDa heat shock protein, mitochondrial	0.84	0.016	−0.567 *	
Q3T149	HSPB1	Heat shock protein beta-1	−0.44	0.038	0.531 *	✓
Q3ZBZ8	STIP1 [b,c]	Stress-induced-phosphoprotein 1	−0.29	0.040	0.484 *	✓
		Oxidative stress (n = 2)				
P35705	PRDX3	Thioredoxin-dependent peroxide reductase	0.32	0.038	−0.488 *	PRDX6/PRDX1/ PRDX2
Q5E946	PARK7	Protein/nucleic acid deglycase DJ-1	−0.47	0.046	0.501 *	✓
		Other pathways (n = 6)				
P11116	LGALS1	Galectin-1	−0.56	0.006	0.639 **	✓
E1BE77	TRIM72	Tripartite motif containing 72	−0.45	0.008	0.620 **	✓
P19879	OGN [b]	Mimecan	−0.64	0.017	0.589 *	
Q2HJF7	CAMK2D	Calcium/calmodulin-dependent protein kinase	−0.20	0.038	0.510 *	
Q6EWQ7	EIF5A	Eukaryotic translation initiation factor 5A-1	−0.45	0.046	0.574 *	
Q3SYR3	APOBEC2	Probable C->U-editing enzyme APOBEC-2	−0.66	0.047	0.530 *	

[a] Significance of the correlations: * $p < 0.05$; ** $p < 0.01$; *** $p < 0.001$. [b] Proteins identified as Quantitative Trait Loci (QTL) of shear force using ProteQTL tool included in ProteINSIDE (http://www.proteinside.org/, accessed on 28 November 2020) from the Animal QTL Database (https://www.animalgenome.org/QTLdb/, accessed on 28 November 2020). [c] Proteins identified as QTL of sensory tenderness.

The regression model built for WBSF is presented in Table 3. The model explained 79% of the variability in WBSF ($p < 0.01$), including the abundance of three proteins: MYOZ3 (Myozenin 3), BIN1 (Bridging Integrator-1), and OGN (Mimecan), which were all positively correlated with WBSF (negatively with tenderness). It should be highlighted that MYOZ3 alone explained 52% of the variability. In this model, the correlation of MYOZ3 with WBSF values is depicted in Figure 1.

Table 3. Best regression equation of WBSF based on the list of the significant differential proteins from Table 2.

R-Squared [a]	S.E	Entered Independent Variable [b]	Partial R-Squared	Regression Coefficient	t-Value	*p*-Value
	0.125	MYOZ3	0.52	0.486	3.875	0.002
0.79 **	0.116	BIN1	0.17	0.454	3.907	0.002
	0.121	OGN	0.1	0.347	2.868	0.012

[a] Significance of the models: ** $p < 0.01$. [b] Variables are shown in order of their entrance, in a stepwise manner, in the regression model.

From the list of the putative protein biomarkers, 30 were negatively correlated with tenderness (positively with WBSF), from which MYOZ3, CFL2, and BIN1 were the most highly significantly correlated proteins. In addition, 4 proteins (COL1A1, ADSL, HSPD1, and PRDX3) were positively correlated with tenderness (negatively with WBSF; Figure 2a and Table 2). From the correlation analyses, Myozenin (MYOZ3) was strongly and significantly correlated with WBSF (Figure 1). No significant correlation was found between WBSF with KLHL41 and VCL (Table 2).

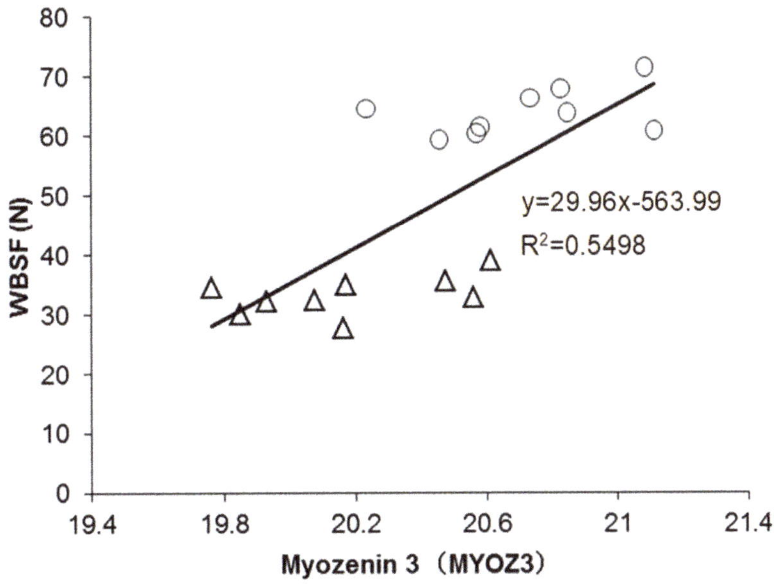

Figure 1. Example of significant correlations between the abundance of Myozenin 3 (MYOZ3) and WBSF values. The tender samples are shown by triangles (Δ) and the tough samples by circles (○). The R-square of the correlation is given.

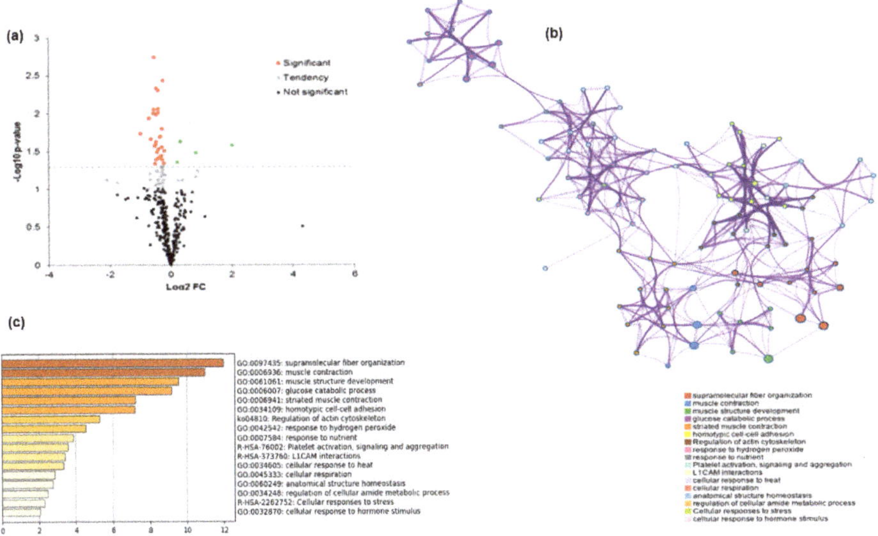

Figure 2. Bioinformatics and statistical analyses of the proteins identified to be differential between the tough and tender *Longissimus thoracis* muscle steaks. (**a**) Volcano plot of the differential proteins in terms of their abundance, with a total of 34 proteins that were significantly different between the two tenderness groups shown in red (negative direction with tenderness) and green colour (positive direction with tenderness). The other proteins that had a tendency or were not significant were in grey and black colour, respectively. (**b**) Networks of pathways and process enrichment cluster analysis based on the 34 differential proteins using Metascape®(https://metascape.org/, accessed on 28 November 2020). (**c**) Functional enrichment analysis based on the list of significant 17 Gene Ontology (GO) terms ranked by their *p*-value.

3.2. Partial Least Squares to Explain the Variability of WBSF Values

Based on the VIP filter (Table 4), the WBSF PLS regression model retained 32 proteins, of which 16 proteins (MYOZ3, FHOD1, CFL2, PDLIM7, TMOD4, LGALS1, TPI1, ADSL, ALDOC, TRIM72, BIN1, MYLPF, CORO6, HSPA8, OGN, EIF5A) had a VIP > 1.0. The other 16 proteins (HSPD1, MYL1, ALDOA, WDR1, PKM, COL1A1, HSPB1, APOBEC2, ACTN3, ACTB, PDLIM3, CAMK2D, PARK7, PRDX3, STIP1, RDX) had a VIP between 0.8 and 1.0; KLHL41 and VCL were the only two proteins whose VIP values were under 0.8. Combined with the results of the correlation analyses, the 32 proteins were identified as related to WBSF regardless of the statistical method. In addition, MYOZ3 was the first ranked protein, with the highest VIP.

Table 4. Partial Least Squares (PLS) prediction of beef tenderness (WBSF) using the list of the 34 putative protein biomarkers based on their variable importance in the projection (VIP).

Proteins	VIP	Direction (+ or −)
MYOZ3: Myozenin 3	1.291	−
FHOD1: Formin homology 2 domain containing 1	1.243	−
CFL2: Cofilin-2	1.168	−
PDLIM7: PDZ and LIM domain protein 7	1.119	−
TMOD4: Tropomodulin-4	1.115	−
LGALS1: Galectin-1	1.113	−
TPI1: Triosephosphate isomerase	1.099	−
ADSL: Adenylosuccinate lyase	1.095	+
ALDOC: Fructose-bisphosphate aldolase	1.082	−
TRIM72: Tripartite motif containing 72	1.080	−
BIN1: Bridging Integrator-1	1.073	−
MYLPF: Myosin regulatory light chain 2, skeletal muscle isoform	1.067	−
CORO6: Coronin	1.056	−
HSPA8: Heat shock cognate 71 kDa protein	1.027	−
OGN: Mimecan	1.025	−
EIF5A: Eukaryotic translation initiation factor 5A-1	1.000	−
HSPD1: 60 kDa heat shock protein, mitochondrial	0.987	+
MYL1: Myosin light chain 1/3, skeletal muscle isoform	0.965	−
ALDOA: Fructose-bisphosphate aldolase	0.965	−
WDR1: WD repeat-containing protein 1	0.943	−
PKM: Pyruvate kinase	0.940	−
COL1A1: Collagen alpha-1(I) chain	0.934	+
HSPB1: Heat shock protein beta-1	0.924	−
APOBEC2: Probable C->U-editing enzyme APOBEC-2	0.923	−
ACTN3: Alpha-actinin-3	0.922	−
ACTB: Actin, cytoplasmic 1	0.913	−
PDLIM3: PDZ and LIM domain protein 3	0.892	−
CAMK2D: Calcium/calmodulin-dependent protein kinase type II subunit delta	0.889	−
PARK7: Protein/nucleic acid deglycase DJ-1	0.873	−
PRDX3: Thioredoxin-dependent peroxide reductase, mitochondrial	0.850	+
STIP1: Stress-induced-phosphoprotein 1	0.843	−
RDX: Radixin	0.830	−
KLHL41: KBTBD10 protein	0.771	−
VCL: Vinculin	0.708	−

3.3. Protein-Protein Interactions (PPI)

The protein-protein interaction network highlighted the importance of the structural and contractile pathways in beef tenderisation (Figure 3). In the network, ACTB (Actin) had the most interactions with other pathways, including energy metabolism and heat shock proteins, while ACTN3 (Alpha-actinin-3) and MYLPF (Myosin regulatory light chain 2) had more involvement within the muscle structure pathway. The next most dominant pathways were cellular processes, binding, apoptosis, and transport proteins, which showed multiple interactions with the energy metabolism (TPI1, ALDOA, and PKM), heat shock proteins (HSPD1), and muscle contraction (PDLIM3). It should be noted that the proteins in the heat shock pathway had a close interaction with each other.

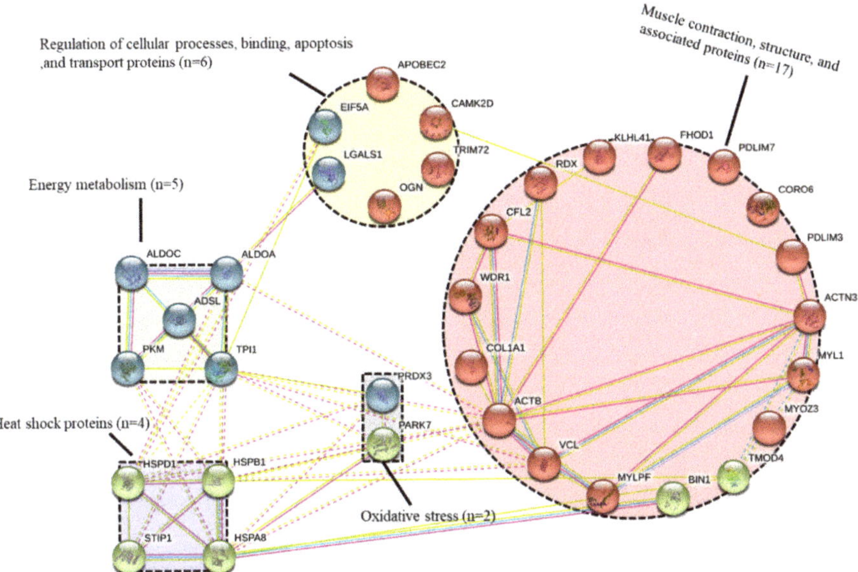

Figure 3. Protein-Protein interaction network built using the 34 differentially abundant proteins. The interaction map was generated from the web-based search STRING database (https://string-db.org/, accessed on 28 November 2020). Default settings of confidence of 0.6 and 4 criteria for linkage: co-occurrence, experimental evidence, existing databases, and text mining were used.

3.4. Pathway and Process Enrichment Analysis

The Gene Ontology (GO) results are given in Table 5. Canonical glycolysis (GO:0061621), glycolytic process (GO:0006096), and muscle contraction (GO:0006936) were the top three Gene-Ontology (GO)-enriched terms identified from the list of the 34 differential proteins (Table 5), while Cellular Component (CC), cytosol (GO:0005829), extracellular exosome (GO:0070062), and cytoplasm (GO:0005737) were the most important three CC terms. It should be noted that a considerable number of proteins were classified as proteins binding (GO:0005515) in molecular function (Table 5). From the Metascape analysis, 17 top and significantly enriched terms were validated and allowed to construct process enrichment networks of the pathways (Figure 2b,c). The top six enriched term clusters were highlighted, including supramolecular fibre organisation (GO:0097435), muscle contraction (GO:0006936), muscle structure development (GO:0061061), glucose catabolic process (GO:0006007), striated muscle contraction (GO:0006941), and homotypic cell-cell adhesion (GO:0034109). The most dominant pathway was supramolecular fibre organisation and muscle contraction, which was consistent with the PPI data confirming their pivotal role in beef tenderisation of young Limousin bull beef (Figure 3).

Table 5. Top20 Gene Ontology (GO) terms computed using the list of the 34 putative protein biomarkers.

GO	Function	Gene Name	GO Frequency within the Dataset (%)	GO Frequency within the Genome (%)	*p*-Values
Biological Process (BP)					
GO:0061621	canonical glycolysis	ALDOC TPI1 PKM ALDOA	11.76	14.81	2.13×10^{-9}
GO:0006096	glycolytic process	ALDOC ALDOA PKM TPI1	11.76	10.26	5.4×10^{-9}
GO:0006936	muscle contraction	TRIM72 MYL1 VCL MYLPF TMOD4	14.71	2.35	2.86×10^{-8}
GO:0043312	neutrophil degranulation	VCL HSPA8 ALDOA ALDOC PKM	14.71	1.03	1.28×10^{-6}
GO:0070527	platelet aggregation	ACTB VCL HSPB1	8.82	7.14	2.06×10^{-6}
GO:0006094	gluconeogenesis	ALDOC TPI1 ALDOA	8.82	6.82	2.27×10^{-6}
GO:0006986	response to unfolded protein	HSPB1 HSPA8 HSPD1	8.82	6.25	2.62×10^{-6}
GO:0035633	maintenance of blood-brain barrier	VCL ACTB	5.88	66.67	5.99×10^{-6}
GO:0030388	fructose 1,6-bisphosphate metabolic process	ALDOA ALDOC	5.88	28.57	1.98×10^{-5}
GO:0030042	actin filament depolymerisation	WDR1 CFL2	5.88	25	2.34×10^{-5}
GO:0043297	apical junction assembly	VCL WDR1	5.88	25	2.34×10^{-5}
GO:0002576	platelet degranulation	ALDOA WDR1 VCL	8.82	2.44	3.11×10^{-5}
GO:0030836	positive regulation of actin filament depolymerisation	CFL2 WDR1	5.88	15.38	4.82×10^{-5}
GO:0006000	fructose metabolic process	ALDOA ALDOC	5.88	13.33	5.97×10^{-5}
GO:0007015	actin filament organisation	ALDOA CORO6 TMOD4	8.82	1.54	9.6×10^{-5}
GO:0042026	protein refolding	HSPD1 HSPA8	5.88	9.52	9.98×10^{-5}
GO:0030239	myofibril assembly	KLHL41 TMOD4	5.88	7.14	0.000162
GO:0034333	adherens junction assembly	ACTB VCL	5.88	5.88	0.000221
GO:0043066	negative regulation of apoptotic process	HSPD1 HSPB1 PRDX3 PARK7	11.76	0.49	0.000221
GO:0086091	regulation of heart rate by cardiac conduction	BIN1 CAMK2D	5.88	5.56	0.000239
Cellular Component (CC)					
GO:0005829	cytosol	VCL TPI1 MYLPF HSPA8 STIP1 EIF5A CAMK2D ADSL MYL1 ACTN3 ALDOC HSPD1 BIN1 PRDX3 PARK7 PDLIM3 ALDOA HSPB1 WDR1 PKM FHOD1 PDLIM7 ACTB KLHL41	70.59	0.5	3.03×10^{-21}
GO:0070062	extracellular exosome	ALDOC HSPD1 ACTN3 ACTB PARK7 TPI1 PKM VCL RDX CFL2 LGALS1 OGN ALDOA HSPA8 HSPB1 WDR1	47.06	0.58	1.67×10^{-15}
GO:0005737	cytoplasm	PARK7 FHOD1 COL1A1 EIF5A PKM BIN1 HSPD1 HSPA8 KLHL41 PRDX3 TRIM72 APOBEC2 CAMK2D ACTB HSPB1 CFL2 LGALS1	50	0.41	3×10^{-14}

Table 5. Cont.

GO	Function	Gene Name	GO Frequency within the Dataset (%)	GO Frequency within the Genome (%)	p-Values
GO:0005615	extracellular space	HSPD1 RDX ALDOA CFL2 HSPA8 TPI1 OGN COL1A1 HSPB1 LGALS1 ACTB	32.35	0.77	3×10^{-12}
GO:0005925	focal adhesion	VCL HSPB1 RDX HSPA8 ACTN3 ACTB PDLIM7	20.59	1.82	2.33×10^{-10}
GO:0015629	actin cytoskeleton	CFL2 MYOZ3 BIN1 PDLIM7 ACTB ALDOA	17.65	3.14	2.54×10^{-10}
GO:0005856	cytoskeleton	VCL KLHL41 HSPB1 BIN1 TMOD4 ACTB FHOD1 ALDOC	23.53	1.08	3.05×10^{-10}
GO:0030018	Z disc	PDLIM7 MYOZ3 BIN1 CFL2 PDLIM3	14.71	4.2	2.57×10^{-9}
GO:1904813	ficolin-1-rich granule lumen	ALDOC HSPA8 PKM ALDOA VCL	14.71	4.03	2.92×10^{-9}
GO:0005634	nucleus	STIP1 CAMK2D EIF5A BIN1 FHOD1 ALDOA PKM HSPB1 APOBEC2 ACTB HSPA8 PARK7 TPI1	38.24	0.26	4.61×10^{-9}
GO:0005576	extracellular region	PKM ALDOC WDR1 LGALS1 ALDOA HSPA8 VCL COL1A1 OGN	26.47	0.49	7.91×10^{-9}
GO:0034774	secretory granule lumen	ALDOC PKM VCL ALDOA HSPA8	14.71	1.56	2×10^{-7}
GO:0031674	I band	ALDOA CFL2 BIN1	8.82	13.64	3.96×10^{-7}
GO:0005912	adherens junction	PARK7 ACTB PDLIM3 PDLIM7 VCL	14.71	1.04	1.28×10^{-6}
GO:0030864	cortical actin cytoskeleton	RDX CFL2 WDR1	8.82	8.57	1.28×10^{-6}
GO:0001725	stress fibre	PDLIM3 PDLIM7 FHOD1	8.82	6.25	2.62×10^{-6}
GO:0030424	axon	PARK7 HSPA8 BIN1 ACTB	11.76	1.65	3.52×10^{-6}
GO:0005654	nucleoplasm	PARK7 CAMK2D ACTB KLHL41 HSPA8 FHOD1 PDLIM7	20.59	0.24	3.81×10^{-5}
GO:0101031	chaperone complex	STIP1 HSPA8	5.88	15.38	4.82×10^{-5}
GO:0005886	plasma membrane	VCL HSPA8 BIN1 HSPD1 ACTB RDX WDR1 KLHL41	23.53	0.18	5.47×10^{-5}
Molecular Function (MF)					
GO:0005515	protein binding	PRDX3 PARK7 ALDOA OGN MYOZ3 ACTN3 FHOD1 CAMK2D PKM HSPD1 TRIM72 COL1A1 CFL2 TMOD4 STIP1 ALDOC CORO6 BIN1 LGALS1 VCL EIF5A HSPA8 HSPB1 PDLIM3 TPI1 KLHL41 RDX ACTB PDLIM7	85.29	0.45	9.5×10^{-24}
GO:0042802	identical protein binding	ALDOA PRDX3 TRIM72 PARK7 ACTB CAMK2D ADSL BIN1 HSPB1 APOBEC2 ACTN3 FHOD1 COL1A1	38.24	0.92	2.98×10^{-15}
GO:0003723	RNA binding	HSPB1 HSPD1 LGALS1 EIF5A APOBEC2 HSPA8 PKM ALDOA STIP1 RDX	29.41	0.62	2.33×10^{-10}

Table 5. *Cont.*

GO	Function	Gene Name	GO Frequency within the Dataset (%)	GO Frequency within the Genome (%)	*p*-Values
GO:0045296	cadherin binding	PARK7 HSPA8 VCL RDX ALDOA PKM	17.65	2.03	2.44×10^{-9}
GO:0051015	actin filament binding	BIN1 FHOD1 WDR1 CFL2 CORO6	14.71	3.57	4.61×10^{-9}
GO:0003779	actin binding	PDLIM3 VCL ALDOA PDLIM7 MYOZ3 RDX	17.65	1.48	1.01×10^{-8}
GO:0008307	structural constituent of muscle	ACTN3 MYL1 MYLPF	8.82	6.52	2.48×10^{-6}
GO:0004332	fructose-bisphosphate aldolase activity	ALDOA ALDOC	5.88	66.67	5.99×10^{-6}
GO:0031625	ubiquitin protein ligase binding	HSPA8 HSPD1 VCL TPI1	11.76	1.36	6.77×10^{-6}
GO:0051087	chaperone binding	BIN1 HSPD1 HSPA8	8.82	3.3	1.44×10^{-5}
GO:0048156	tau protein binding	BIN1 ACTB	5.88	20	3.26×10^{-5}
GO:0023026	MHC class II protein complex binding	PKM HSPA8	5.88	11.76	7.34×10^{-5}
GO:0051371	muscle alpha-actinin binding	PDLIM3 PDLIM7	5.88	11.11	7.98×10^{-5}
GO:0044183	protein folding chaperone	HSPA8 HSPB1	5.88	7.14	0.000162
GO:0042803	protein homodimerisation activity	PARK7 CAMK2D TPI1 HSPB1	11.76	0.53	0.000173
GO:0003697	single-stranded DNA binding	HSPD1 PARK7	5.88	2.06	0.001335
GO:0019901	protein kinase binding	HSPB1 ACTB PRDX3	8.82	0.54	0.001393
GO:0044325	ion channel binding	ACTN3 CAMK2D	5.88	1.82	0.001617
GO:0002020	protease binding	COL1A1 BIN1	5.88	1.61	0.001708
GO:0070626	(S)-2-(5-amino-1-(5-phospho-D-ribosyl)imidazole-4-carboxamido)succinate AMP-lyase (fumarate-forming) activity	ADSL	2.94	100	0.001708

4. Discussion

The beef industry is consistently confronted with challenges in supplying beef with consistent eating qualities. Tenderness is one of the most important palatability traits of beef that affects the repurchase decisions of consumers. The pathways underpinning beef tenderness determination are complex and not fully elucidated, although a recent integromics meta-analysis by Gagaoua et al. [13] on the molecular signatures shed light on some of them. Thus, it was valuable to identify putative protein biomarkers of beef tenderness from two tenderness groups with a strong difference in shear force: tender (33.21 N) vs. tough (63.96 N; Table 1).

This study on Irish Limousin-cross cattle allowed us to get more insights and validate the association of certain proteins with tenderness and propose new ones that will further increase our knowledge and progress in the pipeline of beef tenderness discovery of biomarkers. This study allowed us also to (i) propose preliminary explanatory models of tenderness using multiple regression and partial least squares; (ii) compare the list of putative protein biomarkers identified in this trial with previous studies to verify the robustness of the discovered proteins; and finally, (iii) increase our knowledge on the biological pathways involved in the variation of beef tenderness evaluated in this study using WBSF at an end-point cooking temperature of 71 °C. The relationship between tenderness and the list of candidate proteins was discussed in the following sections.

4.1. The Best Explanatory Proteins in the Regression Model of WBSF

The best regression model built with MYOZ3, BIN1, and OGN proteins explained 79% of the observed variability in WBSF ($p < 0.01$). MYOZ3 is mainly expressed in skeletal muscle and enriched in fast-twitch muscle fibres. MYOZ3 belongs to the myozenin family, of which three other members were previously proposed as tenderness biomarkers [13]. MYOZ3 acts as an intracellular binding protein to link with Z-disc proteins such as alpha-actinin and gamma-filamin and transmit calcineurin signalling to the sarcomere [27]. Due to the capacity to bind multiple proteins, the relationship between MYOZ3 and meat quality, specifically tenderness, could be through regulating the Z-disc structure and signal transduction, influencing muscle fibre differentiation [28]. Consistent with our findings, a previous study reported a negative association between MYOZ3 and the shear force of *M. longissimus thoracis* in heifers [29].

BIN1, also known as Bridging Integrator-1, was identified in the present study for the first time to have a potential association with beef tenderness. BIN1 plays an important role in the regulation of endocytosis and has other roles as a central regulator of cell proliferation and apoptosis [30]. While no evidence was present in the literature on a specific relationship with tenderness, BIN1 was associated with another important beef production attribute, residual feed intake [31], which was previously associated with meat quality. OGN, which is also called mimecan, belongs to a secreted protein family of small leucine-rich proteoglycans located in the extracellular matrix [32]. OGN was negatively correlated with beef tenderness, and both MYOZ3 and OGN genes were located within a Quantitative Trait Loci (QTL) for shear force on chromosome 8 (Table 2). Interestingly, when protein profiles were compared between Japanese Black cattle and Holstein cattle, a higher abundance of OGN protein (mimecan) was found in the Holstein breed known to have lower fat content [33]. An important function of OGN is in collagen fibrillogenesis [32]. For this reason, it could be hypothesised that the greater abundance of OGN protein observed for tougher beef animals may be related to a higher abundance of connective tissue content in the muscle of tough beef [34], although we did not measure the connective tissue content in the present study.

4.2. Dominant Pathway Related to WBSF of Young Limousin-Sired Bulls

Muscle contraction and structure were identified as the most important pathway associated with WBSF in this study. Most of the proteins from this pathway were localised in the sarcomere. Of these, compared with the database of beef tenderness biomarkers of Gagaoua [13], 13 were already identified, and 4 proteins (BIN1, FHOD1, CORO6, and RDX) were reported for the first time in this study.

Myosin and actin were critically important to textural changes in muscle that occurred post-mortem during meat ageing through the weakening of the actin/myosin complex in the myofibril [22]. As the major component of the thick filaments of the myofibril, molecular myosin consisted of two heavy and four light chains. This study revealed, for example, MYLPF (Myosin regulatory light chain 2) and MYL1 (Myosin light chain 1/3) to be negative biomarkers of beef tenderness. Myosin light chains were wrapped around the head/rod junction of the myosin heavy chain in skeletal muscle myosin [35]. The MYLPF and MYL1 proteins were demonstrated [13] to correlate with beef tenderness; however, the direction of their relationships with this trait lacks consistency across studies; this phenomenon was well known to vary depending on the breed and muscle type [5,25,36], and was suggested to be related to post-translational modifications of the proteins [37]. Myosin light chain proteins were highly expressed in fast-twitch fibres. It was noteworthy that phosphorylation of MYL might play an essential role in proteolysis and onset of apoptosis in post-mortem muscle, which was favourable to the degradation of large molecules and final tenderisation of aged meat [38]. As for the second, most abundant myofibrillar protein in muscle, actin was interrelated with the apoptosis of the cytoskeleton during meat tenderisation [39]. In our study, ACTB, ACTN3, and TMOD4 were negatively correlated with beef tenderness, which is consistent with previous studies [9,29].

The collagen alpha-1(I) chain is an abundant connective tissue protein with an important function of support in the muscle tissue and bone in the body, and is encoded by *COL1A1* gene. Several studies showed a close relationship between collagen content and variation in meat tenderness [40,41]. Interestingly, in a previous study by Bjarnadóttir et al. [42], COL1A1 and COL1A2 were found to have lower abundance in tender beef muscle, which was opposite to our findings. However, there was also evidence of a positive relationship between COL1A1 abundance and intramuscular fat content, which could have an effect in promoting beef tenderness [43]. Thus, there was no consistent conclusion regarding the direct influence of COL1A1 on meat quality.

PDLIM3 and PDLIM7 are two members of the PDZ and LIM domain (PDLIM) family, participating in multifunctional protein-protein interaction, cytoskeleton, and signal transduction pathways [44]. PDLIM family proteins contain a PDZ domain in the N-terminal portion and the LIM domain in the C-terminal portion [45]. PDLIM1 and PDLIM7 were previously identified as negative biomarkers of beef tenderness [13], which was consistent with our results stating that PDLIM3 and PDLIM7 were positively correlated with WBSF and the negative relationship of this protein family with beef quality.

Of the putative biomarkers identified for the first time in this study, FHOD1 is an actin regulator which played an important role in the stabilisation of filamentous (F)-actin bundles by selectively covering and binding their barbed ends to actin filaments, thus protecting actin filaments in cytoskeletal structures [46]. Likewise, CORO6 is also an actin-binding protein that is mainly expressed in the heart and skeletal muscle [47]. RDX (Radixin) is referred to as a member of ERM (Ezrin/Radixin/Moesin) proteins which help maintain cytoskeletal organisation by binding specific membrane proteins to the actin cytoskeleton [48]. FHOD1, CORO6, and RDX were all positively correlated with WBSF (and negatively with tenderness), which could be related to their protective effect on the integrity of the cytoskeleton.

4.3. Candidate Protein Biomarkers of WBSF from the Energy Metabolism Pathway

Energy metabolism comprises a series of interconnected pathways that can function in the presence or absence of oxygen to generate adenosine triphosphate (ATP), which is an end-product of the processes of oxidative phosphorylation [49]. Of the five proteins identified from this pathway (TPI1, ALDOA, PKM, ADSL, and ALDOC), the first four proteins have been previously identified as putative biomarkers of beef tenderness. In this study, these proteins were all negatively correlated with beef tenderness except ADSL. TPI1 can catalyse the conversion of dihydroxyacetone phosphate to D-glyceraldehyde 3-phosphate, meanwhile, maintaining the equilibration of the triosephosphates produced by aldolase (ALDOA) [50]. Aldolase is an enzyme that catalyses the reversible conversion of fructose-1,6-bisphosphate to glyceraldehyde 3-phosphate and dihydroxyacetone phosphate [51]. ALDOA and ALDOC are two different isoformes of aldolase. In the literature, ALDOA was both positively and negatively correlated with beef tenderness depending on the gender and muscle fibre type [13], while ALDOC was first identified as a putative negative biomarker of tenderness in the present study. The relationship between aldolase and tenderness could be explained by its participation in muscle glycolysis, which, if variable, could alter the profile and extent of pH decline, thereby further influencing the integrity of the Z-line with consequences for beef tenderness [52].

Pyruvate kinase, also known as PKM, is an enzyme that catalyses the dephosphorylation of phosphoenolpyruvate to pyruvate, generating ATP and regulating cell metabolism during glycolysis [53]. PKM1 and PKM2 are the two predominant isoforms of PKM in skeletal muscles. PKM was listed in the repertoire of beef tenderness biomarkers in *longissimus* muscle [13], and our findings provided further corroboration for its role in beef tenderness determination.

4.4. Heat Shock Proteins (HSPs) as Important Indicators of WBSF

Heat shock proteins (HSPs) are a family of proteins that have as their main function the protection of the organism itself and its cellular structures in response to exposure to stressful conditions [54]. The current study showed a differential abundance of HSPA8, HSPD1, HSPB1, and STIP1, three of which were already discovered to play a role in the variability of tenderness. Interestingly, three of the HSPs identified here were from three different subfamilies of HSPs, i.e., the small HSPs (HSPB1), HSP70s (HSPA8), and HSP60s (HSPD1). Among those four proteins, HSPA8, HSPB1, and STIP1 showed higher abundance in tough meat while HSPD1 was in the opposite direction.

As one important member of the large HSP70 family, HSPA8 was identified in six previous studies to be related to beef tenderness, but the mechanistic connection with tenderness was not clear because the protein was sometimes positively correlated and sometimes negatively correlated with beef tenderness [13]. The impact of HSP70 proteins on meat tenderness was thought to be mainly because they obstruct pro-apoptotic factors such as Bcl-2 in apoptotic pathways [13,55]. HSPA8 is grouped in the response to unfolded protein (GO:0006986) and protein refolding (GO:0042026) in Table 5. In this sense, HSPA8 played an important role in response to cellular stress [56]. Moreover, this protective role might be based on its interaction with structural proteins or by regulating cell signalling pathways (Figure 3). STIP1, known as a stress-induced-phosphoprotein, is a co-chaperone whose negative relationship with tenderness, found here, was consistent with the findings of Picard and Gagaoua [8].

As for the small HSPs, HSPB1 was identified as a robust biomarker of beef tenderness (referring to the database by Gagaoua et al. [13]), and it was, from that integromics study, in the top five biomarkers of beef tenderness from a list of 124 proteins. Extrinsic stressors, such as pre-slaughter or post-mortem management conditions, were sources of the intensive production of sHSPs in the muscle, which, like the larger HSPs, also play a regulatory role in delaying the apoptosis onset, the protection of myofibrillar proteins from proteolysis, and other cellular homeostasis roles [15,39]. The positive and negative relationships identified between sHSPs and tenderness might be due to interactions of factors such as animal type/breed, gender, muscle type, and pre-slaughter conditions [12,39,57]. In this study, HSPB1 was negatively correlated with beef tenderness, which would be consistent with its protective function against proteolysis in skeletal muscle.

It was notable that HSPD1, which is a member of the HSP60 family, was identified to be correlated with beef tenderness for the first time in the present study. Under stress conditions, the HSP60 family of proteins inside the mitochondrial matrix usually acts as molecular chaperones, collaborating with the co-chaperone Hsp10 to promote the correct folding of imported proteins and proper assembly of unfolded polypeptides [58]. A positive relationship between HSPD1 and tenderness might be hypothesised by its function in the energy metabolism pathway to maintain energy supply during proteolysis of myofibril proteins (Figure 3). As with HSPB1, HSPD1 was also associated with beef colour, which deepened our knowledge of the influential role of HSP proteins in post-mortem muscle events and consequences on meat quality [14].

4.5. Putative Biomarkers of Tenderness Related to Oxidative Stress

After slaughter, the lipid and protein fractions of muscle are targeted by various reactive oxygen species (ROS), causing structural alteration or denaturation of proteins [59]. In the context of oxidative stress, meat tenderness can be affected by cellular antioxidants, which include both enzymatic and non-enzymatic scavenger agents engaged in protecting the muscle proteins from damage by ROS, thereby further maintaining cell homeostasis. Meanwhile, it is noteworthy that meat tenderness could also be influenced by ROS damage produced by mitochondria, which play an important role in supplying energy during the conversion of muscle into meat [60,61].

In this study, two important proteins from the oxidative pathway, i.e., PARK7 and PRDX3 were identified and validated in comparison to the previous list of robust beef tenderness biomarkers [13]. PARK7, also named DJ-1, was secreted from the cytosol to mitochondria to remove the mitochondrial H_2O_2 and maintain the integrity of the organelle in response to oxidative stress [62]. Consistent with the results of most previous studies, PARK7 level was negatively correlated with beef tenderness [8,63,64]. A mechanism could be deduced where PARK7 had an inhibitory effect on the pro-apoptotic factors and caspases (proteolytic enzymes) by interacting with other proteins from energy metabolism and HSPs pathways, as depicted by the PPI network (Figure 3), thus contributing to beef toughness by slowing or limiting post-mortem apoptosis of muscle cells [13,65].

PRDX3 is a member of the peroxiredoxins (Prxs), a ubiquitous family with six subgroups, and which, in bovine, contains six members [66]. PRDX3 is exclusively located in mitochondria with an oligomeric ring structure [67]. It should be highlighted that PRDX3 was previously identified to be positively related to beef tenderness in *Semimembranosus* muscle [68], which was consistent with our findings for *Longissimus thoracis* muscle. Regarding the possible mechanism, it could be assumed that the antioxidant enzyme PRDX3 could prevent the accumulation of ROS, protecting the function of proteases and the operation of the electron transport system, and thus, leading to apoptosis promotion and meat tenderisation. Other members of peroxiredoxins were also found to be associated with several beef quality traits, including PRDX1 [69] and PRDX2 [70] with tenderness and PRDX6 with tenderness [9], pH decline [71], and beef colour [25].

4.6. Proteins from Other Pathways

LGALS1 and TRIM72 were negative biomarkers of beef tenderness in this study, which was consistent with previous reports [8,9,42]. LGALS1 (Galectin-1) belongs to a family of β-galactoside-binding proteins, which may act as promoters of apoptosis and have an impact on cell proliferation and skeletal muscle differentiation [42]. However, the mechanism behind the association between Galectin-1 and meat tenderness was still obscure due to its complex functions under different conditions. As a signalling protein expressed in skeletal muscle, Tripartite motif-containing 72 (TRIM72) was considered as a sensor of oxidation on membrane damage [72]. TRIM72 may act as a scavenger of the harmful agents accumulated under the apoptotic process, leading to a limitation of apoptosis and tough meat [9]. In line with our findings, there was also a higher abundance of TRIM72 reported in tough beef, hence showing its negative role in the apoptotic pathway [68]. In addition, TRIM72 was first identified to correlate with beef colour, which confirmed the anti-oxidative properties of this protein, allowing it to be suggested as a relevant marker for multiple beef quality traits [63].

5. Conclusions

The results of this study allowed us to validate 23 putative biomarkers on Irish cattle (Limousin-sired bulls) and to propose 11 new proteins that increased our knowledge on the main biological pathways underpinning beef tenderness variation in the *Longissimus thoracis* muscle of young bulls. The network and gene ontology analyses allowed us to better characterise the enriched molecular pathways. This study also suggested a regression model with an R-squared of 79% using three proteins-MYOZ3, BIN1, and OGN-to explain the relationship between the abundance of these protein biomarkers and WBSF values. Further analyses would assess the robustness of the list of putative biomarkers identified in this study using accurate methods and new populations.

Supplementary Materials: The following are available online at https://www.mdpi.com/article/10.3390/foods10050952/s1, Table S1. Individual Warner-Bratzler shear force (WBSF) values of the *Longissimus thoracis* muscles from the 18 animals used in this trial.

Author Contributions: Conceptualisation, R.M.H., M.G., and A.M.M.; methodology, M.G., R.M.H., and Y.Z.; sampling, Y.Z., R.M.H., and J.C.; lab work, Y.Z. and M.G.; data curation, M.G., D.V., J.C. and Y.Z.; writing-original draft preparation, Y.Z. and M.G.; writing-review and editing, M.G., R.M.H., A.L.K., A.M.M., and T.S.; supervision, R.M.H., A.M.M., A.L.K., and M.G.; project administration,

R.M.H., A.M.M., and A.L.K.; funding acquisition, R.M.H., A.M.M. and T.S. All authors have read and agreed to the published version of the manuscript.

Funding: This research was funded by Teagasc and the Walsh Scholarship programme (Walsh Scholarship 2016022), the BreedQuality project (11/SF/311), which is supported by The Irish Department of Agriculture, Food, and the Marine (DAFM) under the National Development Plan 2007–2013. Mohammed Gagaoua is a Marie Skłodowska–Curie Career-FIT Fellow under project number MF20180029, grant agreement No. 713654.

Institutional Review Board Statement: Not applicable.

Informed Consent Statement: Not applicable.

Data Availability Statement: Data sharing is not applicable to this article.

Acknowledgments: The authors convey special thanks to David Sheehan and Dilip K. Rai for their help to Yao Zhu and for support in funding acquisition. We acknowledge the Irish Cattle Breeders Federation for access to samples.

Conflicts of Interest: The authors declare no conflict of interest.

References

1. Kerth, C.R.; Miller, R.K. Beef Flavor: A Review from Chemistry to Consumer. *J. Sci. Food Agric.* **2015**, *95*, 2783–2798. [CrossRef]
2. Purslow, P.P.; Gagaoua, M.; Warner, R.D. Insights on Meat Quality from Combining Traditional Studies and Proteomics. *Meat Sci.* **2021**, *174*, 108423. [CrossRef] [PubMed]
3. Henchion, M.; McCarthy, M.; Resconi, V.C.; Troy, D. Meat Consumption: Trends and Quality Matters. *Meat Sci.* **2014**, *98*, 561–568. [CrossRef] [PubMed]
4. Cuvelier, C.; Clinquart, A.; Hocquette, J.F.; Cabaraux, J.F.; Dufrasne, I.; Istasse, L.; Hornick, J.L. Comparison of Composition and Quality Traits of Meat from Young Finishing Bulls from Belgian Blue, Limousin and Aberdeen Angus Breeds. *Meat Sci.* **2006**, *74*, 522–531. [CrossRef] [PubMed]
5. Gagaoua, M.; Terlouw, C.; Richardson, I.; Hocquette, J.-F.; Picard, B. The Associations between Proteomic Biomarkers and Beef Tenderness Depend on the End-Point Cooking Temperature, the Country Origin of the Panelists and Breed. *Meat Sci.* **2019**, *157*, 107871. [CrossRef] [PubMed]
6. Cafferky, J.; Hamill, R.M.; Allen, P.; O'Doherty, J.V.; Cromie, A.; Sweeney, T. Effect of Breed and Gender on Meat Quality of M. Longissimus Thoracis et Lumborum Muscle from Crossbred Beef Bulls and Steers. *Foods* **2019**, *8*, 173. [CrossRef]
7. Nian, Y.; Kerry, J.P.; Prendiville, R.; Allen, P. The Eating Quality of Beef from Young Dairy Bulls Derived from Two Breed Types at Three Ages from Two Different Production Systems. *Ir. J. Agric. Food Res.* **2017**, *56*, 31–44. [CrossRef]
8. Picard, B.; Gagaoua, M. Muscle Fiber Properties in Cattle and Their Relationships with Meat Qualities: An Overview. *J. Agric. Food Chem.* **2020**, *68*, 6021–6039. [CrossRef] [PubMed]
9. Gagaoua, M.; Monteils, V.; Picard, B. Data from the Farmgate-to-Meat Continuum Including Omics-Based Biomarkers to Better Understand the Variability of Beef Tenderness: An Integromics Approach. *J. Agric. Food Chem.* **2018**, *66*, 13552–13563. [CrossRef]
10. Terlouw, E.M.C.; Picard, B.; Deiss, V.; Berri, C.; Hocquette, J.-F.; Lebret, B.; Lefèvre, F.; Hamill, R.; Gagaoua, M. Understanding the Determination of Meat Quality Using Biochemical Characteristics of the Muscle: Stress at Slaughter and Other Missing Keys. *Foods* **2021**, *10*, 84. [CrossRef]
11. Gagaoua, M.; Monteils, V.; Couvreur, S.; Picard, B. Identification of Biomarkers Associated with the Rearing Practices, Carcass Characteristics, and Beef Quality: An Integrative Approach. *J. Agric. Food Chem.* **2017**, *65*, 8264–8278. [CrossRef] [PubMed]
12. Picard, B.; Gagaoua, M. Meta-Proteomics for the Discovery of Protein Biomarkers of Beef Tenderness: An Overview of Integrated Studies. *Food Res. Int.* **2020**, *127*, 108739. [CrossRef] [PubMed]
13. Gagaoua, M.; Terlouw, E.M.C.; Mullen, A.M.; Franco, D.; Warner, R.D.; Lorenzo, J.M.; Purslow, P.P.; Gerrard, D.; Hopkins, D.L.; Troy, D.; et al. Molecular Signatures of Beef Tenderness: Underlying Mechanisms Based on Integromics of Protein Biomarkers from Multi-Platform Proteomics Studies. *Meat Sci.* **2021**, *172*, 108311. [CrossRef]
14. Gagaoua, M.; Bonnet, M.; Picard, B. Protein Array-Based Approach to Evaluate Biomarkers of Beef Tenderness and Marbling in Cows: Understanding of the Underlying Mechanisms and Prediction. *Foods* **2020**, *9*, 1180. [CrossRef]
15. Picard, B.; Gagaoua, M. Chapter 11—Proteomic Investigations of Beef Tenderness. In *Proteomics in Food Science*; Colgrave, M.L., Ed.; Academic Press: Cambridge, MA, USA, 2017; pp. 177–197. ISBN 978-0-12-804007-2.
16. Munekata, P.E.; Pateiro, M.; López-Pedrouso, M.; Gagaoua, M.; Lorenzo, J.M. Foodomics in Meat Quality. *Curr. Opin. Food Sci.* **2021**, *38*, 79–85. [CrossRef]
17. Gagaoua, M.; Hughes, J.; Terlouw, E.M.C.; Warner, R.D.; Purslow, P.P.; Lorenzo, J.M.; Picard, B. Proteomic Biomarkers of Beef Colour. *Trends Food Sci. Technol.* **2020**, *101*, 234–252. [CrossRef]
18. Zhu, Y.; Mullen, A.M.; Rai, D.K.; Kelly, A.L.; Sheehan, D.; Cafferky, J.; Hamill, R.M. Assessment of RNAlater®as a Potential Method to Preserve Bovine Muscle Proteins Compared with Dry Ice in a Proteomic Study. *Foods* **2019**, *8*, 60. [CrossRef]

19. Bouley, J.; Chambon, C.; Picard, B. Mapping of Bovine Skeletal Muscle Proteins Using Two-Dimensional Gel Electrophoresis and Mass Spectrometry. *Proteomics* **2004**, *4*, 1811–1824. [CrossRef]
20. Bradford, M.M. A Rapid and Sensitive Method for the Quantitation of Microgram Quantities of Protein Utilizing the Principle of Protein-Dye Binding. *Anal. Biochem.* **1976**, *72*, 248–254. [CrossRef]
21. Zhu, Y.; Gagaoua, M.; Mullen, A.M.; Viala, D.; Rai, D.K.; Kelly, A.L.; Sheehan, D.; Hamill, R.M. Shotgun Proteomics for the Preliminary Identification of Biomarkers of Beef Sensory Tenderness, Juiciness and Chewiness from Plasma and Muscle of Young Limousin-Sired Bulls. *Meat Sci.* **2021**, *176*, 108488. [CrossRef] [PubMed]
22. Gagaoua, M.; Troy, D.; Mullen, A.M. The Extent and Rate of the Appearance of the Major 110 and 30 KDa Proteolytic Fragments during Post-Mortem Aging of Beef Depend on the Glycolysing Rate of the Muscle and Aging Time: An LC–MS/MS Approach to Decipher Their Proteome and Associated Pathways. *J. Agric. Food Chem.* **2021**, *69*, 602–614. [CrossRef] [PubMed]
23. Kaspric, N.; Picard, B.; Reichstadt, M.; Tournayre, J.; Bonnet, M. ProteINSIDE to Easily Investigate Proteomics Data from Ruminants: Application to Mine Proteome of Adipose and Muscle Tissues in Bovine Foetuses. *PLoS ONE* **2015**, *10*, e0128086. [CrossRef]
24. Zhou, Y.; Zhou, B.; Pache, L.; Chang, M.; Khodabakhshi, A.H.; Tanaseichuk, O.; Benner, C.; Chanda, S.K. Metascape Provides a Biologist-Oriented Resource for the Analysis of Systems-Level Datasets. *Nat. Commun.* **2019**, *10*, 1523. [CrossRef]
25. Gagaoua, M.; Terlouw, E.M.C.; Picard, B. The Study of Protein Biomarkers to Understand the Biochemical Processes Underlying Beef Color Development in Young Bulls. *Meat Sci.* **2017**, *134*, 18–27. [CrossRef] [PubMed]
26. Huffman, K.L.; Miller, M.F.; Hoover, L.C.; Wu, C.K.; Brittin, H.C.; Ramsey, C.B. Effect of Beef Tenderness on Consumer Satisfaction with Steaks Consumed in the Home and Restaurant. *J. Anim. Sci.* **1996**, *74*, 91–97. [CrossRef]
27. Ye, M.; Ye, F.; He, L.; Luo, B.; Yang, F.; Cui, C.; Zhao, X.; Yin, H.; Li, D.; Xu, H.; et al. Transcriptomic Analysis of Chicken Myozenin 3 Regulation Reveals Its Potential Role in Cell Proliferation. *PLoS ONE* **2017**, *12*, e0189476. [CrossRef]
28. Ye, M.; Ye, F.; He, L.; Liu, Y.; Zhao, X.; Yin, H.; Li, D.; Xu, H.; Zhu, Q.; Wang, Y. Molecular Cloning, Expression Profiling, and Marker Validation of the Chicken Myoz3 Gene. *Biomed. Res. Int.* **2017**, *2017*, 5930918. [CrossRef] [PubMed]
29. Boudon, S.; Ounaissi, D.; Viala, D.; Monteils, V.; Picard, B.; Cassar-Malek, I. Label Free Shotgun Proteomics for the Identification of Protein Biomarkers for Beef Tenderness in Muscle and Plasma of Heifers. *J. Proteom.* **2020**, *217*, 103685. [CrossRef] [PubMed]
30. Elliott, K.; Sakamuro, D.; Basu, A.; Du, W.; Wunner, W.; Staller, P.; Gaubatz, S.; Zhang, H.; Prochownik, E.; Eilers, M.; et al. Bin1 Functionally Interacts with Myc and Inhibits Cell Proliferation via Multiple Mechanisms. *Oncogene* **1999**, *18*, 3564–3573. [CrossRef]
31. Karisa, B.K.; Thomson, J.; Wang, Z.; Stothard, P.; Moore, S.S.; Plastow, G.S. Candidate Genes and Single Nucleotide Polymorphisms Associated with Variation in Residual Feed Intake in Beef Cattle1. *J. Anim. Sci.* **2013**, *91*, 3502–3513. [CrossRef] [PubMed]
32. Hu, S.-M.; Li, F.; Yu, H.-M.; Li, R.-Y.; Ma, Q.-Y.; Ye, T.-J.; Lu, Z.-Y.; Chen, J.-L.; Song, H.-D. The Mimecan Gene Expressed in Human Pituitary and Regulated by Pituitary Transcription Factor-1 as a Marker for Diagnosing Pituitary Tumors. *J. Clin. Endocrinol. Metab.* **2005**, *90*, 6657–6664. [CrossRef] [PubMed]
33. Ohsaki, H.; Okada, M.; Sasazaki, S.; Hinenoya, T.; Sawa, T.; Iwanaga, S.; Tsuruta, H.; Mukai, F.; Mannen, H. Proteomic Comparison between Japanese Black and Holstein Cattle by Two-Dimensional Gel Electrophoresis and Identification of Proteins. *Asian-Australas. J. Anim. Sci.* **2007**, *20*, 638–644. [CrossRef]
34. Tasheva, E.S.; Koester, A.; Paulsen, A.Q.; Garrett, A.S.; Boyle, D.L.; Davidson, H.J.; Song, M.; Fox, N.; Conrad, G.W. Mimecan/Osteoglycin-Deficient Mice Have Collagen Fibril Abnormalities. *Mol. Vis.* **2002**, *8*, 407–415. [PubMed]
35. Lowey, S.; Waller, G.S.; Trybus, K.M. Skeletal Muscle Myosin Light Chains Are Essential for Physiological Speeds of Shortening. *Nature* **1993**, *365*, 454–456. [CrossRef] [PubMed]
36. Picard, B.; Gagaoua, M.; Micol, D.; Cassar-Malek, I.; Hocquette, J.-F.; Terlouw, C.E.M. Inverse Relationships between Biomarkers and Beef Tenderness According to Contractile and Metabolic Properties of the Muscle. *J. Agric. Food Chem.* **2014**, *62*, 9808–9818. [CrossRef]
37. Mato, A.; Rodríguez-Vázquez, R.; López-Pedrouso, M.; Bravo, S.; Franco, D.; Zapata, C. The First Evidence of Global Meat Phosphoproteome Changes in Response to Pre-Slaughter Stress. *Bmc Genom.* **2019**, *20*, 590. [CrossRef] [PubMed]
38. De Rodrigues, R.T.S.; Chizzotti, M.L.; Vital, C.E.; Baracat-Pereira, M.C.; Barros, E.; Busato, K.C.; Gomes, R.A.; Ladeira, M.M.; Martins, T.D.S. Differences in Beef Quality between Angus (*Bos Taurus Taurus*) and Nellore (*Bos Taurus Indicus*) Cattle through a Proteomic and Phosphoproteomic Approach. *PLoS ONE* **2017**, *12*, e0170294. [CrossRef]
39. Ouali, A.; Gagaoua, M.; Boudida, Y.; Becila, S.; Boudjellal, A.; Herrera-Mendez, C.H.; Sentandreu, M.A. Biomarkers of Meat Tenderness: Present Knowledge and Perspectives in Regards to Our Current Understanding of the Mechanisms Involved. *Meat Sci.* **2013**, *95*, 854–870. [CrossRef] [PubMed]
40. Weston, A.R.; Rogers, R.W.; Althen, T.G. Review: The Role of Collagen in Meat Tenderness. *Prof. Anim. Sci.* **2002**, *18*, 107–111. [CrossRef]
41. Lepetit, J. Collagen Contribution to Meat Toughness: Theoretical Aspects. *Meat Sci.* **2008**, *80*, 960–967. [CrossRef]
42. Bjarnadóttir, S.G.; Hollung, K.; Høy, M.; Bendixen, E.; Codrea, M.C.; Veiseth-Kent, E. Changes in Protein Abundance between Tender and Tough Meat from Bovine Longissimus Thoracis Muscle Assessed by Isobaric Tag for Relative and Absolute Quantitation (ITRAQ) and 2-Dimensional Gel Electrophoresis Analysis1. *J. Anim. Sci.* **2012**, *90*, 2035–2043. [CrossRef]
43. Liao, H.; Zhang, X.H.; Qi, Y.X.; Wang, Y.Q.; Pang, Y.Z.; Liu, Z.B.Z.P. The Relationships of Collagen and ADAMTS2 Expressionlevels with Meat Quality Traits in Cattle. *Indian J. Anim. Res.* **2016**, *52*, 167–172. [CrossRef]

44. Cui, L.; Cheng, Z.; Hu, K.; Pang, Y.; Liu, Y.; Qian, T.; Quan, L.; Dai, Y.; Pang, Y.; Ye, X.; et al. Prognostic Value of the PDLIM Family in Acute Myeloid Leukemia. *Am. J. Transl Res.* **2019**, *11*, 6124–6131.

45. Ríos, H.; Paganelli, A.R.; Fosser, N.S. The Role of PDLIM1, a PDZ-LIM Domain Protein, at the Ribbon Synapses in the Chicken Retina. *J. Comp. Neurol.* **2020**, *528*, 1820–1832. [CrossRef]

46. Schönichen, A.; Mannherz, H.G.; Behrmann, E.; Mazur, A.J.; Kühn, S.; Silván, U.; Schoenenberger, C.-A.; Fackler, O.T.; Raunser, S.; Dehmelt, L.; et al. FHOD1 Is a Combined Actin Filament Capping and Bundling Factor That Selectively Associates with Actin Arcs and Stress Fibers. *J. Cell Sci.* **2013**, *126*, 1891–1901. [CrossRef]

47. Hemerich, D.; Pei, J.; Harakalova, M.; van Setten, J.; Boymans, S.; Boukens, B.J.; Efimov, I.R.; Michels, M.; van der Velden, J.; Vink, A.; et al. Integrative Functional Annotation of 52 Genetic Loci Influencing Myocardial Mass Identifies Candidate Regulatory Variants and Target Genes. *Circ. Genom Precis Med.* **2019**, *12*, e002328. [CrossRef]

48. Hansen, M.D.H.; Kwiatkowski, A.V. Chapter One—Control of Actin Dynamics by Allosteric Regulation of Actin Binding Proteins. In *International Review of Cell and Molecular Biology*; Jeon, K.W., Ed.; Academic Press: Cambridge, MA, USA, 2013; Volume 303, pp. 1–25.

49. Ferguson, D.M.; Gerrard, D.E. Regulation of Post-Mortem Glycolysis in Ruminant Muscle. *Anim. Prod. Sci.* **2014**, *54*, 464–481. [CrossRef]

50. Orosz, F.; Oláh, J.; Ovádi, J. Triosephosphate Isomerase Deficiency: New Insights into an Enigmatic Disease. *Biochim. Biophys. Acta (BBA) Mol. Basis Dis.* **2009**, *1792*, 1168–1174. [CrossRef]

51. Esposito, G.; Vitagliano, L.; Costanzo, P.; Borrelli, L.; Barone, R.; Pavone, L.; Izzo, P.; Zagari, A.; Salvatore, F. Human Aldolase A Natural Mutants: Relationship between Flexibility of the C-Terminal Region and Enzyme Function. *Biochem. J.* **2004**, *380*, 51–56. [CrossRef]

52. Hughes, J.; Clarke, F.; Li, Y.; Purslow, P.; Warner, R. Differences in Light Scattering between Pale and Dark Beef Longissimus Thoracis Muscles Are Primarily Caused by Differences in the Myofilament Lattice, Myofibril and Muscle Fibre Transverse Spacings. *Meat Sci.* **2019**, *149*, 96–106. [CrossRef]

53. Israelsen, W.J.; Vander Heiden, M.G. Pyruvate Kinase: Function, Regulation and Role in Cancer. *Semin. Cell Dev. Biol.* **2015**, *43*, 43–51. [CrossRef]

54. Daugaard, M.; Rohde, M.; Jäättelä, M. The Heat Shock Protein 70 Family: Highly Homologous Proteins with Overlapping and Distinct Functions. *Febs Lett.* **2007**, *581*, 3702–3710. [CrossRef]

55. Jiang, B.; Liang, P.; Deng, G.; Tu, Z.; Liu, M.; Xiao, X. Increased Stability of Bcl-2 in HSP70-Mediated Protection against Apoptosis Induced by Oxidative Stress. *Cell Stress Chaperones* **2011**, *16*, 143–152. [CrossRef]

56. Mayer, M.P. Hsp70 Chaperone Dynamics and Molecular Mechanism. *Trends Biochem. Sci.* **2013**, *38*, 507–514. [CrossRef]

57. Morzel, M.; Terlouw, C.; Chambon, C.; Micol, D.; Picard, B. Muscle Proteome and Meat Eating Qualities of Longissimus Thoracis of "Blonde d'Aquitaine" Young Bulls: A Central Role of HSP27 Isoforms. *Meat Sci.* **2008**, *78*, 297–304. [CrossRef]

58. Caruso Bavisotto, C.; Alberti, G.; Vitale, A.M.; Paladino, L.; Campanella, C.; Rappa, F.; Gorska, M.; Conway de Macario, E.; Cappello, F.; Macario, A.J.L.; et al. Hsp60 Post-Translational Modifications: Functional and Pathological Consequences. *Front. Mol. Biosci.* **2020**, *7*. [CrossRef]

59. McDonagh, B.; Sheehan, D. Redox Proteomics in the Blue Mussel Mytilus Edulis: Carbonylation Is Not a Pre-Requisite for Ubiquitination in Acute Free Radical-Mediated Oxidative Stress. *Aquat. Toxicol.* **2006**, *79*, 325–333. [CrossRef]

60. Sierra, V.; Oliván, M. Role of Mitochondria on Muscle Cell Death and Meat Tenderization. *Recent Pat. Endocr. Metab. Immune Drug Discov.* **2013**. [CrossRef]

61. Lana, A.; Zolla, L. Apoptosis or Autophagy, That Is the Question: Two Ways for Muscle Sacrifice towards Meat. *Trends Food Sci. Technol.* **2015**, *46*, 231–241. [CrossRef]

62. Thomas, K.J.; McCoy, M.K.; Blackinton, J.; Beilina, A.; van der Brug, M.; Sandebring, A.; Miller, D.; Maric, D.; Cedazo-Minguez, A.; Cookson, M.R. DJ-1 Acts in Parallel to the PINK1/Parkin Pathway to Control Mitochondrial Function and Autophagy. *Hum. Mol. Genet.* **2011**, *20*, 40–50. [CrossRef]

63. Gagaoua, M.; Bonnet, M.; Ellies-Oury, M.-P.; De Koning, L.; Picard, B. Reverse Phase Protein Arrays for the Identification/Validation of Biomarkers of Beef Texture and Their Use for Early Classification of Carcasses. *Food Chem.* **2018**, *250*, 245–252. [CrossRef]

64. Jia, X.; Veiseth-Kent, E.; Grove, H.; Kuziora, P.; Aass, L.; Hildrum, K.I.; Hollung, K. Peroxiredoxin-6–a Potential Protein Marker for Meat Tenderness in Bovine Longissimus Thoracis Muscle. *J. Anim. Sci.* **2009**, *87*, 2391–2399. [CrossRef]

65. Fan, J.; Ren, H.; Jia, N.; Fei, E.; Zhou, T.; Jiang, P.; Wu, M.; Wang, G. DJ-1 Decreases Bax Expression through Repressing P53 Transcriptional Activity. *J. Biol. Chem.* **2008**, *283*, 4022–4030. [CrossRef] [PubMed]

66. Fisher, A.B. Peroxiredoxin 6: A Bifunctional Enzyme with Glutathione Peroxidase and Phospholipase A2 Activities. *Antioxid. Redox Signal.* **2010**, *15*, 831–844. [CrossRef]

67. Cao, Z.; Bhella, D.; Lindsay, J.G. Reconstitution of the Mitochondrial PrxIII Antioxidant Defence Pathway: General Properties and Factors Affecting PrxIII Activity and Oligomeric State. *J. Mol. Biol.* **2007**, *372*, 1022–1033. [CrossRef] [PubMed]

68. Grabež, V.; Kathri, M.; Phung, V.; Moe, K.M.; Slinde, E.; Skaugen, M.; Saarem, K.; Egelandsdal, B. Protein Expression and Oxygen Consumption Rate of Early Postmortem Mitochondria Relate to Meat Tenderness. *J. Anim. Sci.* **2015**, *93*, 1967–1979. [CrossRef]

69. Polati, R.; Menini, M.; Robotti, E.; Millioni, R.; Marengo, E.; Novelli, E.; Balzan, S.; Cecconi, D. Proteomic Changes Involved in Tenderization of Bovine Longissimus Dorsi Muscle during Prolonged Ageing. *Food Chem.* **2012**, *135*, 2052–2069. [CrossRef]

70. Malheiros, J.M.; Braga, C.P.; Grove, R.A.; Ribeiro, F.A.; Calkins, C.R.; Adamec, J.; Chardulo, L.A.L. Influence of Oxidative Damage to Proteins on Meat Tenderness Using a Proteomics Approach. *Meat Sci.* **2019**, *148*, 64–71. [CrossRef]
71. Gagaoua, M.; Claudia Terlouw, E.M.; Boudjellal, A.; Picard, B. Coherent Correlation Networks among Protein Biomarkers of Beef Tenderness: What They Reveal. *J. Proteom.* **2015**, *128*, 365–374. [CrossRef] [PubMed]
72. Cai, C.; Masumiya, H.; Weisleder, N.; Matsuda, N.; Nishi, M.; Hwang, M.; Ko, J.-K.; Lin, P.; Thornton, A.; Zhao, X.; et al. MG53 Nucleates Assembly of Cell Membrane Repair Machinery. *Nat. Cell Biol.* **2009**, *11*, 56–64. [CrossRef] [PubMed]

 *foods*

Article

Protein Array-Based Approach to Evaluate Biomarkers of Beef Tenderness and Marbling in Cows: Understanding of the Underlying Mechanisms and Prediction

Mohammed Gagaoua *,†, **Muriel Bonnet and Brigitte Picard** *

National Research Institute for Agriculture, Food and Environment, Université Clermont Auvergne, VetAgro Sup, UMR Herbivores, F-63122 Saint Genès Champanelle, France; muriel.bonnet@inrae.fr
* Correspondence: gmber2001@yahoo.fr (M.G.); brigitte.picard@inrae.fr (B.P.)
† Present Address: Food Quality and Sensory Science Department, Teagasc Ashtown Food Research Centre, Ashtown, Dublin 15, Ireland.

Received: 1 August 2020; Accepted: 21 August 2020; Published: 26 August 2020

Abstract: This study evaluated the potential of a panel of 20 protein biomarkers, quantified by Reverse Phase Protein Array (RPPA), to explain and predict two important meat quality traits, these being beef tenderness assessed by Warner–Bratzler shear force (WBSF) and the intramuscular fat (IMF) content (also termed marbling), in a large database of 188 Protected Designation of Origin (PDO) Maine-Anjou cows. Thus, the main objective was to move forward in the progression of biomarker-discovery for beef qualities by evaluating, at the same time for the two quality traits, a list of candidate proteins so far identified by proteomics and belonging to five interconnected biological pathways: (i) energy metabolic enzymes, (ii) heat shock proteins (HSPs), (iii) oxidative stress, (iv) structural proteins and (v) cell death and protein binding. Therefore, three statistical approaches were applied, these being Pearson correlations, unsupervised learning for the clustering of WBSF and IMF into quality classes, and Partial Least Squares regressions (PLS-R) to relate the phenotypes with the 20 biomarkers. Irrespective of the statistical method and quality trait, seven biomarkers were related with both WBSF and IMF, including three small HSPs (CRYAB, HSP20 and HSP27), two metabolic enzymes from the oxidative pathway (MDH1: Malate dehydrogenase and ALDH1A1: Retinal dehydrogenase 1), the structural protein MYH1 (Myosin heavy chain-IIx) and the multifunctional protein FHL1 (four and a half LIM domains 1). Further, three more proteins were retained for tenderness whatever the statistical method, among which two were structural proteins (MYL1: Myosin light chain 1/3 and TNNT1: Troponin T, slow skeletal muscle) and one was glycolytic enzyme (ENO3: β-enolase 3). For IMF, two proteins were, in this trial, specific for marbling whatever the statistical method: TRIM72 (Tripartite motif protein 72, negative) and PRDX6 (Peroxiredoxin 6, positive). From the 20 proteins, this trial allowed us to qualify 10 and 9 proteins respectively as strongly related with beef tenderness and marbling in PDO Maine-Anjou cows.

Keywords: meat tenderness; fat; proteomics; proteins; enzymes; quantification; biological pathways; chemometrics

---

1. Introduction

During the last two decades, OMICs techniques, especially proteomics, have been applied by meat scientists to understand the modifications occurring in post-mortem muscle in an attempt to explain the variation in several meat quality traits [1–3]. Further, proteomics allowed the identification of putative biomarkers (for review: [4–6]), with the objective of predicting the potential quality and also of proposing a molecular test for the beef industry [7]. Overall, two-dimensional gel electrophoresis

Foods **2020**, *9*, 1180; doi:10.3390/foods9091180　　　　　www.mdpi.com/journal/foods

combined with mass spectrometry was efficiently used to map and characterize bovine muscle proteins [8]. The main protein biomarkers so far identified belong to myriad interconnected pathways, such as structure and contraction, heat stock proteins, energy metabolism including the glycolytic and oxidative pathways, oxidative stress, transport, binding and signaling, apoptosis, and proteolysis, including endogenous muscle inhibitors such as serpins, cystatins and calpastatins [4,5,9,10].

Among the most investigated beef qualities using proteomics, tenderness has gained the most attention [11,12]. Beef tenderness is considered worldwide to be one of the most critical quality attributes for consumers and for re-purchase decisions. However, with beef tenderness being a multifactorial trait, it is highly variable, and is impacted by both intrinsic and extrinsic factors measurable along the continuum from farm-to-fork [13,14]. At the carcass and muscle levels, the intramuscular fat (IMF) content, also termed marbling, impacts tenderness variation and is considered as an important driver of beef palatability [15,16]. Proteomics was also applied to identify potential biomarkers of IMF and to address differences in adiposity [17] under several factors, such as breed [18], rearing practices [19], individual variability [20–22] and distinct stages of IMF development [23,24].

Based on the work by Rifai et al. [25], we recently detailed the process of meat quality biomarker discovery that should be followed so as to identify, evaluate and validate protein biomarkers at the research/industry scale [4,26]. This process is composed of six main steps that are discovery/identification, qualification, verification, research assay optimization, industrial validation and commercialization. In fact, the products of the discovery phase are lists of 10 to 100 proteins with different abundances between two compared situations or conditions (i.e., tender versus tough, or lean versus fat, etc.). Thanks to the current developments in the high-throughput techniques, several research groups have been able to move forward in the process of beef tenderness and IMF (marbling) biomarkers discovery by assaying fast techniques for the qualification of biomarkers in a few animals, which entails the confirmation of the differential abundances of the proteins using a method different from the one used for the discovery step. Thus, some shortlisted biomarkers were tested using immune-based techniques such as western blotting [27–29], Dot-Blot [30–34] and Reverse Phase Protein Array (RPPA) [35,36], or using label-free gel mass spectrometry tools, namely Selected Reaction Monitoring (SRM) and the sequential window acquisition of all theoretical spectra (SWATH) [37]. More recently, a combination of RPPA and Parallel Reaction Monitoring (PRM) [26] was used to both qualify and verify the reliability of a set of 10 proteins for predicting tenderness and marbling. Therefore, this study was designed to check the ability of a previous list of 20 protein candidate markers quantified by the RPPA technique to discriminate both beef tenderness and IMF in a large database of PDO Maine-Anjou cows, comprising 188 animals. It further aimed to qualify the most robust candidates that are predictors of both tenderness and marbling whatever the statistical method. The data of this trial will further increase our understanding of the role played by the qualified proteins in these two important beef qualities, to propose in the future generic biomarker-based tools for the early sorting of carcasses to meet consumer and industrial expectations.

2. Materials and Methods

2.1. Experimental Designs, Cows Handling and Slaughtering

This trial was conducted using two replicate groups of 110 and 78 animals for a total of 188 Protected Designation of Origin (PDO) Maine-Anjou cows representative of the "Rouge des Prés" breed (for details on the experimental designs refer to Gagaoua, et al. [38] and Picard et al. [19]). The Rouge des Prés breed has, since 2004, been approved to be used in France for PDO meat production, hence it has a special economic importance for the valorization of local breeds as it is a dual-purpose cattle used for both beef and milk. PDO is the name of a geographical region or specific area that is recognized by official rules to produce certain foods with special characteristics related to location. The PDO regulation covers agricultural products and foodstuffs that are produced, processed and prepared in a given geographical area using recognized know-how in this specific zone. The certification label

PDO and other labels are required by the European Commission for assuring the authenticity of food products.

The main characteristics of the PDO Maine-Anjou breed are an age at slaughter lower than 10 years, having calved at least once and a minimal carcass weight of 380 kg. In this trial, all the cows had these qualities, with average age at slaughter and carcass weight of 67.4 ± 13.9 months and 445 ± 676 kg, respectively. The cows originated from the north-western part of France and were collected from the same cooperative of livestock farmers located in the department of Maine-et-Loire. All the animals were slaughtered in industrial slaughterhouses (n= 110 at Elivia, Lion d'Angers, France and n = 78 at Charal, Sablé sur Sarthes, France) following the same protocol. The cows had free access to water before their slaughter but food was deprived for 24 h. The exsanguination from the jugular vein was performed after electrical stunning using a captive-bolt pistol. Slaughtering was performed in compliance with French welfare and by respecting EU regulations (Council Regulation (EC) No. 1099/2009). The carcasses were dressed according to standard commercial practices and between 30–50 min post exsanguination the carcasses were split in half then chilled for 24 h at 2–4 °C. None of the carcasses were electrically stimulated. Ultimate pH was recorded at 24 h post-mortem for all the carcasses using a pH meter equipped with a glass electrode, and none of them had pHu > 6.0 (the benchmark used to sort DFD carcasses).

2.2. Muscle and Meat Steaks Sampling

Longissimus thoracis (LT) muscle samples, known as a mixed fast oxido-glycolytic muscle, were taken from the 5th rib of the left-hand side of each carcass 24 h post-mortem. This sampling time is in line with the outcome of the discovery and validation of protein biomarkers of bee quality traits aiming at a management/prediction of the potential quality of the carcasses early post-mortem. Therefore, in this study and for three reasons we used muscle samples taken at 24 h post-mortem. First, this is in line with previous proteomics investigations from our team for the PDO Maine-Anjou breed and other laboratories for other animal types. Second, the sampling at industrial level was only possible at this time due to the company facilities and technical considerations. Third, at this time, we expect that the dynamic properties of the muscle expressed by changes in few but important proteins belonging to the energy metabolic or heat shock proteins pathways are less variable. Thus, the first part, free of connective tissue, was subsequently frozen in liquid nitrogen and stored at −80 °C until protein extractions for the quantification of the protein biomarkers using the RPPA technique. The second part of the sample was cut into 1–2 cm cross-section pieces, vacuum packed, and stored at −20 °C for intramuscular fat (IMF) content determination. The third part was cut into 20 mm thick steaks and placed in sealed plastic bags under vacuum and aged for 14 days at 4 °C (the usual and standard condition of ageing). Each loin sample was then frozen and stored at −20 °C awaiting Warner–Bratzler shear force (WBSF) measurements.

2.3. Intramuscular Fat Content Determination

The amount of IMF on each muscle sample was determined using a Dionex ASE 200 Accelerated Solvent Extractor (Dionex Corporation, Sunnyvale, CA, USA) as previously described [38]. Briefly, 1 ± 0.001 g of meat powder was placed in a 22 mL extraction cell initially prepared with a cellulose filter and silicon balls. Then, petroleum ether at a temperature and pressure of 125 °C and 103 bars respectively was used for the extraction. The slurry containing both fat and petroleum ether was collected and transferred into a previously weighed evaporation vial (±0.001 g). After 15 min of evaporation, the vial was placed in a drying oven at 105 °C for 17 h and then weighed (±0.001 g) to determine the amount of IMF in the sample. The results were expressed as the % of IMF in fresh meat.

2.4. Meat Tenderness Measurement by Warner–Bratzler Shear Force (WBSF)

For objective beef tenderness determination, Warner–Bratzler shear force (WBSF), known as a negative proxy of sensory tenderness, was measured according to Lepetit and Culioli [39] using an

INSTRON 5944 testing machine. Briefly, the frozen steaks firstly thawed for 48 h at 4 °C, then were placed for 4 h in a thermostated bath at 18 °C before cooking on a double grooved plate griddle (SOFRACA, Morangis, France) set at 300 °C until the end-point temperature of 55 °C, which is the usual cooking temperature in France [40]. A temperature probe (Type K-Thermocouple, HI 98704, HANNA Instruments, Newark, NJ, USA) at the geometric center of the steak was used to control the end-point cooking temperature. After cooking, each steak sample was used to prepare five cores (1 cm × 1 cm × 4 cm) parallel to the longitudinal orientation of the muscle fiber. WBSF was assessed 2 or 3 times per core in order to obtain around 10 repetitions per sample. A 1 kN load cell and a 60 mm/min crosshead speed were used (universal testing machine, MTS, Synergie 200H). The force at the rupture during shear compression testing was expressed in N/cm^2.

2.5. Protein Extraction and Quantification

The muscle proteins were extracted by homogenization of the samples in "Precellys 24" tissue homogenizer (Bertin technologies, Saint Quentin-en-Yvelines, France) following the previously described protocol [41]. Briefly, 80 mg of frozen muscle were ground using 1.4 mm ceramic beads in an extraction buffer containing 50 mM Tris (pH 6.8), 5% glycerol, 2% SDS, 2 mM DTT, 2.5 mM EDTA, 2.5 mM EGTA, 1 × HALT Phosphatase inhibitor, Protease inhibitor cocktail complete MINI EDTA-free, 2 mM Na_3VO_4 and 10 mM NaF. The extracts were then boiled for 10 min at 100 °C, sonicated to reduce viscosity and centrifuged for 10 min at 25,000× *g*. The supernatants were collected and stored at −80 °C until use for protein assay and biomarkers quantification by RPPA.

Protein concentrations were determined with a commercial protein assay (Pierce BCA reducing agent compatible kit, ref. 23252, Thermo Scientific, Waltham, MA, USA) with bovine serum albumin (BSA) as standard.

2.6. Reverse Phase Protein Array (RPPA) for Protein Biomarkers Quantification

The relative abundances of 20 protein biomarkers of tenderness and/or IMF (Table 1) were quantified by the Reverse Phase Protein Array [42] following exactly the same protocol recently detailed by our group on bovine muscle [26,35,36,41]. The 20 proteins belong to five biological pathways that are:

(***i***) ***Energy metabolic enzymes (n = 7)***: Malate dehydrogenase (MDH1), β-enolase 3 (ENO3), Retinal dehydrogenase 1 (ALDH1A1), Triosephosphate isomerase (TPI1), Phosphoglycerate kinase 1 (PGK1), Fructose-bisphosphate aldolase (ALDOA) and Glycogen phosphorylase (PYGB);

(***ii***) ***Heat shock proteins (n = 5)***: αB-crystallin (CRYAB), Hsp20 (HSPB6), Hsp27 (HSPB1), Hsp40 (DNAJA1) and Hsp70-1A (HSPA1A);

(***iii***) ***Oxidative stress proteins (n = 1)***: Peroxiredoxin6 (PRDX6);

(***iv***) ***Structural proteins (n = 5)***: MLC-1F (MYL1), Myosin heavy chain-IIx (MYH1), Troponin T, slow skeletal muscle (TNNT1), Titin (TTN) and Tubulin alpha-4A chain (TUBA4A);

(***v***) ***Cell death and protein binding (n = 2)***: Tripartite motif protein 72 (TRIM72) and Four and a half LIM domains 1 (FHL1).

After protein quantification by RPPA and for the determination of the relative abundance of each protein, the raw data were normalized using NormaCurve following the method described by Troncale et al. [43]. This is a SuperCurve-based method that simultaneously quantifies and normalizes RPPA data for fluorescent background per spot, the total protein stain and the potential spatial bias on the slide. Then, each RPPA slide was median-centered and scaled (divided by median absolute deviation). Further corrections to the sample loadings effects were performed individually for each array by correcting the dependency of the data for individual arrays with the median value of each sample over all 20 arrays using a linear regression.

Table 1. List of the 20 protein biomarkers quantified using the Reverse Phase Protein Array (RPPA) technique [1].

Protein Biomarkers Name (*Gene*)	Uniprot ID	Monoclonal (Mo) or Polyclonal (Po) Antibodies References	Antibody Dilutions
Energy metabolic enzymes			
Malate dehydrogenase (*MDH1*)	Q3T145	Mo. anti-pig Rockland 100-601-145	1/1000
β-enolase 3 (*ENO3*)	Q3ZC09	Mo. anti-human Abnova Eno3 (M01), clone 5D1	1/30,000
Retinal dehydrogenase 1 (*ALDH1A1*)	P48644	Po. anti-bovine Abcam ab23375	1/500
Triosephosphate isomerase (*TPI1*)	Q5E956	Po. anti-human Novus NBP1-31470	1/50,000
Phosphoglycerate kinase 1 (*PGK1*)	Q3T0P6	Po. anti-human Abcam ab90787	1/5000
Fructose-bisphosphate aldolase (*ALDOA*)	A6QLL8	Po. anti-human Sigma AV48130	1/4000
Glycogen phosphorylase (*PYGB*)	Q3B7M9	Po. anti-human Santa Cruz SC-46347	1/250
Heat shock proteins			
Alpha-crystallin B chain (*CRYAB*)	P02510	Mo. anti-bovine Assay Designs SPA-222	1/1000
Heat shock protein beta-6, Hsp20 (*HSPB6*)	Q148F8	Mo. anti-human Santa Cruz HSP20-11:SC51955	1/500
Heat shock protein beta-1, Hsp27 (*HSPB1*)	Q3T149	Mo. anti-human Santa Cruz HSP27 (F-4):SC13132	1/3000
DnaJ homolog subfamily A member 1, Hsp40 (*DNAJA1*)	Q5E954	Mo. anti-human Santa Cruz HSP40-4 (SPM251):SC-56400	1/250
Heat shock 70 kDa protein 1A, Hsp70-1A (*HSPA1A*)	Q27975	Mo. anti-human RD Systems MAB1663	1/1000
Oxidative stress proteins			
Peroxiredoxin-6 (*PRDX6*)	O77834	Mo. anti-human Abnova PRDX6 (M01), clone 3A10-2A11	1/500
Structural proteins			
Myosin light chain 1/3 (*MYL1*)	A0JNJ5	Po. anti-human Abnova MYL1 (A01)	1/1000
Myosin heavy chain-IIx (*MYH1*)	Q9BE40	Mo anti-bovine Biocytex 8F4	1/500
Troponin T, slow skeletal muscle (*TNNT1*)	Q8MKH6	Po. anti-human Sigma SAB2102501	1/4000
Titin (*TTN*)	Q8WZ42	Mo. anti-human Novocastra NCL-TITIN	1/100
Tubulin alpha-4A chain (*TUBA4A*)	P81948	Mo anti-human Sigma T6074	1/1000
Cell death and protein binding			
Tripartite motif protein 72 (*TRIM72*)	E1BE77	Po. anti-human Sigma SAB2102571	1/2000
Four and a half LIM domains 1 (*FHL1*)	F1MR86	Po. anti-human Sigma AV34378	1/5000

[1] The suppliers and conditions for each primary antibody after western blotting validation are given as in Picard et al. [36].

2.7. Statistical Analyses

The statistical analyses were carried out using XLSTAT 2018.3 (AddinSoft, Paris, France). Raw data were scrutinized for data entry errors and outliers using Smirnov–Grubb's outlier test at a significance level of 5%. Then, all the data were normalized for replicates (experiment) and the factor related to the rearing practices of the animals [19,44]. This step was based on Z-scores, which represent the number of standard deviations for each observation relative to the mean of the corresponding replicate/condition. Therefore, after this transformation, the data had a mean of 0 and standard deviation of 1. Following this first step, three main statistical approaches were applied to the whole database to predict/explain each quality trait and evaluate the potential of each biomarker to contribute to its associated variation.

Correlations: Pearson correlation coefficients, based on the Z-scores, at the level of 5% were computed between WBSF values and IMF content with the 20 protein biomarkers.

Clustering into WBSF and IMF classes: three unsupervised learning methods, which were (i) hierarchical cluster analysis (HCA), (ii) *k*-means and (iii) partitioning around medoids (PAM), were tested as previously described [45] to create meat quality classes of WBSF and IMF. For both quality traits, *k*-means gave the best results based on the average silhouette width (*Si*) criterion (Euclidean distance), allowing in the two cases three clusters or classes (*k* = 3) that we named tender, medium and tough for WBSF, and for IMF (marbling) fat, medium and lean.

The value of *Si* ranges from −1 to +1, with observations that have a positive large *Si* being very well clustered. Those close to 0 (low *Si*) means that the observation lies between two clusters and those with negative *Si* are partitioned in the wrong cluster [46]. Afterward, variance analyses (ANOVA) were used to compare the protein abundances among the classes for each beef quality trait. Significant differences were performed using Tukey's test at a significance level of 5% and were presented using heatmaps. Subsequently, principal component analyses (PCA) for each beef quality trait were carried out using the significant differential proteins ($p < 0.05$) to illustrate in a more complete picture the separation of the classes, and thus of the individuals and the distribution of the variables.

Partial Least Squares regressions (PLS-R): to deepen our understanding of the mechanisms and identify the most robust explanatory protein biomarkers, Partial Least Squares regressions (PLS-R) were performed per beef quality trait to generate explanatory models using the optimal number of components in each case [47]. This is an appropriate tool to include all the 20 biomarkers in one model and identify those that had a biological and relevant significance using the criterion of the variable's importance in the projection (VIP). This filter method based on VIP scores estimates the importance of each protein in the projection used in a PLS model. A protein with a VIP > 1.0 is considered important, thus highly influential in a given model, 0.8 < VIP < 1.0 is considered moderately influential, and any protein with VIP < 0.8 is less influential thus considered weak and rejected. For the selection of the variables, the jack-knife method was included in the PLS-R as a selective parameter. In this study, all proteins for which the VIP scores were above a threshold of 1.0 and 0.8 (highly and moderately influential proteins) were considered and then compared with those selected from correlation and variance analyses to be used in the future for validation on meat from PDO Maine-Anjou cows.

3. Results

3.1. Pearson Correlation Analyses between the Biomarkers and Meat Quality Traits

The correlation coefficients computed between WBSF values and IMF content with the 20 protein biomarkers and for all the 188 cows are given in Table 2.

For WBSF, 11 proteins were significantly ($p < 0.05$) correlated with this texture trait. Four proteins were from the energy metabolism pathway: MDH1, ENO3, PGK1 (all negative) and ALDH1A1 (positive). These were followed by three small heat shock proteins (sHSP): CRYAB, HSP20 and HSP27 (all positive). Three were structural proteins: MYL1 and MYH1 (both negative) and TNNT1 (positive). One protein from the last pathway, which was FHL1, was positively correlated with WBSF.

For IMF content, 10 proteins were significantly ($p < 0.05$) correlated, of which 2 were energy metabolic enzymes that are MDH1 (negative) and ALDH1A1 (positive), 3 were sHSP (CRYAB, HSP20 and HSP27) that were all positive, and 2 were structural proteins (MYH1 (negative) and TNNT1 (positive)). The PRDX6 from the oxidative pathway was positive, and two proteins from the last pathway, TRIM72 and FHL1, were respectively negatively and positively correlated with IMF content (Table 2).

From the above, eight proteins, MDH1, ALDH1A1, CRYAB, HSP20, HSP27, MYH1, TNNT1 and FHL1, were common to the two traits. Interestingly, sHSPs seemed from this analysis to be important biomarkers for both WBSF and IMF.

Table 2. Pearson correlation analyses between the 20 protein biomarkers with WBSF values and Intramuscular fat (IMF).

Protein Biomarkers [1]	WBSF	IMF
Energy metabolic enzymes		
MDH1	**−0.29 *****	**−0.18 ***
ENO3	−0.21 **	-
ALDH1A1	**+0.34 *****	**+0.38 *****
TPI1	-	-
PGK1	−0.15 *	-
ALDOA	-	-
PYGB	-	-
Heat shock proteins		
CRYAB	**+0.32 *****	**+0.37 *****
HSP20 (HSPB6)	**+0.21 ****	**+0.31 *****
HSP27 (HSPB1)	**+0.28 *****	**+0.22 ****
HSP40 (DNAJA1)	-	-
HSP70-1A	-	-
Oxidative stress proteins		
PRDX6	-	+0.20 *
Structural proteins		
MYL1	−0.18 *	-
MYH1	**−0.26 *****	**−0.26 ****
TNNT1	**+0.22 ****	**+0.15 ***
TTN	-	-
TUBA4A	-	-
Cell death and protein binding		
TRIM72	-	−0.33 ***
FHL1	**+0.18 ***	**+0.22 ****

[1] The correlation coefficients (q = 8) in bold font highlight the protein biomarkers common to WBSF and IMF. WBSF: Warner–Bratzler shear force; IMF: Intramuscular fat. Significance: * $p < 0.05$; ** $p < 0.01$; *** $p < 0.001$.

3.2. Discriminant Biomarkers of WBSF and Marbling

In the attempt to evaluate discriminant biomarkers for both WBSF and IMF classes, the *k*-means algorithm, as the best clustering method, allowed the identification of three classes for each beef quality trait.

For WBSF, the *k*-means clustering of the 188 steaks of the PDO Maine-Anjou cows categorized them into tender ($n = 93$), medium ($n = 71$) and tough ($n = 24$) samples (Figure 1). The tender class has a mean value of 32.96 ± 3.99 N/cm^2, a coefficient of variation (CV) of 12% and WBSF values ranging between 23.05 and 38.76 N/cm^2. The medium tenderness class has a mean value of 44.74 ± 3.69 N/cm^2, a CV of 8% and WBSF values ranging between 39.00 and 52.22 N/cm^2. The tough class has a mean value of 61.18 ± 7.87 N/cm^2, a CV of 13% and WBSF values ranging between 53.03 and 81.49 N/cm^2. Comparison of the protein abundances of the 20 biomarkers, based on variance analysis among the three clusters, highlighted that 11 proteins were significantly different at the level of 5% (Figure 1a). Among them, three proteins were from the energy metabolism pathway, MDH1 and ENO3 being highly abundant in the tender class compared to the tough samples, and the inverse was found for ALDH1A1 ($p < 0.001$). The three sHSPs (CRYAB, HSP20 and HSP27) were all highly abundant in the tough compared to the tender class. Among the structural proteins, three proteins were discriminant, which were MYL1 and MYH1 (highly abundant in tender meat), and TNNT1 was found to be more abundant in the tough class. Finally, TRIM72 and FHL1 from the last pathway were respectively highly and less abundant in the tender class, compared to the tough class (Figure 1a). The main significant discriminant proteins were MDH1, ALDH1A1, CRYAB and HSP27 ($p < 0.001$), followed by MYH1, TNNT1 and FHL1 ($p < 0.01$), and then ENO3, HSP20, MYL1 and TRIM72 ($p < 0.05$). The projection of these WBSF discriminant proteins on a PCA allowed acceptable separation of the three WBSF classes (Figure 1b). The first two principal components (PC) explained around 50% of the WBSF variability, with most variation being explained by the first PC (34.1%). The cows characterized by tender meat, thus being in the tender class, were all loaded on the left, the medium were in the center, and the tough were on the right. A total of six proteins characterize the tough class and the remaining five proteins were higher in the tender class (Figure 1b).

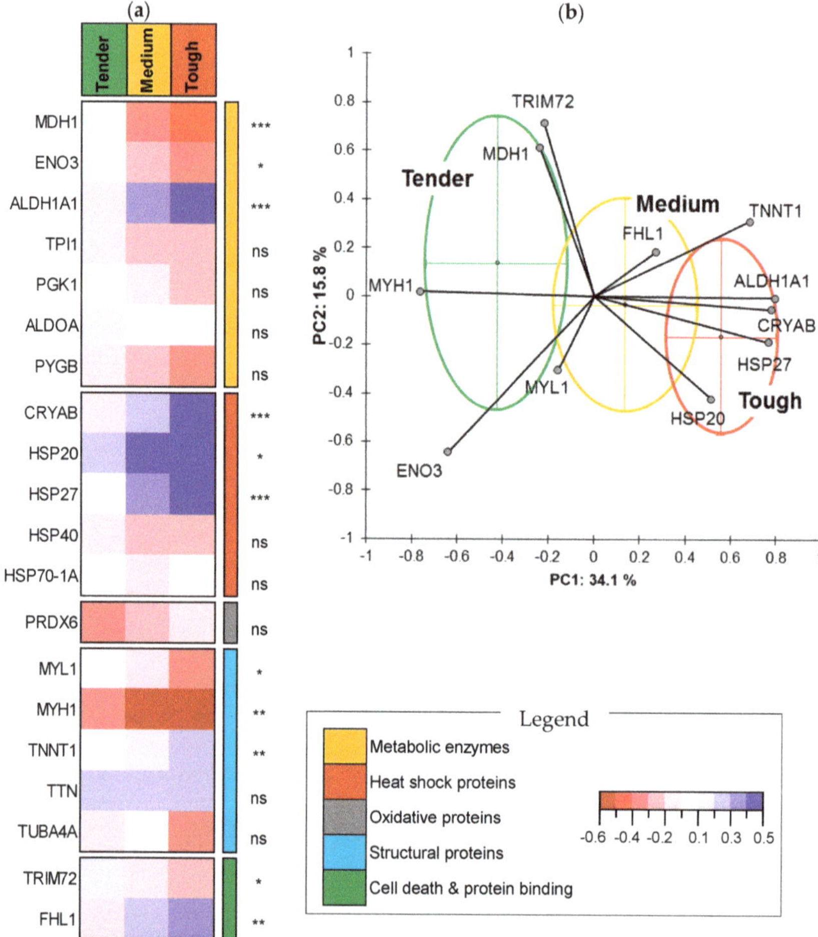

Figure 1. Protein biomarkers differing among the tenderness classes (Tender (n = 93), Medium (n = 71) and Tough (n = 24)). (**a**) Heatmap comparing the protein abundances among the three WBSF (tenderness) classes. Significance—ns: not significant; *: $p < 0.05$; **: $p < 0.01$; ***: $p < 0.001$. The proteins are given by their biological family following the legend. (**b**) Principal component analysis highlighting the distribution of the individuals of each tenderness class based on the 11 discriminant protein biomarkers. Individuals belonging to the same class are encircled in clusters using the corresponding schematic colors. The descriptive statistics of the three tenderness classes are as follows—**Tender class:** mean value of 32.96 ± 3.99 N/cm^2 (CV, 12%), Min = 23.05 and Max = 38.76 N/cm^2. **Medium class:** mean value of 44.74 ± 3.69 N/cm^2 (CV, 8%), Min = 39.00 and Max = 52.22 N/cm^2. **Tough class:** mean value of 61.18 ± 7.87 N/cm^2 (CV, 13%), Min = 53.03 and Max = 81.49 N/cm^2.

For IMF, the 188 steaks of the PDO Maine-Anjou cows were categorized into three marbling classes, as they were for WBSF, (Figure 2) namely fat (n = 28), medium (n = 69) and lean (n = 87). The fat class has a mean value of 7.72 ± 1.58%, a CV of 20% and values ranging between 6.34 and

13.82%. The medium fat class has a mean value of 4.72 ± 0.63%, a CV of 13% and IMF values ranging between 3.76 and 6.11%. The lean class has a mean value of 2.72 ± 0.62%, a CV of 23% and IMF values ranging between 0.45 and 3.69%. Eleven proteins were significantly different ($p < 0.05$) among the three marbling classes (Figure 2a). The main discriminant biomarkers were from the HSP superfamily, with a total of four proteins (CRYAB, HSP20, HSP27 and HSP40) that were all highly abundant in the fat class compared to the others (Figure 2a). This is followed by the energy metabolism pathway, with three proteins: MDH1 (high in lean class), ALDH1A1 (high in fat class) and PYGB (high in medium class). PRDX6 from the oxidative stress pathway and MYH1 from the structural pathway were respectively high and low in the fat class (Figure 2a,b). From the last family group, the two proteins TRIM72 and FHL1 were both different in their abundance among the marbling classes, being respectively low and high in the fat class. FHL1 was in this trial more abundant in the intermediate fat group. The main most significant discriminant proteins were ALDH1A1, CRYAB and TRIM72 ($p < 0.001$), followed by HSP20, MYH1 and FHL1 ($p < 0.01$), and then MDH1, PYGB, HSP27, HSP40 and PRDX6 ($p < 0.05$). The projection of these 11 discriminant proteins on a PCA allowed for separating efficiently the three marbling classes, especially the lean and fat samples (Figure 2b). The first two PC explained around 44% of the variation, with 30.5% in the first PC.

From the above, the clustering analysis allowed us to observe that eight proteins (MDH1, ALDH1A1, CRYAB, HSP20, HSP27, MYH1, TRIM72 and FHL1) are common for the two traits, and clearly delineate the beef quality classes. As for the correlation analyses, sHSPs seemed also to be important biomarkers for both WBSF and IMF.

3.3. Partial Least Squares for the Prediction of WBSF and Marbling Using the Panel of 20 Protein Biomarkers

The investigation using PLS-R of the relationships between the 20 protein biomarkers and the two beef quality traits evaluated on the 188 PDO Maine-Anjou cows generated explanatory models with the main drivers of their variation (Figure 3a,b). The WBSF model retained 10 proteins (Figure 3a). Among them, eight proteins had VIP values > 1.0 (ALDH1A1, CRYAB, MDH1, HSP27, MYH1, TNNT1, ENO3 and HSP20) and 2 hade 0.8 < VIP < 1.0 (FHL1 and MYL1). Three proteins for the energy metabolism, small HSP and structural proteins pathways, respectively, were retained. From the cell death and binding protein family, one protein, FHL1, was retained for WBSF.

For IMF, the PLS-R generated an explanatory model with nine proteins (Figure 3a). Among the retained proteins, seven had a VIP value > 1.0 (ALDH1A1, CRYAB, TRIM72, HSP20, MYH1, HSP27 and FHL1) and two had 0.8 < VIP < 1.0 (PRDX6 and MDH1).

For both PLS-R explanatory models, seven proteins were common (ALDH1A1, CRYAB, MDH1, HSP20, HSP27, MYH1 and FHL1) in explaining WBSF and IMF variations. Interestingly, for both quality traits models, the first ranked proteins with VIP values > 1.5 were ALDH1A1 and CRYAB (Figure 3a,b).

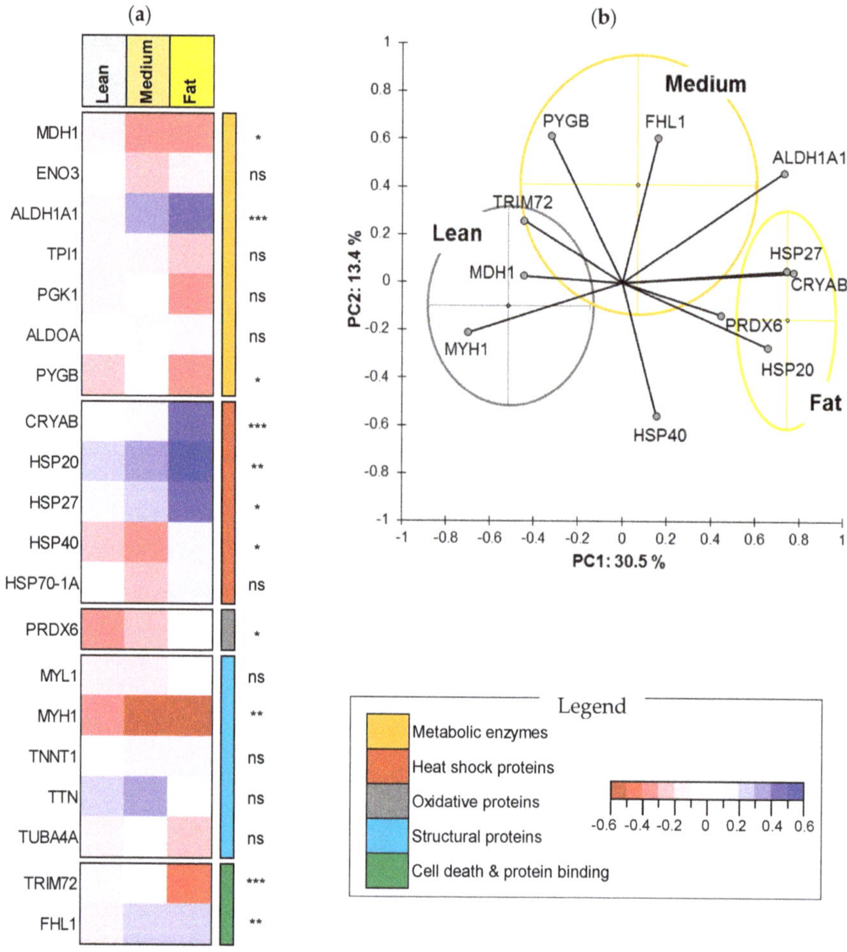

Figure 2. Protein biomarkers differing among the marbling classes (Fat ($n = 28$), Medium ($n = 69$) and Lean ($n = 87$)). (**a**) Heatmap comparing the protein abundances among the three IMF (marbling) classes. Significance—ns: not significant; *: $p < 0.05$; **: $p < 0.01$; ***: $p < 0.001$. The proteins are given by their biological family following the legend. (**b**) Principal component analysis highlighting the distribution of the individuals of each marbling class based on the 11 discriminant protein biomarkers. Individuals belonging to the same class are encircled in clusters using the corresponding schematic colors. The descriptive statistics of the three marbling classes are as follows—**Fat class:** mean value of 7.72 ± 1.58% (CV, 20%), Min = 6.34 and Max = 13.82%. **Medium class:** 4.72 ± 0.63% (CV, 13%), Min = 3.76 and Max = 6.11%. **Lean class:** 2.72 ± 0.62% (CV, 23%), Min = 0.45 and Max = 3.69%.

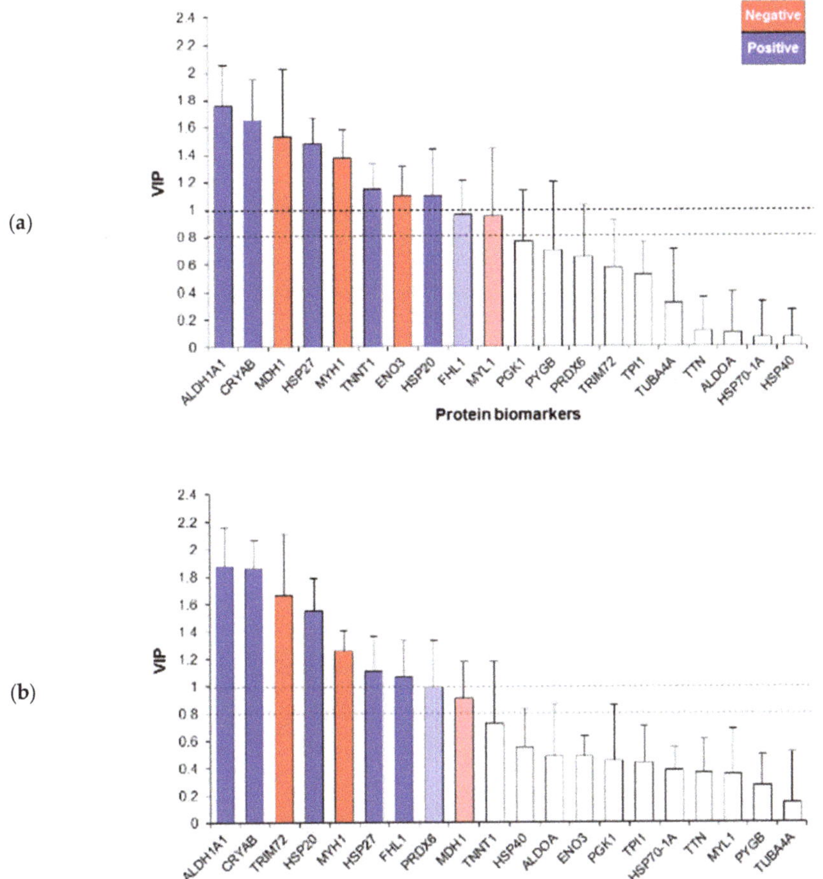

Figure 3. Partial least squares highlighting the protein biomarkers retained to explain (**a**) tenderness evaluated by WBSF and (**b**) IMF content (marbling) based on their variable importance in the projection (VIP). The proteins retained in positive and negative directions are shown in blue and red colors, respectively. For (**a**) WBSF, a total of 10 proteins were retained from which 8 had VIP > 1.0, and for (**b**) IMF, 9 proteins were retained, from which 7 had VIP > 1.0. A total of 7 proteins were common (MDH1, ALDH1A1, CRYAB, HSP20, HSP27, MYH1 and FHL1) in the two models to explain both WBSF and IMF variation.

3.4. Summary of the Putative Common Protein Biomarkers from the Three Statistical Methods

The summary of the proteins retained from the 20 panel biomarkers, based on the results of the three statistical methods presented above to explain/predict WBSF and IMF content, is given in Figure 4. Overall, irrespective of the statistical method and quality trait and based on the variable importance of the proteins in the models, their accuracy in discriminating the quality groups and their significant associations with WBSF and IMF, seven biomarkers were common, including three small HSPs (CRYAB, HSP20 and HSP27), two energy metabolic enzymes from the oxidative pathway (MDH1 and ALDH1A1), the structural protein MYH1 and the multifunctional protein FHL1. Further, for WBSF three more proteins were retained whatever the statistical method, among which two were structural proteins (MYL1 and TNNT1) and one was a metabolic enzyme (ENO3). For IMF, two proteins were retained whatever the statistical method, these being TRIM72 and PRDX6.

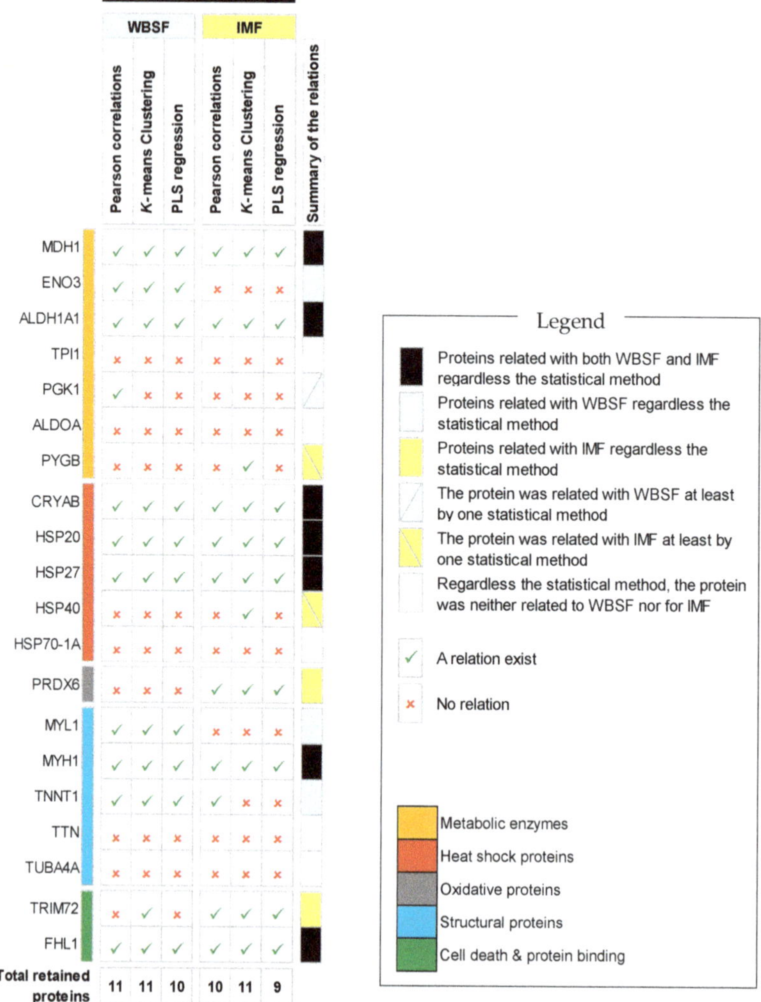

Figure 4. Summary of the evaluation of the 20 protein biomarkers quantified by RPPA using the three statistical methods (Pearson correlations, *k*-means clustering and Partial Least Squares regressions (PLS-R)) to explain/predict WBSF and IMF content on the *Longissimus thoracis* muscle of the 188 PDO Maine-Anjou cows.

4. Discussion

This trial aimed to evaluate the potential of 20 protein biomarkers, previously identified by proteomics to be potential markers of beef tenderness and marbling [4,11,12,17,19,22,26,34,35,48–50] and belonging to five interconnected biological pathways ((i) energy metabolic enzymes, (ii) heat shock proteins (HSPs), (iii) oxidative stress, (iv) structural proteins and (v) cell death and protein binding) to explain/predict two important beef quality traits for both the consumers and the meat sector (meat tenderness measured by WBSF and the marbling evaluated by the percentage of IMF content). The list of proteins was selected for qualification in this trial based on two main criteria: (i) association of the protein with the quality traits from previous studies, and (ii) validation of an antibody against the protein for its quantification by RPPA. Therefore, we used the immunological

RPPA technique for their semi-quantification, and applied three statistical methods to explain the variability of each quality trait, being Pearson correlations to assess the type of associations with the quality traits, unsupervised learning to perform a clustering of the quality traits and determine the main protein splitters, and Partial Least Squares regressions (PLS-R) to propose the first overall regressions models and identify the main predictor proteins based on their importance in the model. This firstly correlates the relative abundances of the candidate protein markers with WBSF and IMF, and secondly describes the consistencies and differences for the two traits.

This study is the first to use such a large database of 188 cows to perform this qualification/evaluation step on the selected 20 protein biomarkers, hence allowing us to move forward in the progression of biomarkers discovery for beef qualities [4] by evaluating at the same time two beef quality traits. From the evaluated list of proteins, we revealed (regardless of statistical method) that certain biomarkers are robust for both beef qualities of PDO Maine-Anjou cows. A robust biomarker is the protein that is identified in this study to be related with both beef quality traits whatever the statistical method. In the following sections, the proteins common to the two traits are discussed together, and those that were trait-dependent are presented separately. The biological pathways behind the associations identified between the two phenotypes and the proteins were further briefly presented. The relationships between the robust evaluated protein biomarkers and the two beef quality traits showed that the most tender meat of PDO Maine-Anjou cows had higher abundances of glycolytic enzymes, such as ENO3, and of fast contractile proteins such as MYL1 and MYH1, while they had lower abundances of slow contractile proteins such as TNNT1, and lower abundances of small HSPs with higher abundances of FHL1. These relationships are consistent with each other and are in accordance with the results of Couvreur et al. [51] based on the contractile and metabolic properties of the *Longissimus thoracis* muscle. Furthermore, the contractile properties of the *Longissimus thoracis* muscle from PDO Maine-Anjou cows have been associated with a specific fiber type composition, as the muscle contains a very low proportion of IIX fibers and a higher proportion of IIA fibers compared to the French beef breeds [38,51–53]. Consequently, its glycolytic metabolism is very low [51,52]. This demonstrates that for a slow oxidative type of muscle, the most tender are the less slow oxidative and the most fast glycolytic, as observed for Aberdeen Angus [54] or Chianina breeds [49].

4.1. Common Biomarkers Explaining the Variation in WBSF (Tenderness) and IMF (Marbling)

The data of the present trial showed that the common proteins between tenderness and IMF are usually inversely associated, except for FHL1 which was positive. WBSF was weakly and negatively correlated with IMF (r = +0.16; p = 0.032). The fatter meat samples are those containing higher proportions of proteins of the slow oxidative type, such as TNNT1 and small HSPs, and low proportions of the fast glycolytic type, such as MYH1, which is consistent with previous results about biomarkers of adiposity from the same breed [22,26]. From these data, it seems that the fatter meat samples do not lead to the tenderest beef. Thus, irrespective of the quality trait and the statistical method, seven proteins were robustly related with the WBSF and IMF content of PDO Maine-Anjou cows (Figure 4).

Without any surprise and in line with the results presented above, the superfamily of heat shock proteins, specifically the small HSP members, seemed to be just as good and important biomarkers for both beef qualities. Indeed, CRYAB, HSP20 and HSP27 were positively related in this trial with WBSF (negatively with tenderness), and so with toughness and IMF content. An inverse relationship between the abundance of small HSPs and tenderness was reported in the *Longissimus thoracis* muscle of the Angus breed [54]. It is worthwhile to note that the *Longissimus thoracis* muscles of the PDO Maine-Anjou breed investigated in this study, as well as those of Aberdeen Angus, were described as having more oxidative metabolisms with high amounts of fat content [18,38,52,54]. The slow oxidative type fibers are known to contain high levels of sHSP proteins [52], as also evidenced by Golenhofen et al. [55] in rats, showing more HSPs in slow oxidative muscles.

The involvement of these chaperones in post-mortem muscle, and thus in its conversion into meat, was reported by many earlier studies (for review: [5,11,12,56]). This is in agreement with the onset of apoptosis, the first phase in the conversion of muscle into meat, involving major biochemical and structural changes that influence not only the meat tenderization process but also the homeostasis of the post-mortem muscle [9,57]. Small HSP proteins were thought to stabilize and ensure the correct folding of newly synthesized proteins, or help refold proteins altered by cell stress to protect them against metabolic disorders and ischemia [58]. For example, small HSPs can bind to myofibrils [59–61], thereby protecting skeletal muscle through structural protein complexes, which partly explain the high abundance observed in the tough meat class. Accordingly and during the process of apoptosis, CRYAB is able to negatively regulate Cytochrome c and Caspase-8, hence inhibiting the executor caspase 3 [62,63]. For instance, any increase in the level of CRYAB (maybe also the other small HSPs) leads to less muscle structure degradation, which would produce tough meat [48]. In support of their important roles in meat tenderization, small HSPs were identified by several previous proteomics and under different conditions as good biomarkers for the two qualities investigated in this trial [6,17], or other sensory qualities of beef such as color [5] and pH decline [52,64,65]. CRYAB, which was ranked second in the PLS-R models, is a well-known beef tenderness biomarker [4,29,50,66–69], also recently identified [22] using gel-based and gel-free proteomics methods for IMF of the same breed using PRM and antibody-based proteomics [26]. HSP27 was reported by proteomics in several studies for beef tenderness [4,27,29,35,66,67,70], and as a consistent biomarker of adiposity (marbling) in the few proteomics that investigated this trait [18,20,22,24,71]. An absolute quantification of HSP27 in PDO Maine-Anjou cows confirmed a higher abundance in the less tender *Longissimus thoracis* muscle [26], as observed in this study. For HSP20, this study is the first to identify it as a biomarker for marbling, but it is already connected by several authors to beef tenderness [4,34,35,49,66–68]. Worthy of note is that in humans and among the HSP protein members, HSP20 was found by DeLany et al. [72] to be the most up-regulated chaperone during the differentiation of human adipose-derived stem cells into mature adipocytes.

In line with the above, ALDH1A1, a metabolic enzyme that is ranked first in the PLS-R models regardless of the quality trait, and is also significantly correlated with both traits and discriminates groups of tenderness and IMF, appeared as the most robust biomarker in this study with CRYAB. ALDH1A1 has been already identified as a biomarker of adiposity of PDO Maine-Anjou [22,26]. The authors proposed that any increase in this protein would mediate an increase in CRYAB and maybe in other small HSPs thanks to the retinoic acid content [73]. Further, the same authors postulated a mechanism in PDO Maine-Anjou [22,26], suggesting that a higher abundance of CRYAB resulting from a high proportion of slow fibers may sustain the elevated oxidative metabolism of marbled muscle, which may partly explain the important role that ALDH1A1 and CRYAB play in the determination of both WBSF and IMF content. ALDH1A1 was identified as a good biomarker of beef tenderness in different breeds and muscles [35,68,74,75], and was also connected to several color parameters [5], especially redness (a^*) [41] and metmyoglobin-reducing activity [76]. These last parameters are associated with the oxidative metabolism resulting from the fiber composition of the *Longissimus thoracis* muscle from Rouge des Près cows (with a very low proportion of fast glycolytic fibers [52]). It is worthy of mention that ALDH1A1 is able to protect cells against the cytotoxic effects of various aldehydes [77], which are generated in the cytosol by lipid peroxidation [78] and not cleared due to the cessation of blood flow.

The second metabolic enzyme, the cytosolic MDH1, is a member of the malate dehydrogenase enzymes very important for both gluconeogenesis and the Krebs cycle, and therefore plays a crucial role in energy and cellular metabolism [79]. In adipocytes, including muscular adipocytes, MDH1 is involved in the reduced nicotinamide adenine dinucleotide phosphate (NADPH) supply for de novo fatty acid synthesis, and is considered as a lipogenic enzyme. It was retained in this study in agreement with earlier studies. For instance, it was reported (as a biomarker of beef tenderness in bulls) to be positive in two studies [28,54] and negative in another [4]. MDH1 was also shown as a good predictor of *Longissimus thoracis* tenderness variability (in a few animals) when quantified by PRM [26]. Moreover, malic enzyme activity (MDH1) and IMF quantitative values in the *Rectus abdominis* and *Semitendinosus* muscles from Limousin, Angus and Japanese Black cross Angus steers were shown to be positively correlated [80]. Similar positive correlations between MDH1 and IMF were also observed when the abundance of MDH1 was assayed by absolute PRM quantification in both *Longissimus thoracis* and *Semitendinosus* muscles, but not by RPPA [26]. In the present study, using a larger number of animals and using the RPPA technique, an expected link between MDH1 and IMF was revealed by three mathematical methods. However, an unexpected negative correlation between MDH1 abundance and IMF probably arose from the relative and normalized abundance of MDH1 in the present study, making difficult the comparison between the current and previous results. This warrants further investigation in order to understand and describe the link between the MDH1 protein's abundance and IMF depending on the methodological specificities of both the protein quantification method and data processing.

MYH1 is the only structural protein that was related to both WBSF and IMF content whatever the statistical method (Figure 4). Changes in the cytoskeletal proteins have been shown and investigated for decades to play a role in meat tenderization [81], and are also evidenced by proteomics [4,6,11,82]. Further, myosin fibers play pivotal roles from fetal life to the slaughter of cattle (for review: [83]). MYH1 is the gene that encodes the fast glycolytic IIX fibers. In this study, the low abundance of MYH1 in the lean marbling class is in agreement with the earlier data reported for breeds of French origin, characterized by high proportions of fast glycolytic IIX fibers [84]. Cytoskeletal proteins were proposed to participate in IMF deposition [22,85,86], and MYH1 has already been identified by proteomics as a negative biomarker of adiposity [71]. MYH1 was revealed in several studies and under different factors as a good biomarker of beef tenderness [4,34,35,54,67,87,88]. As reported by Picard et al. [54] and Gagaoua et al. [34], the direction of its relationships with tenderness depends on muscle type, breed, end-point cooking temperature and origin of the panelists. In fact, in muscles with a low proportion of fast glycolytic fibers, such as the *Longissimus thoracis* muscle in breeds like Rouge des Prés or Aberdeen Angus, MYH1 was already reported to be positively related with tenderness. On the contrary, in French beef breeds in which the *Longissimus thoracis* muscle contains a high proportion of IIX fibers, the relationship with tenderness is negative. The identification of MYH1 as an important biomarker agrees with the theory stating that muscles with high proportions of fast fiber types are more susceptible to early post-mortem proteolytic degradation [89]. This can be further explained by their susceptibility to post-mortem glycolysis, hence leading to more tender than tough meat [83].

The seventh and last protein retained in this trial for the two beef quality traits was FHL1, known as a multifunctional protein regulating metabolism, cell proliferation, gene transcription and apoptosis [90]. This protein can further interact with metabolic enzymes as a response to the oxidative stress in muscle as well as hypoxia [90], thereby explaining its tendency to be projected with proteins characterizing slow oxidative properties, such as ALDH1A1 and TNNT1. To our knowledge, FHL1 was reported as a biomarker of beef tenderness in the *Longissimus thoracis* muscle in four previous studies [35,91–93], and of marbling in three studies [22,26,94]. We suggest that it is mainly via the regulation of calcium homeostasis [95] that FHL1 plays a role in beef tenderness determination. Indeed, Ca^{2+} ions contribute to the regulation of the energy metabolism pathways [96], as they affect the enzymatic speed of several crucial metabolic enzymes [11,12]. This is consistent with our results showing modifications of the abundances of the proteins ALDH1A1, MDH1 and ENO3. Calcium concentrations have also been implicated as initiators of apoptosis via some signaling pathways in skeletal muscle [12,97]. For example, apoptosis was documented to affect the integrity of the skeletal muscle, through the modification of Ca^{2+} flux during ageing and its consequences on protein proteolysis involving ultra-structural modifications. This can be supported in this trial by the robust association of TNNT1, MYL1 and MYH1, which are all proteins of structure. Further, FHL1 belongs to this last group of proteins and was described as an activator of the myostatin signaling pathway in skeletal muscle by promoting muscle atrophy [98]. This would explain the variation in the different muscles fibers and the involvement of MYH1 (further details, see [83]). For marbling, the mechanism would also be partly through the activation of myostatin signaling [98]. It is worthy of note that in beef, FHL1 was further related negatively with the lightness (L^*) and positively with the redness (a^*) color parameters [41].

4.2. Biomarkers Specific to WBSF

Irrespective of the statistical method, three proteins were found to be robust and specific to WBSF (Figure 4), these being two structural proteins (MYL1 and TNNT1) and one glycolytic enzyme (ENO3). The positive association of MYL1 with tenderness (negative with WBSF) is in agreement with several earlier studies on cows [4,38], steers [68,99] and young bulls [54,66,70]. MYL1 is considered as an indicator of proteolysis [81] and is prone to phosphorylation, a reaction playing a pivotal role in muscle to meat conversion [82,100]. It is worthy to consider that this phosphorylation is induced through sarcoplasmic reticulum Ca^{2+} release in a concentration-dependent manner. TNNT1 is a slow isoform of the troponins complex that is involved in the regulation of muscle contraction [81]. The release of TNNT members, including TNNT1, has been extensively studied and they have been considered as important substrates of the endogenous muscle proteolytic systems [11,12,82]. TNNT members were thought to be easily degraded by calpains during the aging period of muscle. In fact, several proteomics trials identified TNNT1 as a good biomarker of beef tenderness [4,66,99,101]. Its negative association with tenderness in this trial agrees with what we know from the literature [102], as recently evidenced by a meta-proteomics on different muscles and genders [4], highlighting inverse relationships. In the present study, the identification of TNNT1 is consistent with the positive relation observed between the fast glycolytic isoform of MYH1 and the glycolytic enzyme ENO3, and tenderness.

ENO3 is the last protein we identified as robustly related to tenderness. Its identification as a positive biomarker of beef tenderness in this study on PDO Maine-Anjou cows and several previous proteomics on beef [4,35,67,87,92,103] points to its importance, as it is especially described as being a key moonlighting enzyme associated with hypoxic conditions and stress [104]. Enolase is a cytosolic enzyme responsible for the conversion of 2-phosphoglycerate into phosphoenolpyruvate, thereby playing an important role in pH decline and post-mortem metabolism [41]. Indeed, ENO3 induces glucose metabolism under hypoxic conditions [105], hence enacting a cellular stress response to the deprivation of oxygen supply and glucose levels.

4.3. Biomarkers Specific to IMF

Two proteins, Peroxiredoxin 6 (PRDX6) and Tripartite motif protein 72 (TRIM72), appeared to be robustly related as biomarkers of marbling (Figure 4). PRDX6 was already identified as a good biomarker of beef tenderness [35,49,50,68,92], color [5] and pH decline [33]. PRDX6 is a bi-functional protein with both phospholipase A2 (PLA2) and glutathione peroxidase activities, which is expressed in nearly all tissues and protects cells against oxidative stress [106]. Earlier studies reported the high abundance of PRDX6 in Aberdeen Angus (young bulls and cross sired steers), known as a marbled breed compared to young Limousin bulls [32] and Belgian Blue sired steer [18]. PRDX6 might further play a role through its PLA_2 activity via the catalysis of the hydrolysis of the acyl group at the sn-2 position of glycerophospholipids, with a specific link to phosphatidylcholine to release free fatty acids and a lysophospholipid [106], hence partly explaining its relation with IMF. Further, using PRDX6 knockout mice, Arriga et al. [107] demonstrated that PRDX6 modulates the link between glycemic and lipogenic metabolisms, thereby playing a pivotal role in fat deposition.

TRIM72, also termed MG53 (Mitsugumin 53), is a signaling protein that acts as a sensor of oxidation on membrane damage [108], including playing a crucial role in the muscle membrane repair process. Its negative relationship with IMF would be partly explained by its implication in the clearance of harmful agents collected under the cell death process and lipid oxidation. TRIM72 was reported as a negative biomarker of beef tenderness [4,35,74], and thus one can suggest that a reduced cell death phase in tough meat occurred. TRIM72 was further identified as a biomarker of beef color Lightness (L^*) [41]. The identification of TRIM72 as a biomarker of marbling in PDO Maine-Anjou would be supported by its involvement in the negative feedback regulation of myogenesis, by targeting the insulin receptor substrate-1 [109]. Further investigations are warranted in order to clarify the exact role of TRIM72 in the muscle to meat conversion, including the role it plays in marbling.

5. Conclusions

This study allowed us to qualify, on *Longissimus thoracis* muscle and using the immune-based RPPA technique and different statistical methods, the potential of a list of 20 protein biomarkers to explain the variation of two important beef quality traits of PDO Maine-Anjou cows. These were tenderness, measured by the instrumental method WBSF, and the marbling of the carcasses as evaluated by IMF content. This study is the first to propose 10 and 9 proteins as robust candidate biomarkers of WBSF and IMF, respectively, in Rouge des Prés cows regardless the statistical method. Seven proteins are of specific interest, as they are related to both traits. They are, in the order of importance, ALDH1A1, CRYAB, HSP27, HSP20, MYH1, FHL1 and MDH1. These proteins belong to the superfamily of heat shock proteins, energy metabolism (especially the oxidative pathway), and the structural proteins (including FHL1) playing roles in cell death, metabolism and the regulation of calcium homeostasis. Our findings further highlight that similarities exist in the biological pathways underpinning tenderness and marbling determination.

Author Contributions: Conceptualization, M.G., M.B. and B.P.; software, M.G.; validation, M.G., M.B., and B.P.; formal analysis, M.G.; investigation, M.G., M.B. and B.P.; resources, M.B. and B.P.; data curation, M.G., M.B. and B.P.; writing—original draft preparation, M.G.; writing—review and editing, M.G., M.B. and B.P.; visualization, M.G.; funding acquisition, M.B. and B.P. All authors have read and agreed to the published version of the manuscript.

Funding: This work was supported by the regional council of Pays de Loire (Quafima 1 & 2).

Acknowledgments: The authors thank the SICA Rouge des Prés, especially A. Valais and G. Aminot for animal handling and sampling. We further thank S. Couvreur and G. Le Bec (Ecole Supérieure d'Agriculture (ESA) Angers) for muscle sampling, as well as IMF content and shear force assays. The authors thank Nicole Dunoyer for extraction of proteins and convey special thanks to the Institut Curie RPPA platform for acquisition of RPPA data under the supervision of Leanne De Koning and the support of Aurélie Cartier and Bérengère Ouine.

Conflicts of Interest: The authors declared no conflict of interest.

References

1. Jia, X.; Hollung, K.; Therkildsen, M.; Hildrum, K.I.; Bendixen, E. Proteome analysis of early post-mortem changes in two bovine muscle types: M. longissimus dorsi and M. semitendinosis. *Proteomics* **2006**, *6*, 936–944. [CrossRef]
2. Lametsch, R.; Karlsson, A.; Rosenvold, K.; Andersen, H.J.; Roepstorff, P.; Bendixen, E. Postmortem proteome changes of porcine muscle related to tenderness. *J. Agric. Food Chem.* **2003**, *51*, 6992–6997. [CrossRef] [PubMed]
3. Huang, H.; Lametsch, R. Application of Proteomics for Analysis of Protein Modifications in Postmortem Meat. In *Proteomics in Foods: Principles and Applications*; Toldrá, F., Nollet, L.M.L., Eds.; Springer: Boston, MA, USA, 2013; pp. 111–125. [CrossRef]
4. Picard, B.; Gagaoua, M. Meta-proteomics for the discovery of protein biomarkers of beef tenderness: An overview of integrated studies. *Food Res. Int.* **2020**, *127*, 108739. [CrossRef] [PubMed]
5. Gagaoua, M.; Hughes, J.; Terlouw, E.M.C.; Warner, R.D.; Purslow, P.P.; Lorenzo, J.M.; Picard, B. Proteomic biomarkers of beef colour. *Trends Food Sci. Technol.* **2020**, *101*, 234–252. [CrossRef]
6. Picard, B.; Gagaoua, M.; Hollung, K. Chapter 12—Gene and Protein Expression as a Tool to Explain/Predict Meat (and Fish) Quality. In *New Aspects of Meat Quality: From Genes to Ethics*; Purslow, P., Ed.; Woodhead Publishing: Sawston, UK; Cambridge, UK, 2017; pp. 321–354. [CrossRef]
7. Te Pas, M.F.; Hoekman, A.J.; Smits, M.A. Biomarkers as management tools for industries in the pork production chain. *J. Chain Netw. Sci.* **2011**, *11*, 155–166. [CrossRef]
8. Bouley, J.; Chambon, C.; Picard, B. Mapping of bovine skeletal muscle proteins using two-dimensional gel electrophoresis and mass spectrometry. *Proteomics* **2004**, *4*, 1811–1824. [CrossRef] [PubMed]

9. Gagaoua, M.; Hafid, K.; Boudida, Y.; Becila, S.; Ouali, A.; Picard, B.; Boudjellal, A.; Sentandreu, M.A. Caspases and Thrombin Activity Regulation by Specific Serpin Inhibitors in Bovine Skeletal Muscle. *Appl. Biochem. Biotechnol.* **2015**, *177*, 279–303. [CrossRef] [PubMed]

10. Gagaoua, M.; Dunoyer, N.; Picard, B. Differences in the levels of bov-SERPINA3 in *Rectus Abdominis* Muscle Explain the Variability of Beef Tenderness. In Proceedings of the 63rd International Congress of Meat Science and Technology, Cork, Ireland, 13–18 August 2017; pp. 832–833.

11. Picard, B.; Gagaoua, M. Chapter 11—Proteomic Investigations of Beef Tenderness. In *Proteomics in Food Science: From Farm to Fork*; Colgrave, M.L., Ed.; Academic Press: London, UK, 2017; pp. 177–197. [CrossRef]

12. Ouali, A.; Gagaoua, M.; Boudida, Y.; Becila, S.; Boudjellal, A.; Herrera-Mendez, C.H.; Sentandreu, M.A. Biomarkers of meat tenderness: Present knowledge and perspectives in regards to our current understanding of the mechanisms involved. *Meat Sci.* **2013**, *95*, 854–870. [CrossRef]

13. Gagaoua, M.; Picard, B.; Monteils, V. Assessment of cattle inter-individual cluster variability: The potential of continuum data from the farm-to-fork for ultimate beef tenderness management. *J. Sci. Food Agric.* **2019**, *99*, 4129–4141. [CrossRef]

14. Gagaoua, M.; Monteils, V.; Picard, B. Data from the farmgate-to-meat continuum including omics-based biomarkers to better understand the variability of beef tenderness: An integromics approach. *J. Agric. Food Chem.* **2018**, *66*, 13552–13563. [CrossRef]

15. Hocquette, J.F.; Gondret, F.; Baeza, E.; Medale, F.; Jurie, C.; Pethick, D.W. Intramuscular fat content in meat-producing animals: Development, genetic and nutritional control, and identification of putative markers. *Animal* **2010**, *4*, 303–319. [CrossRef] [PubMed]

16. Wood, J.D.; Enser, M.; Fisher, A.V.; Nute, G.R.; Sheard, P.R.; Richardson, R.I.; Hughes, S.I.; Whittington, F.M. Fat deposition, fatty acid composition and meat quality: A review. *Meat Sci.* **2008**, *78*, 343–358. [CrossRef] [PubMed]

17. Ceciliani, F.; Lecchi, C.; Bazile, J.; Bonnet, M. Proteomics Research in the Adipose Tissue. In *Proteomics in Domestic Animals: From Farm to Systems Biology*; de Almeida, A.M., Eckersall, D., Miller, I., Eds.; Springer: Cham, Switzerland, 2018; pp. 233–254. [CrossRef]

18. Keady, S.M.; Kenny, D.A.; Ohlendieck, K.; Doyle, S.; Keane, M.G.; Waters, S.M. Proteomic profiling of bovine M. longissimus lumborum from Crossbred Aberdeen Angus and Belgian Blue sired steers varying in genetic merit for carcass weight. *J. Anim. Sci.* **2013**, *91*, 654–665. [CrossRef]

19. Picard, B.; Gagaoua, M.; Al Jammas, M.; Bonnet, M. Beef tenderness and intramuscular fat proteomic biomarkers: Effect of gender and rearing practices. *J. Proteom.* **2019**, *200*, 1–10. [CrossRef]

20. Mao, Y.; Hopkins, D.L.; Zhang, Y.; Li, P.; Zhu, L.; Dong, P.; Liang, R.; Dai, J.; Wang, X.; Luo, X. Beef quality with different intramuscular fat content and proteomic analysis using isobaric tag for relative and absolute quantitation of differentially expressed proteins. *Meat Sci.* **2016**, *118*, 96–102. [CrossRef]

21. Shen, Y.N.; Kim, S.H.; Yoon, D.H.; Lee, H.G.; Kang, H.S.; Seo, K.S. Proteome Analysis of Bovine Longissimus dorsi Muscle Associated with the Marbling Score. *Asian Australas. J. Anim. Sci.* **2012**, *25*, 1083–1088. [CrossRef] [PubMed]

22. Bazile, J.; Picard, B.; Chambon, C.; Valais, A.; Bonnet, M. Pathways and biomarkers of marbling and carcass fat deposition in bovine revealed by a combination of gel-based and gel-free proteomic analyses. *Meat Sci.* **2019**, *156*, 146–155. [CrossRef]

23. Kim, N.K.; Lee, S.H.; Cho, Y.M.; Son, E.S.; Kim, K.Y.; Lee, C.S.; Yoon, D.; Im, S.K.; Oh, S.J.; Park, E.W. Proteome analysis of the m. longissimus dorsi between fattening stages in Hanwoo steer. *BMB Rep.* **2009**, *42*, 433–438. [CrossRef]

24. Zhang, Q.; Lee, H.G.; Han, J.A.; Kim, E.B.; Kang, S.K.; Yin, J.; Baik, M.; Shen, Y.; Kim, S.H.; Seo, K.S.; et al. Differentially expressed proteins during fat accumulation in bovine skeletal muscle. *Meat Sci.* **2010**, *86*, 814–820. [CrossRef]

25. Rifai, N.; Gillette, M.A.; Carr, S.A. Protein biomarker discovery and validation: The long and uncertain path to clinical utility. *Nat. Biotechnol.* **2006**, *24*, 971–983. [CrossRef]

26. Bonnet, M.; Soulat, J.; Bons, J.; Léger, S.; De Koning, L.; Carapito, C.; Picard, B. Quantification of biomarkers for beef meat qualities using a combination of Parallel Reaction Monitoring- and antibody-based proteomics. *Food Chem.* **2020**, *317*, 126376. [CrossRef] [PubMed]

27. Kim, N.K.; Cho, S.; Lee, S.H.; Park, H.R.; Lee, C.S.; Cho, Y.M.; Choy, Y.H.; Yoon, D.; Im, S.K.; Park, E.W. Proteins in longissimus muscle of Korean native cattle and their relationship to meat quality. *Meat Sci.* **2008**, *80*, 1068–1073. [CrossRef] [PubMed]

28. Jia, X.; Veiseth-Kent, E.; Grove, H.; Kuziora, P.; Aass, L.; Hildrum, K.I.; Hollung, K. Peroxiredoxin-6—A potential protein marker for meat tenderness in bovine longissimus thoracis muscle. *J. Anim. Sci.* **2009**, *87*, 2391–2399. [CrossRef] [PubMed]

29. Oh, E.; Lee, B.; Choi, Y.M. Associations of Heat-Shock Protein Expression with Meat Quality and Sensory Quality Characteristics in Highly Marbled Longissimus Thoracis Muscle from Hanwoo Steers Categorized by Warner–Bratzler Shear Force Value. *Foods* **2019**, *8*, 638. [CrossRef] [PubMed]

30. Guillemin, N.; Meunier, B.; Jurie, C.; Cassar-Malek, I.; Hocquette, J.F.; Leveziel, H.; Picard, B. Validation of a Dot-Blot quantitative technique for large scale analysis of beef tenderness biomarkers. *J. Physiol. Pharmacol.* **2009**, *60* (Suppl. 3), 91–97. [PubMed]

31. Guillemin, N.; Jurie, C.; Cassar-Malek, I.; Hocquette, J.F.; Renand, G.; Picard, B. Variations in the abundance of 24 protein biomarkers of beef tenderness according to muscle and animal type. *Animal* **2011**, *5*, 885–894. [CrossRef]

32. Gagaoua, M.; Terlouw, E.M.C.; Picard, B. The study of protein biomarkers to understand the biochemical processes underlying beef color development in young bulls. *Meat Sci.* **2017**, *134*, 18–27. [CrossRef]

33. Gagaoua, M.; Terlouw, E.M.; Micol, D.; Boudjellal, A.; Hocquette, J.F.; Picard, B. Understanding Early Post-Mortem Biochemical Processes Underlying Meat Color and pH Decline in the Longissimus thoracis Muscle of Young Blond d'Aquitaine Bulls Using Protein Biomarkers. *J. Agric. Food Chem.* **2015**, *63*, 6799–6809. [CrossRef]

34. Gagaoua, M.; Terlouw, C.; Richardson, I.; Hocquette, J.F.; Picard, B. The associations between proteomic biomarkers and beef tenderness depend on the end-point cooking temperature, the country origin of the panelists and breed. *Meat Sci.* **2019**, *157*, 107871. [CrossRef]

35. Gagaoua, M.; Bonnet, M.; Ellies-Oury, M.P.; De Koning, L.; Picard, B. Reverse phase protein arrays for the identification/validation of biomarkers of beef texture and their use for early classification of carcasses. *Food Chem.* **2018**, *250*, 245–252. [CrossRef]

36. Picard, B.; Gagaoua, M.; Al-Jammas, M.; De Koning, L.; Valais, A.; Bonnet, M. Beef tenderness and intramuscular fat proteomic biomarkers: Muscle type effect. *Peer J.* **2018**, *6*, e4891. [CrossRef] [PubMed]

37. Wu, W.; Dai, R.T.; Bendixen, E. Comparing SRM and SWATH Methods for Quantitation of Bovine Muscle Proteomes. *J. Agric. Food Chem.* **2019**, *67*, 1608–1618. [CrossRef] [PubMed]

38. Gagaoua, M.; Monteils, V.; Couvreur, S.; Picard, B. Identification of Biomarkers Associated with the Rearing Practices, Carcass Characteristics, and Beef Quality: An Integrative Approach. *J. Agric. Food Chem.* **2017**, *65*, 8264–8278. [CrossRef] [PubMed]

39. Lepetit, J.; Culioli, J. Mechanical properties of meat. *Meat Sci.* **1994**, *36*, 203–237. [CrossRef]

40. Gagaoua, M.; Micol, D.; Picard, B.; Terlouw, C.E.; Moloney, A.P.; Juin, H.; Meteau, K.; Scollan, N.; Richardson, I.; Hocquette, J.F. Inter-laboratory assessment by trained panelists from France and the United Kingdom of beef cooked at two different end-point temperatures. *Meat Sci.* **2016**, *122*, 90–96. [CrossRef] [PubMed]

41. Gagaoua, M.; Bonnet, M.; De Koning, L.; Picard, B. Reverse Phase Protein array for the quantification and validation of protein biomarkers of beef qualities: The case of meat color from Charolais breed. *Meat Sci.* **2018**, *145*, 308–319. [CrossRef] [PubMed]

42. Akbani, R.; Becker, K.F.; Carragher, N.; Goldstein, T.; de Koning, L.; Korf, U.; Liotta, L.; Mills, G.B.; Nishizuka, S.S.; Pawlak, M.; et al. Realizing the promise of reverse phase protein arrays for clinical, translational, and basic research: A workshop report: The RPPA (Reverse Phase Protein Array) society. *Mol. Cell. Proteom. MCP* **2014**, *13*, 1625–1643. [CrossRef] [PubMed]

43. Troncale, S.; Barbet, A.; Coulibaly, L.; Henry, E.; He, B.; Barillot, E.; Dubois, T.; Hupe, P.; de Koning, L. NormaCurve: A SuperCurve-based method that simultaneously quantifies and normalizes reverse phase protein array data. *PLoS ONE* **2012**, *7*, e38686. [CrossRef]

44. Gagaoua, M.; Monteils, V.; Couvreur, S.; Picard, B. Beef Tenderness Prediction by a Combination of Statistical Methods: Chemometrics and Supervised Learning to Manage Integrative Farm-To-Meat Continuum Data. *Foods* **2019**, *8*, 274. [CrossRef]

45. Gagaoua, M.; Picard, B.; Soulat, J.; Monteils, V. Clustering of sensory eating qualities of beef: Consistencies and differences within carcass, muscle, animal characteristics and rearing factors. *Livest. Sci.* **2018**, *214*, 245–258. [CrossRef]

46. Rousseeuw, P.J. Silhouettes: A graphical aid to the interpretation and validation of cluster analysis. *J. Comput. Appl. Math.* **1987**, *20*, 53–65. [CrossRef]

47. Gagaoua, M.; Listrat, A.; Andueza, D.; Gruffat, D.; Normand, J.; Mairesse, G.; Picard, B.; Hocquette, J.-F. What are the drivers of beef sensory quality using metadata of intramuscular connective tissue, fatty acids and muscle fiber characteristics? *Livest. Sci.* **2020**. [CrossRef]

48. Bernard, C.; Cassar-Malek, I.; Le Cunff, M.; Dubroeucq, H.; Renand, G.; Hocquette, J.F. New indicators of beef sensory quality revealed by expression of specific genes. *J. Agric. Food Chem.* **2007**, *55*, 5229–5237. [CrossRef] [PubMed]

49. D'Alessandro, A.; Marrocco, C.; Rinalducci, S.; Mirasole, C.; Failla, S.; Zolla, L. Chianina beef tenderness investigated through integrated Omics. *J. Proteom.* **2012**, *75*, 4381–4398. [CrossRef] [PubMed]

50. Guillemin, N.; Bonnet, M.; Jurie, C.; Picard, B. Functional analysis of beef tenderness. *J. Proteom.* **2011**, *75*, 352–365. [CrossRef] [PubMed]

51. Couvreur, S.; Le Bec, G.; Micol, D.; Picard, B. Relationships Between Cull Beef Cow Characteristics, Finishing Practices and Meat Quality Traits of Longissimus thoracis and Rectus abdominis. *Foods* **2019**, *8*, 141. [CrossRef]

52. Gagaoua, M.; Couvreur, S.; Le Bec, G.; Aminot, G.; Picard, B. Associations among Protein Biomarkers and pH and Color Traits in Longissimus thoracis and Rectus abdominis Muscles in Protected Designation of Origin Maine-Anjou Cull Cows. *J. Agric. Food Chem.* **2017**, *65*, 3569–3580. [CrossRef]

53. Gagaoua, M.; Terlouw, E.M.C.; Micol, D.; Hocquette, J.F.; Moloney, A.P.; Nuernberg, K.; Bauchart, D.; Boudjellal, A.; Scollan, N.D.; Richardson, R.I.; et al. Sensory quality of meat from eight different types of cattle in relation with their biochemical characteristics. *J. Integr. Agric.* **2016**, *15*, 1550–1563. [CrossRef]

54. Picard, B.; Gagaoua, M.; Micol, D.; Cassar-Malek, I.; Hocquette, J.F.; Terlouw, C.E. Inverse relationships between biomarkers and beef tenderness according to contractile and metabolic properties of the muscle. *J. Agric. Food Chem.* **2014**, *62*, 9808–9818. [CrossRef]

55. Golenhofen, N.; Perng, M.D.; Quinlan, R.A.; Drenckhahn, D. Comparison of the small heat shock proteins alphaB-crystallin, MKBP, HSP25, HSP20, and cvHSP in heart and skeletal muscle. *Histochem. Cell Biol.* **2004**, *122*, 415–425. [CrossRef]

56. Lomiwes, D.; Farouk, M.M.; Wiklund, E.; Young, O.A. Small heat shock proteins and their role in meat tenderness: A review. *Meat Sci.* **2014**, *96*, 26–40. [CrossRef] [PubMed]

57. Picard, B.; Kammoun, M.; Gagaoua, M.; Barboiron, C.; Meunier, B.; Chambon, C.; Cassar-Malek, I. Calcium Homeostasis and Muscle Energy Metabolism Are Modified in HspB1-Null Mice. *Proteomes* **2016**, *4*, 17. [CrossRef] [PubMed]

58. Dreiza, C.M.; Komalavilas, P.; Furnish, E.J.; Flynn, C.R.; Sheller, M.R.; Smoke, C.C.; Lopes, L.B.; Brophy, C.M. The small heat shock protein, HSPB6, in muscle function and disease. *Cell Stress Chaperones* **2010**, *15*, 1–11. [CrossRef] [PubMed]

59. Rembold, C.M.; Foster, D.B.; Strauss, J.D.; Wingard, C.J.; Eyk, J.E. cGMP-mediated phosphorylation of heat shock protein 20 may cause smooth muscle relaxation without myosin light chain dephosphorylation in swine carotid artery. *J. Physiol.* **2000**, *524 Pt 3*, 865–878. [CrossRef]

60. Lomiwes, D.; Hurst, S.M.; Dobbie, P.; Frost, D.A.; Hurst, R.D.; Young, O.A.; Farouk, M.M. The protection of bovine skeletal myofibrils from proteolytic damage post mortem by small heat shock proteins. *Meat Sci.* **2014**, *97*, 548–557. [CrossRef]

61. Ma, D.; Kim, Y.H.B. Proteolytic changes of myofibrillar and small heat shock proteins in different bovine muscles during aging: Their relevance to tenderness and water-holding capacity. *Meat Sci.* **2020**, *163*, 108090. [CrossRef]

62. Kamradt, M.C.; Chen, F.; Cryns, V.L. The Small Heat Shock Protein αB-Crystallin Negatively Regulates Cytochrome c- and Caspase-8-dependent Activation of Caspase-3 by Inhibiting Its Autoproteolytic Maturation. *J. Biol. Chem.* **2001**, *276*, 16059–16063. [CrossRef]

63. Kamradt, M.C.; Chen, F.; Sam, S.; Cryns, V.L. The Small Heat Shock Protein αB-crystallin Negatively Regulates Apoptosis during Myogenic Differentiation by Inhibiting Caspase-3 Activation. *J. Biol. Chem.* **2002**, *277*, 38731–38736. [CrossRef]

64. Pulford, D.J.; Fraga Vazquez, S.; Frost, D.F.; Fraser-Smith, E.; Dobbie, P.; Rosenvold, K. The intracellular distribution of small heat shock proteins in post-mortem beef is determined by ultimate pH. *Meat Sci.* **2008**, *79*, 623–630. [CrossRef]

65. Pulford, D.J.; Dobbie, P.; Fraga Vazquez, S.; Fraser-Smith, E.; Frost, D.A.; Morris, C.A. Variation in bull beef quality due to ultimate muscle pH is correlated to endopeptidase and small heat shock protein levels. *Meat Sci.* **2009**, *83*, 1–9. [CrossRef]

66. D'Alessandro, A.; Rinalducci, S.; Marrocco, C.; Zolla, V.; Napolitano, F.; Zolla, L. Love me tender: An Omics window on the bovine meat tenderness network. *J. Proteom.* **2012**, *75*, 4360–4380. [CrossRef] [PubMed]

67. Polati, R.; Menini, M.; Robotti, E.; Millioni, R.; Marengo, E.; Novelli, E.; Balzan, S.; Cecconi, D. Proteomic changes involved in tenderization of bovine Longissimus dorsi muscle during prolonged ageing. *Food Chem.* **2012**, *135*, 2052–2069. [CrossRef] [PubMed]

68. Zapata, I.; Zerby, H.N.; Wick, M. Functional proteomic analysis predicts beef tenderness and the tenderness differential. *J. Agric. Food Chem.* **2009**, *57*, 4956–4963. [CrossRef] [PubMed]

69. Morzel, M.; Terlouw, C.; Chambon, C.; Micol, D.; Picard, B. Muscle proteome and meat eating qualities of Longissimus thoracis of "Blonde d'Aquitaine" young bulls: A central role of HSP27 isoforms. *Meat Sci.* **2008**, *78*, 297–304. [CrossRef] [PubMed]

70. Carvalho, M.E.; Gasparin, G.; Poleti, M.D.; Rosa, A.F.; Balieiro, J.C.; Labate, C.A.; Nassu, R.T.; Tullio, R.R.; Regitano, L.C.; Mourao, G.B.; et al. Heat shock and structural proteins associated with meat tenderness in Nellore beef cattle, a Bos indicus breed. *Meat Sci.* **2014**, *96*, 1318–1324. [CrossRef]

71. Thornton, K.J.; Chapalamadugu, K.C.; Eldredge, E.M.; Murdoch, G.K. Analysis of Longissimus thoracis Protein Expression Associated with Variation in Carcass Quality Grade and Marbling of Beef Cattle Raised in the Pacific Northwestern United States. *J. Agric. Food Chem.* **2017**, *65*, 1434–1442. [CrossRef]

72. DeLany, J.P.; Floyd, Z.E.; Zvonic, S.; Smith, A.; Gravois, A.; Reiners, E.; Wu, X.; Kilroy, G.; Lefevre, M.; Gimble, J.M. Proteomic Analysis of Primary Cultures of Human Adipose-derived Stem Cells. *Modul. Adipogenes.* **2005**, *4*, 731–740. [CrossRef]

73. Takeda, K.; Sriram, S.; Chan, X.H.D.; Ong, W.K.; Yeo, C.R.; Tan, B.; Lee, S.-A.; Kong, K.V.; Hoon, S.; Jiang, H.; et al. Retinoic Acid Mediates Visceral-Specific Adipogenic Defects of Human Adipose-Derived Stem Cells. *Diabetes* **2016**, *65*, 1164–1178. [CrossRef]

74. Grabez, V.; Kathri, M.; Phung, V.; Moe, K.M.; Slinde, E.; Skaugen, M.; Saarem, K.; Egelandsdal, B. Protein expression and oxygen consumption rate of early postmortem mitochondria relate to meat tenderness. *J. Anim. Sci.* **2015**, *93*, 1967–1979. [CrossRef]

75. Jia, X.; Ekman, M.; Grove, H.; Faergestad, E.M.; Aass, L.; Hildrum, K.I.; Hollung, K. Proteome changes in bovine longissimus thoracis muscle during the early postmortem storage period. *J. Proteome Res.* **2007**, *6*, 2720–2731. [CrossRef] [PubMed]

76. Wu, W.; Yu, Q.Q.; Fu, Y.; Tian, X.J.; Jia, F.; Li, X.M.; Dai, R.T. Towards muscle-specific meat color stability of Chinese Luxi yellow cattle: A proteomic insight into post-mortem storage. *J. Proteom.* **2016**, *147*, 108–118. [CrossRef] [PubMed]

77. Vasiliou, V.; Thompson, D.C.; Smith, C.; Fujita, M.; Chen, Y. Aldehyde dehydrogenases: From eye crystallins to metabolic disease and cancer stem cells. *Chem. Biol. Interact.* **2013**, *202*, 2–10. [CrossRef] [PubMed]

78. Domínguez, R.; Pateiro, M.; Gagaoua, M.; Barba, F.J.; Zhang, W.; Lorenzo, J.M. A Comprehensive Review on Lipid Oxidation in Meat and Meat Products. *Antioxidants* **2019**, *8*, 429. [CrossRef] [PubMed]

79. Sasaki, K.; Motoyama, M.; Narita, T.; Hagi, T.; Ojima, K.; Oe, M.; Nakajima, I.; Kitsunai, K.; Saito, Y.; Hatori, H.; et al. Characterization and classification of Japanese consumer perceptions for beef tenderness using descriptive texture characteristics assessed by a trained sensory panel. *Meat Sci.* **2014**, *96*, 994–1002. [CrossRef]

80. Bonnet, M.; Faulconnier, Y.; Leroux, C.; Jurie, C.; Cassar-Malek, I.; Bauchart, D.; Boulesteix, P.; Pethick, D.; Hocquette, J.F.; Chilliard, Y. Glucose-6-phosphate dehydrogenase and leptin are related to marbling differences among Limousin and Angus or Japanese Black × Angus steers1,2. *J. Anim. Sci.* **2007**, *85*, 2882–2894. [CrossRef]

81. Ertbjerg, P.; Puolanne, E. Muscle structure, sarcomere length and influences on meat quality: A review. *Meat Sci.* **2017**, *132*, 139–152. [CrossRef]

82. Lana, A.; Zolla, L. Proteolysis in meat tenderization from the point of view of each single protein: A proteomic perspective. *J. Proteom.* **2016**, *147*, 85–97. [CrossRef]

83. Picard, B.; Gagaoua, M. Muscle Fiber Properties in Cattle and Their Relationships with Meat Qualities: An Overview. *J. Agric. Food Chem.* **2020**, *68*, 6021–6039. [CrossRef]

84. Hocquette, J.-F.; Jurie, C.; Picard, B.; Alberti, P.; Panea, B.; Christensen, M.; Failla, S.; Gigli, S.; Levéziel, H.; Olleta, J.L.; et al. Metabolic and contractile characteristics of Longissimus thoracis muscle of young bulls from 15 European breeds in relationship with body composition. In Proceedings of the International Symposium on Energy and Protein Metabolism and Nutrition (ISEP 2007), Vichy, France, 9–13 September 2007; pp. 111–112.

85. Poleti, M.D.; Regitano, L.C.A.; Souza, G.H.M.F.; Cesar, A.S.M.; Simas, R.C.; Silva-Vignato, B.; Oliveira, G.B.; Andrade, S.C.S.; Cameron, L.C.; Coutinho, L.L. Longissimus dorsi muscle label-free quantitative proteomic reveals biological mechanisms associated with intramuscular fat deposition. *J. Proteom.* **2018**, *179*, 30–41. [CrossRef]

86. Poleti, M.D.; Regitano, L.C.A.; Souza, G.H.M.F.; Cesar, A.S.M.; Simas, R.C.; Silva-Vignato, B.; Montenegro, H.; Pértille, F.; Balieiro, J.C.C.; Cameron, L.C.; et al. Proteome alterations associated with the oleic acid and cis-9, trans-11 conjugated linoleic acid content in bovine skeletal muscle. *J. Proteom.* **2020**, *222*, 103792. [CrossRef]

87. Zhao, C.; Zan, L.; Wang, Y.; Scott Updike, M.; Liu, G.; Bequette, B.J.; Baldwin Vi, R.L.; Song, J. Functional proteomic and interactome analysis of proteins associated with beef tenderness in Angus cattle. *Livest. Sci.* **2014**, *161*, 201–209. [CrossRef]

88. Beldarrain, L.R.; Aldai, N.; Picard, B.; Sentandreu, E.; Navarro, J.L.; Sentandreu, M.A. Use of liquid isoelectric focusing (OFFGEL) on the discovery of meat tenderness biomarkers. *J. Proteom.* **2018**, *183*, 25–33. [CrossRef]

89. Ouali, A. Meat Tenderization: Possible Causes and Mechanisms. A Review. *J. Muscle Foods* **1990**, *1*, 129–165. [CrossRef]

90. Shathasivam, T.; Kislinger, T.; Gramolini, A.O. Genes, proteins and complexes: The multifaceted nature of FHL family proteins in diverse tissues. *J. Cell. Mol. Med.* **2010**, *14*, 2702–2720. [CrossRef]

91. Laville, E.; Sayd, T.; Morzel, M.; Blinet, S.; Chambon, C.; Lepetit, J.; Renand, G.; Hocquette, J.F. Proteome changes during meat aging in tough and tender beef suggest the importance of apoptosis and protein solubility for beef aging and tenderization. *J. Agric. Food Chem.* **2009**, *57*, 10755–10764. [CrossRef]

92. Bjarnadottir, S.G.; Hollung, K.; Hoy, M.; Bendixen, E.; Codrea, M.C.; Veiseth-Kent, E. Changes in protein abundance between tender and tough meat from bovine longissimus thoracis muscle assessed by isobaric Tag for Relative and Absolute Quantitation (iTRAQ) and 2-dimensional gel electrophoresis analysis. *J. Anim. Sci.* **2012**, *90*, 2035–2043. [CrossRef]

93. Boudon, S.; Ounaissi, D.; Viala, D.; Monteils, V.; Picard, B.; Cassar-Malek, I. Label free shotgun proteomics for the identification of protein biomarkers for beef tenderness in muscle and plasma of heifers. *J. Proteom.* **2020**, *217*, 103685. [CrossRef]

94. Wang, Y.H.; Bower, N.I.; Reverter, A.; Tan, S.H.; De Jager, N.; Wang, R.; McWilliam, S.M.; Cafe, L.M.; Greenwood, P.L.; Lehnert, S.A. Gene expression patterns during intramuscular fat development in cattle1. *J. Anim. Sci.* **2009**, *87*, 119–130. [CrossRef]

95. Pillar, N.; Pleniceanu, O.; Fang, M.; Ziv, L.; Lahav, E.; Botchan, S.; Cheng, L.; Dekel, B.; Shomron, N. A rare variant in the FHL1 gene associated with X-linked recessive hypoparathyroidism. *Hum. Genet.* **2017**, *136*, 835–845. [CrossRef]

96. Carafoli, E. Calcium signaling: A tale for all seasons. *Proc. Natl. Acad. Sci. USA* **2002**, *99*, 1115–1122. [CrossRef]

97. Demaurex, N.; Distelhorst, C. Cell biology. Apoptosis–the calcium connection. *Science* **2003**, *300*, 65–67. [CrossRef]

98. Lee, J.Y.; Lori, D.; Wells, D.J.; Kemp, P.R. FHL1 activates myostatin signalling in skeletal muscle and promotes atrophy. *FEBS Open Bio* **2015**, *5*, 753–762. [CrossRef]

99. Malheiros, J.M.; Braga, C.P.; Grove, R.A.; Ribeiro, F.A.; Calkins, C.R.; Adamec, J.; Chardulo, L.A.L. Influence of oxidative damage to proteins on meat tenderness using a proteomics approach. *Meat Sci.* **2019**, *148*, 64–71. [CrossRef]

100. Mato, A.; Rodríguez-Vázquez, R.; López-Pedrouso, M.; Bravo, S.; Franco, D.; Zapata, C. The first evidence of global meat phosphoproteome changes in response to pre-slaughter stress. *BMC Genom.* **2019**, *20*, 590. [CrossRef]

101. Rosa, A.F.; Moncau, C.T.; Poleti, M.D.; Fonseca, L.D.; Balieiro, J.C.C.; Silva, S.L.E.; Eler, J.P. Proteome changes of beef in Nellore cattle with different genotypes for tenderness. *Meat Sci.* **2018**, *138*, 1–9. [CrossRef]

102. Silva, L.H.P.; Rodrigues, R.T.S.; Assis, D.E.F.; Benedeti, P.D.B.; Duarte, M.S.; Chizzotti, M.L. Explaining meat quality of bulls and steers by differential proteome and phosphoproteome analysis of skeletal muscle. *J. Proteom.* **2019**, *199*, 51–66. [CrossRef]

103. Marino, R.; Albenzio, M.; Della Malva, A.; Caroprese, M.; Santillo, A.; Sevi, A. Changes in meat quality traits and sarcoplasmic proteins during aging in three different cattle breeds. *Meat Sci.* **2014**, *98*, 178–186. [CrossRef]

104. Didiasova, M.; Schaefer, L.; Wygrecka, M. When Place Matters: Shuttling of Enolase-1 across Cellular Compartments. *Front. Cell Dev. Biol.* **2019**, *7*, 61. [CrossRef]

105. Sedoris, K.C.; Thomas, S.D.; Miller, D.M. Hypoxia induces differential translation of enolase/MBP-1. *BMC Cancer* **2010**, *10*, 157. [CrossRef]

106. Fisher, A.B. Peroxiredoxin 6 in the repair of peroxidized cell membranes and cell signaling. *Arch. Biochem. Biophys.* **2017**, *617*, 68–83. [CrossRef]

107. Arriga, R.; Pacifici, F.; Capuani, B.; Coppola, A.; Orlandi, A.; Scioli, M.G.; Pastore, D.; Andreadi, A.; Sbraccia, P.; Tesauro, M.; et al. Peroxiredoxin 6 Is a Key Antioxidant Enzyme in Modulating the Link between Glycemic and Lipogenic Metabolism. *Oxidative Med. Cell. Longev.* **2019**, *2019*, 9685607. [CrossRef]

108. Cai, C.; Masumiya, H.; Weisleder, N.; Matsuda, N.; Nishi, M.; Hwang, M.; Ko, J.-K.; Lin, P.; Thornton, A.; Zhao, X.; et al. MG53 nucleates assembly of cell membrane repair machinery. *Nat. Cell Biol.* **2009**, *11*, 56–64. [CrossRef]

109. Jung, S.-Y.; Ko, Y.-G. TRIM72, a novel negative feedback regulator of myogenesis, is transcriptionally activated by the synergism of MyoD (or myogenin) and MEF2. *Biochem. Biophys. Res. Commun.* **2010**, *396*, 238–245. [CrossRef]

*foods*

Article

Proteomic Changes in Sarcoplasmic and Myofibrillar Proteins Associated with Color Stability of Ovine Muscle during Post-Mortem Storage

Xiaoguang Gao [1,2], Dandan Zhao [1], Lin Wang [1], Yue Cui [1], Shijie Wang [1], Meng Lv [1], Fangbo Zang [1] and Ruitong Dai [2,*]

[1] College of Food Science and Biology, Hebei University of Science and Technology, No. 26 Yuxiang Street, Yuhua District, Shijiazhuang 050000, China; gaoxiaoguang23@hotmail.com (X.G.); zdd6364@126.com (D.Z.); w592148022@163.com (L.W.); cuiy1223@hotmail.com (Y.C.); mrshjwang@163.com (S.W.); lmeng1126@163.com (M.L.); 15076005611@163.com (F.Z.)
[2] College of Food Science and Nutritional Engineering, China Agricultural University, No. 17 Qinghua East Road, Haidian District, Beijing 100083, China
* Correspondence: dairuitong@hotmail.com; Tel.: +86-010-62737547

Abstract: The objective of this study was to investigate the proteomic characteristics for the sarcoplasmic and myofibrillar proteomes of *M. longissimus lumborum* (LL) and *M. psoasmajor* (PM) from *Small-tailed Han* Sheep. During post-mortem storage periods (1, 3, and 5 days), proteome analysis was applied to elucidate sarcoplasmic and myofibrillar protein changes in skeletal muscles with different color stability. Proteomic results revealed that the identified differentially abundant proteins were glycolytic enzymes, energy metabolism enzymes, chaperone proteins, and structural proteins. Through Pearson's correlation analysis, a few of those identified proteins (Pyruvate kinase, Adenylate kinase isoenzyme 1, Creatine kinase M-type, and Carbonic anhydrase 3) were closely correlated to representative meat color parameters. Besides, bioinformatics analysis of differentially abundant proteins revealed that the proteins mainly participated in glycolysis and energy metabolism pathways. Some of these proteins may have the potential probability to be predictors of meat discoloration during post-mortem storage. Within the insight of proteomics, these results accumulated some basic theoretical understanding of the molecular mechanisms of meat discoloration.

Keywords: ovine; color stability; enzyme; proteomics; bioinformatics

Citation: Gao, X.; Zhao, D.; Wang, L.; Cui, Y.; Wang, S.; Lv, M.; Zang, F.; Dai, R. Proteomic Changes in Sarcoplasmic and Myofibrillar Proteins Associated with Color Stability of Ovine Muscle during Post-Mortem Storage. *Foods* 2021, 10, 2989. https://doi.org/10.3390/foods10122989

Academic Editors: Mohammed Gagaoua and Brigitte Picard

Received: 1 November 2021
Accepted: 1 December 2021
Published: 3 December 2021

Publisher's Note: MDPI stays neutral with regard to jurisdictional claims in published maps and institutional affiliations.

1. Introduction

Among various sensory characteristics, the color of fresh meat is a critical quality factor, influencing the purchase decision of the consumers [1]. The defects of meat color have always been connected to spoilage and unwholesomeness [2]. Prevention of meat discoloration has always been a challenging task and long-term objective for scientific researchers [3–5]. Consequently, maintaining the "cherry-red" color of fresh meat is very critical for the meat industry.

The relative contents of the reduced deoxymyoglobin (DeoMb, purple), the oxygenated oxymyoglobin (OxyMb, bright red), and the oxidized metmyoglobin (MetMb, brown) determine the color of fresh meat [6]. The reason for meat discoloration is most commonly attributed to the formation of MetMb. Furthermore, the affecting factors of redox forms of myoglobin are formed by many intrinsic and extrinsic ones such as breed, muscle type, temperature, and oxygen partial pressure of storage [7,8]. Because of the distinct metabolic function and biochemical profile, *M. longissimus lumborum* (LL) and *M. psoas major* (PM) were considered to be typical color-stable and color-labile muscles, respectively, by previous researchers [9,10]. Thus, both of the skeletal muscles whose color stabilities (color-stable and color-labile) were opposing can be used as typical experimental subjects to elaborate the underlying biochemical mechanisms of post-mortem color stability [11]. The post-mortem changes in muscles constitute a very complicated biochemical processes influencing meat color stability, as skeletal muscle consist of proteins, lipids, and other biomacromolecules and micro-molecules [11,12]. Proteomics provides an efficient way, in post-genome era, to illustrate the post-mortem changes in meat quality development, including the field of color characteristics [13,14]. In the previous literature, some investigators explored a series of post-mortem proteomic changes in beef and pork [15–18]. However, data of physiological and biochemical mechanisms of ovine color stability during post-mortem storage has rarely been reported [7,19]. The proteomics for the variations of different ovine muscles (color-stable and color-labile) still need to be investigated. Therefore, the potential molecular mechanisms need further research. The objective of this study was to investigate the proteomic characteristics for the sarcoplasmic and myofibrillar proteome of *M. longissimus lumborum* (LL) and *M. psoasmajor* (PM) from *Small-tailed Han* Sheep during post-mortem storage.

2. Materials and Methods

2.1. Sample Preparation

The skeletal muscle samples were harvested from 6 male Small-tailed Han Sheep, slaughtered at 8 months, with a mean carcass weight of 15.6 ± 0.2 kg. There were 2 typical types of skeletal muscles (*M. longissimus lumborum* (LL) and *M. psoas major* (PM)) that were excised (30 min after slaughter) from both sides of each carcass (*n* = 6 carcasses). The whole process of slaughter followed the industrial practice. The animals received humanely non-painful manipulations at the termination of the procedure, without regaining consciousness. Muscles samples were cut along the direction perpendicular to the muscle fibers. For each sample (corresponding to 1 carcass and 1 type of muscle), 6 equal-weight (30 g, trimmed free of connective tissue and fat) pieces were randomly assigned to 12 replications [*n* = 6, (6 muscles per type × 2 sides × 3 sections per muscle)/3 days].

The samples were wrapped individually with polyethylene films (350–400 cm^{3}m^{-2}h^{-1}atm^{-1}, Mitsui Chemical, Japan) and put in a refrigerator for a storage time of 5 days (4 ± 1 °C). Muscles were taken out for analyses at day 1, day 3, and day 5. At each time point (day 1, day 3, and day 5), instrumental color parameters and metmyoglobin reducing activity (MRA) of the muscles were measured. Sarcoplasmic and myofibrillar proteins were extracted for comparing the change in the early stage of storage (day 1) and the late stage of storage (day 5). At the storage time of day 3, proteins were not extracted accordingly. At the end of storage period, myofibrillar proteins were extracted for the comparison of proteomic difference between muscle types (LL and PM).

2.2. Instrumental Color

All the measurements of color parameters were determined on the surface of each sample. The instrumental color parameters were measured by utilizing a Minolta chromameter (CR-400, Minolta Inc., Osaka, Japan), expressing as CIE Lab lightness (*L**), redness (*a**), and yellowness (*b**) with D$_{65}$ standard illuminant. Accordingly, the parameters were calibrated through the white and black reference standards. The observer angle and measurement area were 2° and 8 mm, respectively. Hue angle (h°) and chroma (*C**) were determined by the following formula: h° = [ATAN(*b**/*a**) × (180/π)], *C** = [(*a**^{2} + *b**^{2})$^{0.5}$] [20].

LL and PM samples (5 g of each sample) were collected and homogenized for 10 s in a phosphate buffer (25 mL, pH 6.8, 0 °C, 40 mM) by using a homogenizer (F6-10, Fluko, Shanghai, China). The homogenate solution was placed in 4 °C for an hour and centrifuged (4500× *g*, 4 °C). Supernatant fluid was collected and filtered (Whatman-No.1 filter paper). The absorbance of the filtrate was tested with a spectrometer (wavelength of 503, 525, 557, 582 nm; TU-1810, PERSEE, Beijing, China). The proportion of myoglobin redox forms was calculated as follows [21]:

$$[DeoMb] = C_{DeoMb} \div C_{Mb} = -0.543R_1 + 1.594R_2 + 0.552R_3 - 1.3290$$
$$[OxyMb] = C_{OxyMb} \div C_{Mb} = 0.722R_1 - 1.432R_2 - 1.659R_3 + 2.599 \quad (1)$$
$$[MetMb] = C_{MetMb} \div C_{Mb} = -0.159R_1 - 0.085R_2 + 1.262R_3 - 0.520$$

where $R_1 = A_{582} \div A_{525}$; $R_2 = A_{557} \div A_{525}$; $R_3 = A_{503} \div A_{525}$.

2.3. Metmyoglobin Reducing Activity

The metmyoglobin reducing activity (MRA) was determined through spectrophotography and based on Mikkelsen et al. [22], including the following experimental equipment: spectrometer (TU-1810, PERSEE, Beijing, China), refrigeration centrifuge (3K15, SIGMA, Hamburg, Germany) and homogenizer (F6-10, Fluko, Shanghai, China). Muscle sample (12 g, connective tissue and fat removed) was homogenized (20 mL, 2.0 mM phosphate buffer, pH 7.0) for 30 s. The homogenate was centrifuged ($35,000 \times g$, 4 °C) for 30 min. The supernatant fluid was collected and filtered (Whatman-No.1 filter paper) for further removal of fat.

$K_3Fe(CN)_6$ was used to oxidize OxyMb into MetMb. The solution was dialyzed against phosphate buffer (2.0 mM, pH 7.0, 4 °C, 14,000 Mw cut-off) and centrifuged for 20 min ($15,000 \times g$, 4 °C). The supernatant was adjusted to volume of 20 mL through phosphate buffer (2.0 mM, pH 7.0). The standard assay (pH 6.4, 25 °C) was the mixed solution of NADH (0.1 mL, 2.0 mM, used for initiating the reaction), $K_4[Fe(CN)_6]$ (0.1 mL, 3.0 mM), EDTA (0.1 mL, 5.0 mM), Mb Fe(III) (0.2 mL of 0.75 mM within 2.0 mM phosphate buffer, pH 7.0), phosphate buffer (0.1 mL, 50 mM, pH 7.0), deionized water (0.1 mL) and muscle extract (0.3 mL). The blank contained all chemicals except for NADH (replace NADH with water, no reaction). MRA was followed the changes in absorbance (580 nm) and determined as nmol MetMb reduced ($min^{-1} g^{-1}$).

2.4. Sarcoplasmic and Myofibrillar Protein Extraction

Sarcoplasmic and myofibrillar protein extraction methods were based on Sayd et al. [23] and Kim et al. [24] with a minor modification, respectively. Each muscle sample (3 g) was homogenized with extraction buffer (30 mL, 2 mM EDTA, 40 mM Tris, pH 7.0, 4 °C) containing a protease inhibitors cocktail by utilizing a homogenizer (F6-10, Fluko, Shanghai, China). The sample was centrifuged (10 min, $10,000 \times g$, 4 °C) and the supernatant (sarcoplasmic proteins) was collected for further experiments (stored in −80 °C). The precipitate of each sample was incubated for 40 min in 2 mL of extraction buffer (1% pH 3–10 bio-lyte ampholytes (Bio-Rad, Hercules, CA, USA), 2 M thio-urea, 8 M urea, 2% CHAPS (w/v), 65 mM DTT, with protease inhibitors cocktail). The sample was centrifuged ($40,000 \times g$, 4 °C) for 60 min and the supernatant (myofibrillar proteins) was stored at −80 °C for further experiments.

2.5. Two-Dimensional Electrophoresis and Gel Image Analysis

The sarcoplasmic and myofibrillar proteome (800 µg, respectively) was included in the Bio-Rad buffer (8 M urea, 2% CHAPS (w/v), 50 mM DTT, 0.3% carrier ampholyte (v/v), bromophenol blue). The samples were loaded on the IPG strips (immobilized pH gradient, 24 cm, pH 3–10, Bio-Rad). The PROTEAN II XL and PROTEAN IEF (isoelectric focusing) cell system (Bio-Rad, Hercules, CA, USA) performed in the first and second dimension electrophoreses, respectively.

In the first dimension, IEF was subjected to passive rehydration (16 h) and then rapid voltage ramping (80,000 V h) was applied. In the second dimension, proteins were resolved on 12% SDS-PAGE gels, which were stained in Coomassie Brilliant Blue (48 h). Gel images were scanned by an image scanner (ImageScanner III, GE Healthcare, Branford, CT, USA). ImageMaster 2D Platinum software (6.0 version, GE Healthcare, Branford, CT, USA) was used to analyze the scanning images. Through expressing the relative number of each spot as the ratio of the number of single spots to the total number of valid spots, the detected and matched spots were normalized. The mean values of gels for each sample and spot were calculated in triplicate. A spot was considered as differential proteins when it came to 5% statistical significance ($p < 0.05$) in one-way ANOVA.

2.6. Identification of Protein Spots by Mass Spectrometry

The protein spots were carefully excised from gels and then de-stained for 30 min in wash buffer (100 μL, 25 mM NH_4HCO_3/50% acetonitrile (v/v)). Washed-out gel-spots (dehydrated in 100% acetonitrile) were dried completely with a centrifuge (Vacufuge plus, Eppendorf, Hamburg, Germany) and then incubated in 15 ng/μL trypsin and 25 mM NH_4HCO_3 (37 °C, 16 h). The peptides were incubated in trifluoroacetic acid (20 μL, 0.1% (v/v), 37 °C, 40 min) after digestion. The above extraction procedures were repeated by using 50% acetonitrile/0.1% trifluoroacetic acid (v/v). Sediments were washed in trifluoroacetic acid and then vacuum freeze-dried for further analysis. The proteins were identified by using AUTOFLEX II TOF-TOF mass spectrometer (autoflex™ speed, Bruker Daltonik, Bremen, Germany). Samples in 1 μL of buffer (50% acetonitrile and 5 mg/mL α-CHCA in 0.1% trifluoroacetic acid) were loaded on plate (AnchorChip, 384-MPT).

Each crystallized sample was washed by using 0.1% trifluoroacetic acid for removing salt ions. Protein identification was performed by peptide mass fingerprinting (PMF), searching the Mascot (2.2 version, Matrix Science, London, UK), and matched with a sheep (*Ovis aries*) family in the Uniprot database (https://www.uniprot.org/ (accessed on 8 October 2021)), correspondingly.

2.7. Statistical Analysis

Proportions of MetMb, OxyMb, DeoMb, and instrumental color attributes (L^*, a^*, b^* and MRA) were analyzed separately. Random terms for all models included the carcass and the processing day. For color models, an additional term for side nested within the carcass was added to the random model to account for the repetition of measures [25]. The correlation between meat color attributes and differentially abundant proteins was analyzed by Pearson's correlation. The mixed models, which were prediction models for muscle color traits, were based on Starkey et al.

Data were expressed as mean ± SE (standard error). The GLM (general linear model) procedure was used to analyze the instrumental color and MRA data with the SAS 8.2 software (Institute Inc., Cary, NC, USA). Means and the correlation data were analyzed by one-way ANOVA by using the SPSS 20.0 software (IBM Inc., Chicago, IL, USA) with the 5% ($p < 0.05$) level of statistical significance difference.

Protein–protein interaction (PPI), gene ontology (GO), and Kyoto encyclopedia of genes and genomes (KEGG) were applied to characterize the functional information of the identified proteins. The protein–protein interaction was analyzed by String 10.0. Bioinformatics analysis was performed through DAVID Bioinformatics Resources 6.7. The species of *Ovis aries* was selected.

3. Results and Discussion

3.1. Instrumental Color and MRA

The instrumental color parameters of skeletal muscles (*M. longissimus lumborum* (LL) and *M. psoas major* (PM)) are presented in Table 1. There was an insignificant ($p > 0.05$) decrease trend in L^*-value. Besides, no significant difference ($p > 0.05$) was found between these two kinds of skeletal muscles within storage time points in L^*-value. Although some previous studies [26] showed that the variation in L^*-value were very subtle. Besides being a color parameter, the L^*-value is also considered an indicator of the water-holding capacity of meat [11]. During ageing in ovine muscle, structural changes of proteins were considered one of the causes of the change in lightness (L^*-value) [27].

For the other color parameters, there were decreasing trends in redness (a^*-value), yellowness (b^*-value), and chroma (C^*-value) of both kinds of muscles. Meanwhile, hue angle ($h^°$-value) exhibited increasing trends during 5 days of post-mortem storage. Those results were consistent with studies in the previous literature [26,28,29]. In an associated study [17], the relative content of OxyMb, MetMb, and DeoMb, on the value of the color parameters of minced pork loin, were evaluated. The color parameters were associated with myoglobin redox forms. Our results (Figure 1) were partially in agreement with Karamucki et al. [17]. For instance, an increase in OxyMb contributes most greatly to an increase in redness (*a**-value), yellowness (*b**-value), and chroma (*C**-value). Chroma (*C**-value) was deemed as an indicator for vividness of color [30]. Compared single color coordinate, hue angle (h°-value), showed more realistic perspectives on meat discoloration [8]. In the literature, muscles with unstable color often go along with greater hue angle (h°-value) [10]. In this study, compared with PM muscle, LL showed lower ($p < 0.05$) hue angle (h°-value) values, indicating a greater color stability.

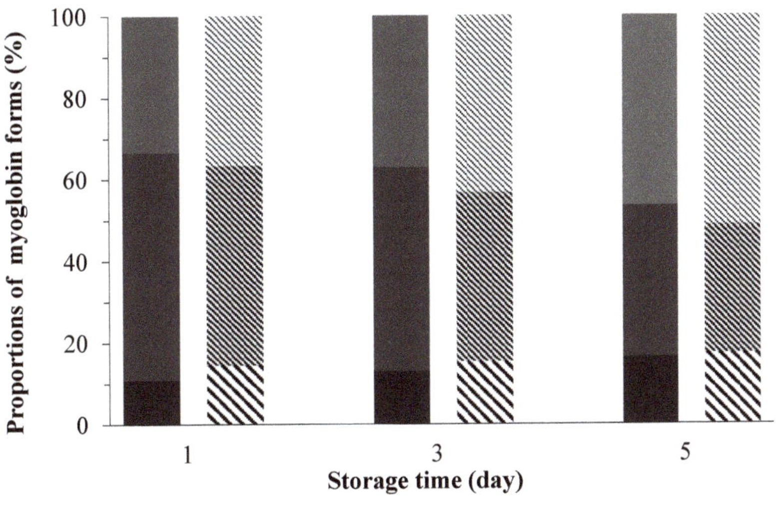

Figure 1. Changes in the relative proportions of MetMb, OxyMb, and DeoMb in ovine *M. longissimus lumborum* (LL) and *M. psoas major* (PM) muscles during postmortem storage (4 ± 1 °C).

Table 1. Instrumental color attributes of ovine *M. longissimus lumborum* (LL) and *M. psoas major* (PM) muscles at 1, 3, and 5 days of postmortem storage (4 °C).

Attribute	Muscle	Storage Time (Day)			SEM
		1	3	5	
*L**	LL	44.25 ax	43.61 ax	37.83 by	0.403
	PM	44.81 bx	46.75 ax	44.70 bx	0.472
*a**	LL	21.41 ax	17.20 bx	13.62 cx	0.35
	PM	16.96 ay	12.98 by	10.59 cy	0.306
*b**	LL	15.57 ay	15.02 ax	12.72 by	0.262
	PM	17.33 ax	15.88 bx	13.28 cx	0.261
*C**	LL	26.47 ax	22.83 bx	18.64 cx	0.423
	PM	24.25 ay	20.51 by	16.99 cy	0.374
h°	LL	36.03 cy	41.12 by	43.06 ay	0.322
	PM	45.62 bx	50.74 ax	51.43 ax	0.372
MRA	LL	0.210 ax	0.145 bx	0.067 cx	0.003
	PM	0.170 ay	0.121 by	0.041 cy	0.004

Means in rows with different superscripts (a–c) are different ($p < 0.05$) and means in column for LL and PM for each storage time and attribute with different superscripts (x–z) are different ($p < 0.05$). (a) SEM, standard error of the mean. (b) C^* (Chroma) = $(a^{*2} + b^{*2})^{0.5}$. (c) h° (Hue angle) = $\mathrm{ATAN}(b^*/a^*) \times (180/\pi)$.

MRA (metmyoglobin reducing activity) decreased with storage time. LL showed greater metmyoglobin-reducing activity than PM ($p < 0.05$, Table 1). Muscles with more color stability had greater ability to reduce metmyoglobin into reduction state. In practice, the loss of reducing activity was influenced by many cofactors, such as the decline of pH, the depletion of substrates, the loss of respiration enzymes or functional properties, and the structural integrity of mitochondria during post-mortem storage of meat [29,31]. MRA gradually decreased, partially due to depletion of the NADH pool and reduction in material consumption [32]. Consequently, LL muscle displayed greater color stability than PM during storage period.

3.2. Sarcoplasmic and Myofibrillar Proteome Analysis

The differentially abundant ($p < 0.05$) proteins, which were identified by two-dimensional electrophoresis image analysis of ovine muscles (*M. longissimus lumborum* (LL, color-stable) and *M. psoas major* (PM, color-labile)), were presented in Figure 2. Differential abundance changes of spots were at least twofold ($p < 0.05$). The identification and related information ware shown in Table 2. The identified proteins were matched with the sheep (*Ovis aries*) family in the Uniprot database. According to physiological function in metabolism, the differentially abundant sarcoplasmic and myofibrillar proteins were classified into three categories: enzymes, chaperone proteins, and structural proteins.

Please note: The early stage of storage (day 1) of myofibrillar proteins gel images were not shown, as there was little differentially abundant proteins between LL and PM muscles at day 1.

Table 2. Differentially abundant proteins in ovine *M. longissimus lumborum* (LL) and *M. psoas major* (PM) muscles at day 1 or day 5 postmortem storage (4 °C).

Spot. [a]	Protein Name [b]	Uniprot ID [b]	Gene Names	Protein Score	Mw/pI [c]	Matched Peptides [d]	Sequence Coverage (%) [d]	Overabundant in Muscle [e]
	Sarcoplasmic proteins (The early stage of storage, day 1, corresponding Figure 2A,B)							
1314	Serum albumin	P14639	ALB	153	71,139/5.80	15	43	LD
1404	Creatine kinase M-type	W5PJ69	CKM	127	43,213/6.66	15	40	LD
1410	Fructose-bisphosphate aldolase	W5PCA0	ALDOB	96	41,555/7.57	12	34	LD
1417	Glyceraldehyde-3-phosphate dehydrogenase	W5PDG3	GAPDH	111	36,110/8.51	11	40	LD
1448	Carbonic anhydrase 3	W5PUC1	CA3	83	29,726/7.70	6	22	LD
1469	Heat shock protein family B	W5P9U1	HSPB11	75	15,737/5.08	5	34	LD
1475	Adenylate kinase isoenzyme 1	C5IJA8	AK1	172	21,750/8.40	15	62	LD
	Sarcoplasmic proteins (The late stage of storage, day 5, corresponding Figure 2C,D)							
1058	Pyruvate kinase	W5QC41	PKM	147	58,551/7.24	20	40	LD
1122	Creatine kinase M-type	W5PJ69	CKM	109	43,213/6.66	14	37	LD
1128	Fructose-bisphosphate aldolase	W5PCA0	ALDOB	87	41,555/7.57	12	36	PM
1134	Glyceraldehyde-3-phosphate dehydrogenase	Q28554	GAPDH	95	36,110/8.51	13	39	LD
1172	Adenylate kinase isoenzyme 1	C5IJA8	AK1	147	21,750/8.40	14	64	LD
	Myofibrillar proteins (The late stage of storage, day 5, corresponding Figure 2E,F)							
2172	Actin, alpha 1	W5NYJ1	ACTA1	70	42,338/5.23	11	25	LD
2242	Enolase 2	W5P5C0	ENO2	117	47,382/7.60	14	37	LD
2323	Fructose-bisphosphate aldolase	W5PCA0	ALDOB	62	41,555/7.57	7	26	LD
2327	Creatine kinase M-type	W5PJ69	CKM	104	43,213/6.66	16	32	LD
2330	Myosin light chain 1	A0A0H3V7A0	MYL1B	73	20,950/4.95	9	58	LD

[a] The numbered spots in gel image (Figure 2). [b] Protein names and accession numbers were taken from the Uniprot database (http://www.uniprot.org (accessed on 8 October 2021)). [c] Theoretical protein mass (Mw; kDa) and isoelectric pH (pI). [d] Number of peptides that matched the protein sequence and total percentage of sequence coverage. [e] Muscle with greater abundance of the protein significance level was indicated ($p < 0.05$).

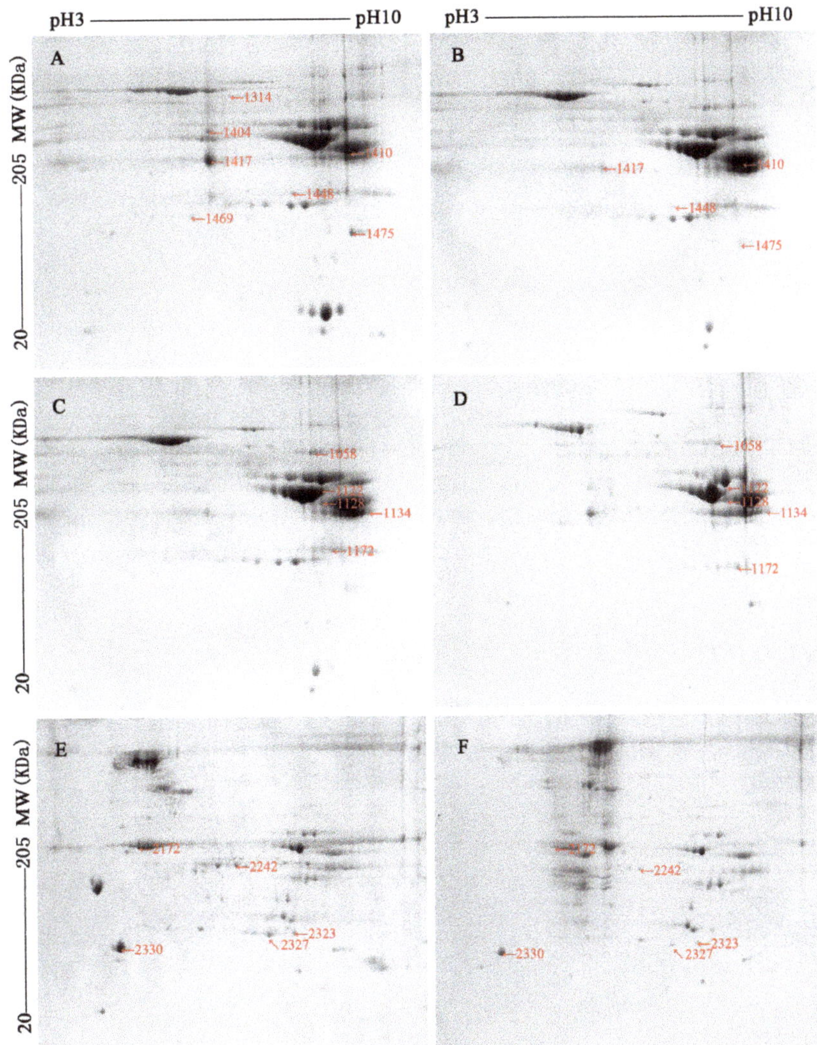

Figure 2. Two-dimensional gel images (pH range between 3 and 10 and molecular weight from about 20 to 205 kDa) of sarcoplasmic ((**A,B**), the early of storage, day 1; (**C,D**), the late stage of storage, day 5) and myofibrillar ((**E,F**), the late stage of storage, day 5) proteome from ovine *M. longissimus lumborum* (LL) and *M. psoas major* (PM) muscles.

3.3. Correlation of Differentially Abundant Proteins with Meat Color Attributes

The correlation data between meat color attributes and differentially abundant sarcoplasmic proteins are presented in Table 3. In this research, the proteomic changes of proteins may affect the color traits of ovine muscle. Therefore, the method of Pearson's Spearman's correlation was used to infer the possible correlation between meat color attributes and differentially abundant sarcoplasmic proteins.

Table 3. Correlation coefficients (Pearson and Spearman) of differentially abundant proteins in ovine *M. longissimus lumborum* (LL) with color attributes (*n* = 6).

Protein Name	Trait [a]	Pearson's Correlation Coefficient [b]	Spearman's Correlation Coefficient [b]
Fructose-bisphosphate aldolase (ALDOA)	a^*	0.126	0.086
	C^*	0.179	0.257
	h°	−0.061	0.086
	MRA	0.145	0.143
Gyceraldehyde-3-phosphate dehydrogenase (GAPDH)	a^*	0.127	0.314
	C^*	0.072	0.371
	h°	−0.139	−0.143
	MRA	0.232	0.486
Carbonic anhydrase 3 (CA3)	a^*	0.702	0.829 *
	C^*	0.857 *	0.771
	h°	−0.543	−0.771
	MRA	0.346	0.200
Adenylate kinase isoenzyme 1 (AK1)	a^*	0.950 **	0.771
	C^*	0.876 *	0.829 *
	h°	−0.956 **	−0.714
	MRA	0.876 *	0.657
Pyruvate kinase (PKM)	a^*	0.575	0.543
	C^*	0.416	0.543
	h°	−0.644	−0.429
	MRA	0.653	0.657
Creatine kinase M-type (CKM)	a^*	0.955 **	0.886 *
	C^*	0.877 *	0.886 *
	h°	−0.957 **	−0.878 *
	MRA	0.889 *	0.771

[a] C^* means chroma; h° means hue angle. [b] * means significant at $p < 0.05$, ** means significant at $p < 0.01$.

During post-mortem storage, six proteins were correlated with MRA and instrumental color data in *M. longissimus lumborum* muscle (Table 3). There was a significant correlation between redness (a^*-value) and adenylate kinase isoenzyme 1 or creatine kinase M-type (r = 0.950, 0.955; $p < 0.01$). Contrarily, a negative relationship was found between hue angle (h°-value) and adenylate kinase isoenzyme 1 or creatine kinase M-type (r = −0.956, 0.957; $p < 0.01$). Chroma (C^*-value) also has been found a positive correlation with carbonic anhydrase 3, adenylate kinase isoenzyme 1 or creatine kinase M-type (r = 0.857, 0.876, 0.877; $p < 0.05$). MRA of LL and PM muscles were positively correlated with adenylate kinase isoenzyme 1 and creatine kinase M-type (r = 0.876, 0.889; $p < 0.05$).

Those proteins (overabundant in color-stable muscles), such as fructose-bisphosphate aldolase [16], glyceraldehyde-3-phosphate dehydrogenase [33], adenylate kinase isoenzyme 1, pyruvate kinase [34], and creatine kinase M-type [16,29], exhibited positive correlations with several meat color parameters (a^*-value, b^*-value, and C^*-value), associating them with stability of meat color. In addition, Joseph et al. [12] observed that some sarcoplasmic proteins (antioxidant) showed positive correlations with a^*-values and MRA. Those results were partially in agreement with present study.

3.4. Functional Roles of Differentially Abundant Proteins and Their Relevance to Color Stability

3.4.1. Glycolytic Enzymes

Four sarcoplasmic proteins (glyceraldehyde-3-phosphate dehydrogenase, fructose-bisphosphate aldolase, pyruvate kinase, enolase 2), which belong to glycolytic enzymes, were over-abundant in LL group (color-stable muscle, Table 2).

Fructose-bisphosphate aldolase catalyzes the reaction that splits the fructose 1,6-bisphosphate into glyceraldehyde 3-phosphate, and dihydroxyacetone phosphate [35]. During the early stage of storage, fructose-bisphosphate aldolase was over-abundant in LL muscle. Nevertheless, an opposite result was observed during late stage of storage (Table 2). Fructose-bisphosphate aldolase was closely related to fast-twitch fiber displaying a higher glycolysis metabolic activity [2]. The down-regulated trend of fructose-bisphosphate aldolase involved in glycolytic metabolism indicated that glycolytic activity was highly active in PM muscle which exhibited lower redness (*a**-value) and color stability [34].

The glycolysis pathway continues with the conversion of glyceraldehyde 3-phosphate into dihydroxyacetone phosphate, the reaction of which is catalyzed by glyceraldehyde 3-phosphate dehydrogenase. The reaction is NAD-dependent, resulting in the production of pyruvate and NADH [34,35]. Pyruvate kinase catalyzes conversion of phosphoenolpyruvate (PEP) and ADP into ATP and pyruvate, which is one of the main rate-limiting enzymes in glycolysis [35,36]. Enolase (also called muscle-specific enolase) catalyzes the conversion of 2-phosphoglycerate to PEP [37]. The three glycolytic enzymes mentioned above were over-abundant in LL group (Table 2). Glyceraldehyde 3-phosphate dehydrogenase and pyruvate kinase exhibited a positive correlation with MRA (Table 3). Some researchers have discussed the differences between color-stable and color-labile muscles in the literature. Several glycolytic enzymes (β-enolase, pyruvate kinase M2, and glyceraldehyde-3-phosphate dehydrogenase), which were over abundant in *Longissimus lumborum* (color-stable muscle), were found to be positively correlated with *a**-values [12,34]. The over-abundance of the enzymes mentioned above could result in higher metabolic activity of glycolytic pathway [34]. This stimulated biological process could accelerate the NADH and pyruvate production. Furthermore, pyruvate is one of the mitochondrial substrates that promote the regeneration of NADH [38]. Besides, NADH supplied by glycolysis is necessary for the NADH–cytochrome b_5 reductase system of mitochondria [39]. The NADH–cytochrome b_5 reductase system was deemed as an important electron transport-mediated reduction system for metmyoglobin, whose mechanism has been expounded by some researchers [7]. Researchers [30] considered NADH a critical component of MRA, whose metabolic function plays a key role in reducing metmyoglobin accumulation [19]. In fact, a direct relationship was found between color stability and MRA, in that muscles with enzyme activity (MRA) in higher had greater color stability [7]. Consequently, NADH may play a key role in MetMb enzymatic reduction, which further influences color stability of meat [2].

In this study, Enolase 2 was overabundant in the LL group (color-stable muscle, Table 2). Enolase, one of the key enzymes in glycolysis, catalyzes the formation of phosphoenolpyruvate from 2-phosphoglycerate, leading to an increased rate of glycolysis [40]. Previous literature findings showed that enolase 1 was related to meat color development and quality variation [23,41,42]. In some proteome research, the profile of muscles of different color stability was compared, revealing the correlation between enolase and attributes of meat color [12,32]. The findings in proteomic analysis indicated that enolase was a key enzyme associated with the color of meat.

3.4.2. Energy Metabolism Enzymes

Adenylate kinase isoenzyme 1 was overabundant and positively correlated with MRA and chroma (*C**-value) in the color-stable LL muscle (Tables 2 and 3). Adenylate kinase (also known as myokinase), a reversible enzyme, plays a role in adenine synthesis and energy metabolism [43]. It provides a way to catalyze the conversion of ADP to ATP and convert AMP to ADP in the cytoplasm [44]. There were different findings within adenylate kinase in the previous associated muscle proteomic studies. Canto et al. [34] noted that the exact mechanism of how adenylate kinase affects color stability remains unclear. Wu et al. [2] found a positive relationship between this enzyme and MRA. It was considered to be a possible potential predictor for meat color stability of M. *longissimuss lumborum*. Thus, the exact function mechanism of adenylate kinase in proteomics of muscle color stability is waiting for further investigation.

Creatine kinase M-type was over-abundant in LL muscle (Table 2), demonstrating positive relationships (r = 0.702, 0.955; *p* < 0.05) with redness (*a**-value) and MRA (Table 3). Creatine kinase M-type was critical in maintenance of ATP-ADP level, reversibly catalyzing the mutual transformation of phosphate between ATP and phosphocreatine [16]. Creatine kinase was degraded during the early post-mortem stages due to the depletion of ATP [42].

Findings from previous literature showed that the denaturation of creatine kinase and several other related proteins were considered to be related to the paler colored (lower *a**-value) *longissimus lumborum* muscle [45]. The present study was in agreement with a previous study, which found that *longissimus lumborum* muscle that creatine kinase M-type was positively correlated ($p < 0.05$) with *a**-values and MRA [16]. Creatine can be used as a natural selective antioxidant because of its ability to scavenge oxygen free radicals [46,47].

Creatine kinase M-type, which is relatively overabundant, can increase creatine levels and minimize myoglobin oxidation, which leads to a lower formation of metmyoglobin and enhanced color stability in color-stable muscles [34].

The functions of carbonic anhydrase 3 were considered as an oxyradical scavenger. In skeletal muscle, this enzyme prevents cells for oxidative damage [48]. In agreement with our results, Yu et al. found that carbonic anhydrase 3 was overabundant in *M. longissimus lumborum* (color-stable muscle) compared with *M. psoas major* (color-labile muscle) [11]. Moreover, we also found this enzyme was correlated with redness (*a**-value) and hue angle (h°-value) (r =0.702, −0.543; $p < 0.05$). In practice, an increased glycolysis and oxidative stress can be indicated by carbonic anhydrase 3 in skeletal muscle [49]. In color-stable muscles, the overabundance of this enzyme can increase glycolysis levels and minimize myoglobin oxidation, possibly resulting in lower redness (*a**-value) and stabilized color attributes [5].

3.4.3. Chaperone Proteins and Structural Proteins

One of the physiological functions of chaperone protein is preventing protein from denaturation [23]. Myoglobin and meat color stability can be compromised by protein denaturation in the process of transformation from muscle to meat [12]. The major chaperone proteins and structural proteins identified were heat shock protein family B, actin (alpha 1), myosin light chain 1, serum albumin, and carbonic anhydrase 3.

The heat shock protein family B (HSP beta-1, also known as HSP27) belongs to chaperone proteins. HSP27 is involved in small heat shock protein family which can prevent structure damage or degradation of proteins in muscle cells and promotes the survival of cell under physiological stress [50,51]. In pig muscle, lower abundance of HSP27 was related to the development of lighter color and PSE zones in semimembranosus [23,52]. Higher abundance of HSP27 was found in beef which was lower in *a** and *L** values compared with their counterparts. [24]

The physiological function of myosin and actin are mainly related to contractile properties of muscle [43]. During muscle contraction, the function of alpha actin is associated with ATP binding. [53]. Actin alpha has been identified as related to the color of meat in some previous proteomic analyses [4,54]. In a similar study, alpha actin and myosin light chain 1, identified in porcine longissimus muscle, were related to *L** value. In the study of Gagaoua et al. [15], actin alpha was correlated with all color coordinates. This protein and several structural proteins were deemed as biomarkers for meat color and of great interest in discriminating between beef color classes. Thus, the functional role of structural proteins in meat color proteomics was important.

3.5. Bioinformatics Analysis of Differentially Abundant Proteins

The analytical method of bioinformatics has been applied to the related research area of meat color investigation to explore the biological processes and molecular functions of proteins [2,41,55]. Bioinformatics analysis of differentially abundant proteins between muscle types (color-stable and color-labile) provides an efficient tool for deeper understanding of the molecular mechanisms of meat discoloration, based on the proteomic results. In this study, Protein–protein interaction (PPI), gene ontology (GO), and the Kyoto encyclopedia of genes and genomes (KEGG) analyses were applied to characterize the functional information of the identified proteins (DAVID Bioinformatics Resources 6.7).

Protein–Protein Interactions

The protein–protein interactions were illustrated in Figure 3 and assessed by String 10.0. There are nodes and edges networks, representing the identified proteins and functional annotation of protein–protein interactions with different colors. The local clustering coefficient and PPI enrichment p-value were 0.49 and 1.81×10^{-8} respectively.

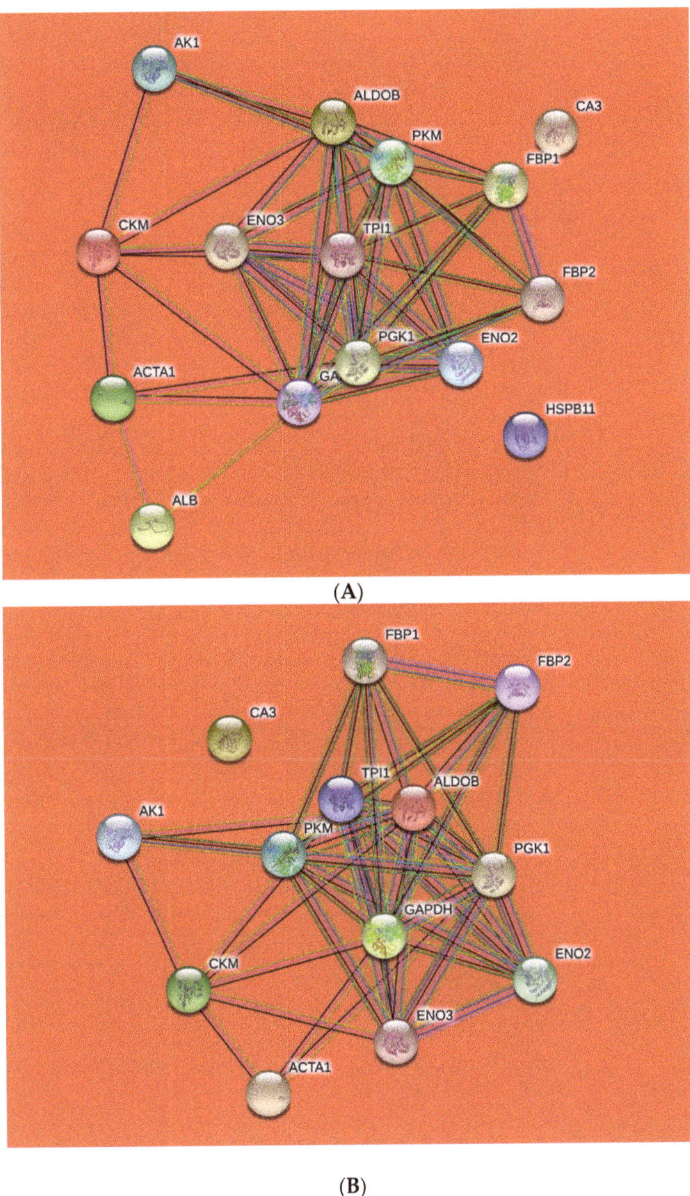

Figure 3. Protein–protein interaction (PPI) network of differentially expressed proteins from ovine *M. longissimus lumborum* (**A**) and *M. psoas major* (**B**) muscles.

The highest number of proteins that interacted strongly with each other were implicated in glycolysis pathway (ALDOB, GAPDH, PKM, and ENO2). Four proteins with close range were involved in energy metabolic process or structural protein (AK1, CKM, CA3, and ACTA1). Furthermore, ALB and HSPB11 were exhibited further distance with the other proteins in the network.

The nodes are differentially abundant proteins (colored nodes—query proteins and first shell of interactors; white nodes—second shell of interactors; empty nodes—proteins of unknown 3D structure; filled nodes—some 3D structure is known or predicted) in Ovis aries database and connection lines are the predicted functional annotations (known interactions: blue—from curated databases; purple—experimentally determined; and predicted interactions: green—gene neighborhood; red—gene fusions; dark blue—gene co-occurrence; and others: yellow—text mining; black—co-expression; light blue—protein homology).

4. Conclusions

In the present research, the post-mortem proteome profiles associated with the typical color-stable (*M. longissimus lumborum*) and color-labile (*M. psoasmajor*) muscles of *Small-tailed Han* Sheep were studied and compared. The results of differential proteome indicated that the identified proteins were glycolytic enzymes, energy metabolism enzymes, chaperone proteins, structural proteins, and chaperone and binding proteins. Several proteins were demonstrated to be related to post-mortem discoloration through the methods of Pearson's correlation and multiple linear regression model analysis between differentially abundant proteins and meat color indices.

Bioinformatics analysis showed that these proteins (pyruvate kinase, fructose-bisphosphate aldolase, enolase 2, and glyceraldehyde-3-phosphate dehydrogenase) mainly involved in the glycolysis pathway. The relationship between post-mortem proteome profiles and discoloration could play a key role in explaining the different color stabilities within muscles. However, further research is needed to confirm if the specific differentially abundant proteins have the potential probability of being predictors for meat discoloration during post-mortem storage.

Author Contributions: Conceptualization, R.D.; Data curation, D.Z.; Formal analysis, X.G.; Funding acquisition, X.G. and R.D.; Investigation, L.W.; Methodology, X.G.; Resources, Y.C.; Software, M.L. and F.Z.; Supervision, R.D.; Validation, S.W.; Writing—review & editing, X.G. All authors have read and agreed to the published version of the manuscript.

Funding: This research was funded by the financial support by Science and Technology research project of Hebei Education Department (Project No.: BJ2018061); and the Ministry of Agriculture of China (China Agriculture Research System-39, CARS-39).

Institutional Review Board Statement: The study was conducted according to the guidelines of the Declaration of Helsinki, and approved by the Institutional Review Board of China Agricultural University Laboratory Animal Welfare and Animal Experimental Ethical Inspection Form (protocol code: AW30211202-4-1, 30 March 2020).

Acknowledgments: This work was supported by the financial support by Science and Technology research project of Hebei Education Department (Project No.: BJ2018061); and the Ministry of Agriculture of China (China Agriculture Research System-39, CARS-39).

Conflicts of Interest: The authors declare that there are no conflict of interest.

References

1. Mancini, R.A.; Hunt, M.C. Current research in meat color. *Meat Sci.* **2005**, *71*, 100–121. [CrossRef]
2. Wu, W.; Yu, Q.; Fu, Y.; Tian, X.; Jia, F.; Li, X.; Dai, R. Towards muscle-specific meat color stability of Chinese Luxi yellow cattle: A proteomic insight into post-mortem storage. *J. Proteom.* **2016**, *147*, 108–118. [CrossRef]
3. Zhang, Y.; Zhang, X.; Wang, T.; Hopkins, D.L.; Mao, Y.; Liang, R.; Yang, G.; Luo, X.; Zhu, L. Implications of step-chilling on meat color investigated using proteome analysis of the sarcoplasmic protein fraction of beef longissimus lumborum muscle. *J. Integr. Agric.* **2018**, *17*, 2118–2125. [CrossRef]

4. Gagaoua, M.; Terlouw, E.M.C.; Picard, B. The study of protein biomarkers to understand the biochemical processes underlying beef color development in young bulls. *Meat Sci.* **2017**, *134*, 18–27. [CrossRef]

5. Schilling, M.W.; Suman, S.P.; Zhang, X.; Nair, M.N.; Desai, M.A.; Cai, K.; Ciaramella, M.A.; Allen, P.J. Proteomic approach to characterize biochemistry of meat quality defects. *Meat Sci.* **2017**, *132*, 131–138. [CrossRef]

6. Sørheim, O.; Westad, F.; Larsen, H.; Alvseike, O. Colour of ground beef as influenced by raw materials, addition of sodium chloride and low oxygen packaging. *Meat Sci.* **2009**, *81*, 467–473. [CrossRef]

7. Bekhit, A.E.; Faustman, C. Metmyoglobin reducing activity: A review. *Meat Sci.* **2005**, *71*, 407–439. [CrossRef]

8. Luciano, G.; Monahan, F.J.; Vasta, V.; Pennisi, P.; Bella, M.; Priolo, A. Lipid and colour stability of meat from lambs fed fresh herbage or concentrate. *Meat Sci.* **2009**, *82*, 193–199. [CrossRef]

9. Jacob, R.H.; D'Antuono, M.F.; Gilmour, A.R.; Warner, R.D. Phenotypic characterisation of colour stability of lamb meat. *Meat Sci.* **2014**, *96*, 1040–1048. [CrossRef]

10. King, D.A.; Shackelford, S.D.; Rodriguez, A.B.; Wheeler, T.L. Effect of time of measurement on the relationship between metmyoglobin reducing activity and oxygen consumption to instrumental measures of beef longissimus color stability. *Meat Sci.* **2011**, *87*, 26–32. [CrossRef]

11. Yu, Q.; Wu, W.; Tian, X.; Jia, F.; Xu, L.; Dai, R.; Li, X. Comparative proteomics to reveal muscle-specific beef color stability of Holstein cattle during post-mortem storage. *Food Chem.* **2017**, *229*, 769–778. [CrossRef]

12. Joseph, P.; Suman, S.P.; Rentfrow, G.; Li, S.; Beach, C.M. Proteomics of muscle-specific beef color stability. *J. Agric. Food Chem.* **2012**, *60*, 3196–3203. [CrossRef]

13. Ramanathan, R.; Suman, S.P.; Faustman, C. Biomolecular Interactions Governing Fresh Meat Color in Post-mortem Skeletal Muscle: A Review. *J. Agric. Food Chem.* **2020**, *68*, 12779–12787. [CrossRef] [PubMed]

14. Gagaoua, M.; Hughes, J.; Terlouw, E.M.C.; Warner, R.D.; Purslow, P.P.; Lorenzo, J.M.; Picard, B. Proteomic biomarkers of beef colour. *Trends Food Sci. Tech.* **2020**, *101*, 234–252. [CrossRef]

15. Gagaoua, M.; Bonnet, M.; De Koning, L.; Picard, B. Reverse Phase Protein array for the quantification and validation of protein biomarkers of beef qualities: The case of meat color from Charolais breed. *Meat Sci.* **2018**, *145*, 308–319. [CrossRef]

16. Nair, M.N.; Suman, S.P.; Chatli, M.K.; Li, S.; Joseph, P.; Beach, C.M.; Rentfrow, G. Proteome basis for intramuscular variation in color stability of beef semimembranosus. *Meat Sci.* **2016**, *113*, 9–16. [PubMed]

17. Karamucki, T.; Jakubowska, M.; Rybarczyk, A.; Gardzielewska, J. The influence of myoglobin on the colour of minced pork loin. *Meat Sci.* **2013**, *94*, 234–238. [CrossRef] [PubMed]

18. Te Pas, M.F.W.; Kruijt, L.; Pierzchala, M.; Crump, R.E.; Boeren, S.; Keuning, E.; Hoving-Bolink, R.; Hortós, M.; Gispert, M.; Arnau, J.; et al. Identification of proteomic biomarkers in M. Longissimus dorsi as potential predictors of pork quality. *Meat Sci.* **2013**, *95*, 679–687. [CrossRef] [PubMed]

19. Faustman, C.; Cassens, R.G. The biochemical basis for meat discoloration in fresh meat: A review. *J. Muscle Foods* **1990**, *1*, 217–243. [CrossRef]

20. Mclaren, K. Colour space, colour scales and colour difference. In *Colour Physics for Industry*; Mcdonald, R., Ed.; Dyers Company Publications Trust: Bradford, UK, 1987; pp. 97–115.

21. Tang, J.; Faustman, C.; Hoagland, T.A. Krzywicki revisited: Equations for spectrophotometric determination of myoglobin redox forms in aqueous meat extracts. *J. Food Sci.* **2004**, *69*, C717–C720. [CrossRef]

22. Mikkelsen, A.; Juncher, D.; Skibsted, L.H. Metmyoglobin reductase activity in porcine m. longissimus dorsi muscle. *Meat Sci.* **1999**, *51*, 155–161. [CrossRef]

23. Sayd, T.; Morzel, M.; Chambon, C.; Franck, M.; Figwer, P.; Larzul, C.; Le Roy, P.; Monin, G.; Chérel, P.; Laville, E. Proteome Analysis of the Sarcoplasmic Fraction of Pig Semimembranosus Muscle: Implications on Meat Color Development. *J. Agric. Food Chem.* **2006**, *54*, 2732–2737. [CrossRef]

24. Kim, N.K.; Cho, S.; Lee, S.H.; Park, H.R.; Lee, C.S.; Cho, Y.M.; Choy, Y.H.; Yoon, D.; Im, S.K.; Park, E.W. Proteins in longissimus muscle of Korean native cattle and their relationship to meat quality. *Meat Sci.* **2008**, *80*, 1068–1073. [CrossRef]

25. Biffin, T.E.; Smith, M.A.; Bush, R.D.; Collins, D.; Hopkins, D.L. The effect of electrical stimulation and tenderstretching on colour and oxidation traits of alpaca (Vicunga pacos) meat. *Meat Sci.* **2019**, *156*, 125–130. [CrossRef]

26. Suman, S.P.; Hunt, M.C.; Nair, M.N.; Rentfrow, G. Improving beef color stability: Practical strategies and underlying mechanisms. *Meat Sci.* **2014**, *98*, 490–504. [CrossRef] [PubMed]

27. Kim, Y.H.B.; Frandsen, M.; Rosenvold, K. Effect of ageing prior to freezing on colour stability of ovine longissimus muscle. *Meat Sci.* **2011**, *88*, 332–337. [CrossRef] [PubMed]

28. Wu, W.; Gao, X.; Dai, Y.; Fu, Y.; Li, X.; Dai, R. Post-mortem changes in sarcoplasmic proteome and its relationship to meat color traits in M. semitendinosus of Chinese Luxi yellow cattle. *Food Res. Int.* **2015**, *72*, 98–105. [CrossRef]

29. Gao, X.; Wu, W.; Ma, C.; Li, X.; Dai, R. Postmortem changes in sarcoplasmic proteins associated with color stability in lamb muscle analyzed by proteomics. *Eur. Food Res. Technol.* **2016**, *242*, 527–535. [CrossRef]

30. Seyfert, M.; Mancini, R.A.; Hunt, M.C.; Tang, J.; Faustman, C.; Garcia, M. Color Stability, Reducing Activity, and Cytochrome c Oxidase Activity of Five Bovine Muscles. *J. Agric. Food Chem.* **2006**, *54*, 8919–8925. [CrossRef]

31. Renerre, M.; Labas, R. Biochemical factors influencing metmyoglobin formation in beef muscles. *Meat Sci.* **1987**, *19*, 151–165. [CrossRef]

32. Yu, Q.; Wu, W.; Tian, X.; Hou, M.; Dai, R.; Li, X. Unraveling proteome changes of Holstein beef M. semitendinosus and its relationship to meat discoloration during post-mortem storage analyzed by label-free mass spectrometry. *J. Proteom.* **2017**, *154*, 85–93. [CrossRef] [PubMed]

33. Żelechowska, E.; Przybylski, W.; Jaworska, D.; Santé-Lhoutellier, V. Technological and sensory pork quality in relation to muscle and drip loss protein profiles. *Eur. Food Res. Technol.* **2012**, *234*, 883–894. [CrossRef]

34. Canto, A.C.; Suman, S.P.; Nair, M.N.; Li, S.; Rentfrow, G.; Beach, C.M.; Silva, T.J.P.; Wheeler, T.L.; Shackelford, S.D.; Grayson, A.; et al. Differential abundance of sarcoplasmic proteome explains animal effect on beef Longissimus lumborum color stability. *Meat Sci.* **2015**, *102*, 90–98. [CrossRef] [PubMed]

35. Robert, K.M.; Daryl, K.G.; Peter, A.M.; Victor, W.R. Glycolysis & the Oxidation of Pyruvate. In *Harper's Illustrated Biochemistry*; Foltin, J., Ed.; McGraw-Hill: New York, NY, USA, 2003; pp. 136–144.

36. Zhou, B.; Shen, Z.; Liu, Y.; Wang, C.; Shen, Q.W. Proteomic analysis reveals that lysine acetylation mediates the effect of antemortem stress on postmortem meat quality development. *Food Chem.* **2019**, *293*, 396–407. [CrossRef] [PubMed]

37. Subramanian, A.; Miller, D.M. Structural Analysis of alpha-Enolase: Mapping the functional domains involved in down-regulation of the c-myc protooncogene. *J. Biol. Chem.* **2000**, *275*, 5958–5965. [CrossRef]

38. Ramanathan, R.; Mancini, R.A. Effects of pyruvate on bovine heart mitochondria-mediated metmyoglobin reduction. *Meat Sci.* **2010**, *86*, 738–741. [CrossRef]

39. Kim, Y.H.; Hunt, M.C.; Mancini, R.A.; Seyfert, M.; Loughin, T.M.; Kropf, D.H.; Smith, J.S. Mechanism for lactate-color stabilization in injection-enhanced beef. *J. Agric. Food Chem.* **2006**, *54*, 7856–7862. [CrossRef]

40. Gagaoua, M.; Terlouw, E.M.C.; Micol, D.; Boudjellal, A.; Hocquette, J.; Picard, B. Understanding Early Post-Mortem Biochemical Processes Underlying Meat Color and pH Decline in the Longissimus thoracis Muscle of Young Blond d'Aquitaine Bulls Using Protein Biomarkers. *J. Agric. Food Chem.* **2015**, *63*, 6799–6809. [CrossRef]

41. Joseph, P.; Nair, M.N.; Suman, S.P. Application of proteomics to characterize and improve color and oxidative stability of muscle foods. *Food Res. Int.* **2015**, *76 Pt 4*, 938–945. [CrossRef]

42. Jia, X.; Ekman, M.; Grove, H.; Færgestad, E.M.; Aass, L.; Hildrum, K.I.; Hollung, K. Proteome Changes in Bovine Longissimus Thoracis Muscle During the Early Postmortem Storage Period. *J. Proteome Res.* **2007**, *6*, 2720–2731. [CrossRef] [PubMed]

43. Desai, M.A.; Joseph, P.; Suman, S.P.; Silva, J.L.; Kim, T.; Schilling, M.W. Proteome basis of red color defect in channel catfish (Ictalurus punctatus) fillets. *LWT-Food Sci. Technol.* **2014**, *57*, 141–148. [CrossRef]

44. Sánchez, B.; Champomier-Vergès, M.; Anglade, P.; Baraige, F.; de Los Reyes-Gavilán, C.G.; Margolles, A.; Zagorec, M. Proteomic analysis of global changes in protein expression during bile salt exposure of Bifidobacterium longum NCIMB 8809. *J. Bacteriol.* **2005**, *187*, 5799–5808. [CrossRef]

45. Choi, Y.M.; Ryu, Y.C.; Lee, S.H.; Go, G.W.; Shin, H.G.; Kim, K.H.; Rhee, M.S.; Kim, B.C. Effects of supercritical carbon dioxide treatment for sterilization purpose on meat quality of porcine longissimus dorsi muscle. *LWT-Food Sci. Technol.* **2008**, *41*, 317–322. [CrossRef]

46. Fang, Y.; Yang, S.; Wu, G. Free radicals, antioxidants, and nutrition. *Nutrition* **2002**, *18*, 872–879. [CrossRef]

47. Sestili, P.; Martinelli, C.; Colombo, E.; Barbieri, E.; Potenza, L.; Sartini, S.; Fimognari, C. Creatine as an antioxidant. *Amino Acids* **2011**, *40*, 1385–1396. [CrossRef]

48. Räisänen, S.R.; Lehenkari, P.; Tasanen, M.; Rahkila, P.; Härkönen, P.L.; Väänänen, H.K. Carbonic anhydrase III protects cells from hydrogen peroxide-induced apoptosis. *FASEB J.* **1999**, *13*, 513–522. [CrossRef]

49. Vasilaki, A.; Simpson, D.; Mcardle, F.; Mclean, L.; Beynon, R.J.; Van Remmen, H.; Richardson, A.G.; Mcardle, A.; Faulkner, J.A.; Jackson, M.J. Formation of 3-nitrotyrosines in carbonic anhydrase III is a sensitive marker of oxidative stress in skeletal muscle. *Proteom.-Clin. Appl.* **2007**, *1*, 362–372. [CrossRef]

50. Beere, H.M. Death versus survival: Functional interaction between the apoptotic and stress-inducible heat shock protein pathways. *J. Clin. Investig.* **2005**, *115*, 2633–2639. [CrossRef] [PubMed]

51. Thompson, H.S.; Maynard, E.B.; Morales, E.R.; Scordilis, S.P. Exercise-induced HSP27, HSP70 and MAPK responses in human skeletal muscle. *Acta Physiol. Scand.* **2003**, *178*, 61–72. [CrossRef] [PubMed]

52. Laville, E.; Sayd, T.; Santé-Lhoutellier, V.; Morzel, M.; Labas, R.; Franck, M.; Chambon, C.; Monin, G. Characterisation of PSE zones in semimembranosus pig muscle. *Meat Sci.* **2005**, *70*, 167–172. [CrossRef] [PubMed]

53. Chen, F.; Lee, Y.; Jiang, Y.; Wang, S.; Peatman, E.; Abernathy, J.; Liu, H.; Liu, S.; Kucuktas, H.; Ke, C.; et al. Identification and characterization of full-length cDNAs in channel catfish (Ictalurus punctatus) and blue catfish (Ictalurus furcatus). *PLoS ONE* **2010**, *5*, e11546. [CrossRef] [PubMed]

54. Polati, R.; Menini, M.; Robotti, E.; Millioni, R.; Marengo, E.; Novelli, E.; Balzan, S.; Cecconi, D. Proteomic changes involved in tenderization of bovine Longissimus dorsi muscle during prolonged ageing. *Food Chem.* **2012**, *135*, 2052–2069. [CrossRef] [PubMed]

55. Wu, S.; Luo, X.; Yang, X.; Hopkins, D.L.; Mao, Y.; Zhang, Y. Understanding the development of color and color stability of dark cutting beef based on mitochondrial proteomics. *Meat Sci.* **2020**, *163*, 108046. [CrossRef] [PubMed]

Communication

Preliminary Results about Lamb Meat Tenderness Based on the Study of Novel Isoforms and Alternative Splicing Regulation Pathways Using Iso-seq, RNA-seq and CTCF ChIP-seq Data

Zehu Yuan [1], Ling Ge [2], Weibo Zhang [2], Xiaoyang Lv [1], Shanhe Wang [2], Xiukai Cao [1] and Wei Sun [1,2,*]

[1] Joint International Research Laboratory of Agriculture and Agri-Product Safety of Ministry of Education, Yangzhou University, Yangzhou 225000, China; yuanzehu@yzu.edu.cn (Z.Y.); dx120170085@yzu.edu.cn (X.L.); cxkai0909@163.com (X.C.)

[2] College of Animal Science and Technology, Yangzhou University, Yangzhou 225000, China; mx120200799@stu.yzu.edu.cn (L.G.); mz120191042@stu.yzu.edu.cn (W.Z.); 007121@yzu.edu.cn (S.W.)

* Correspondence: sunwei@yzu.edu.cn; Tel.: +86-0514-87926332

Abstract: Tenderness is an important indicator of meat quality. Novel isoforms associated with meat tenderness and the role of the CCCTC-binding factor (CTCF) in regulating alternative splicing to produce isoforms in sheep are largely unknown. The current project studied six sheep from two crossbred populations (Dorper × Hu × Hu, DHH and Dorper × Dorper × Hu, DDH) with divergent meat tenderness. Pooled Iso-seq data were used to annotate the sheep genomes. Then, the updated genome annotation and six RNA-seq data were combined to identify differentially expressed isoforms (DEIs) in muscles between DHH and DDH. These data were also combined with peaks detected from CTCF ChIP-seq data to investigate the regulatory role of CTCF for the alternative splicing. As a result, a total of 624 DEIs were identified between DDH and DHH. For example, isoform 7.524.18 transcribed from *CAPN3* may be associated with meat tenderness. In addition, a total of 86 genes were overlapped between genes with transcribed DEIs and genes in differential peaks identified by CTCF ChIP-seq. Among these overlapped genes, *ANKRD23* produces different isoforms which may be regulated by CTCF via methylation. As preliminary research, our results identified novel isoforms associated with meat tenderness and revealed the possible regulating mechanisms of alternative splicing to produce isoforms.

Keywords: sheep; tenderness; novel isoforms; alternative splicing; CTCF; Iso-seq; ChIP-seq

Citation: Yuan, Z.; Ge, L.; Zhang, W.; Lv, X.; Wang, S.; Cao, X.; Sun, W. Preliminary Results about Lamb Meat Tenderness Based on the Study of Novel Isoforms and Alternative Splicing Regulation Pathways Using Iso-seq, RNA-seq and CTCF ChIP-seq Data. *Foods* **2022**, *11*, 1068. https://doi.org/10.3390/foods11081068

Academic Editors: Mohammed Gagaoua and Brigitte Picard

Received: 19 February 2022
Accepted: 4 April 2022
Published: 7 April 2022

Publisher's Note: MDPI stays neutral with regard to jurisdictional claims in published maps and institutional affiliations.

1. Introduction

Meat tenderness is determined by several intrinsic factors (e.g., proteolytic activity, amount of glycogen, fiber type and connective tissue) and extrinsic factors (e.g., animal breed and nutritional conditions) [1]. Among these factors, the calpain system plays an essential role in meat tenderness [2]. In beef, the most important factor that determines tenderness is the proteolytic activity of the calpain system. Calpain3 (CAPN3), a member of the calpain system, is mainly expressed in skeletal muscle, especially in type II fibers [3–5], and it may play an important role in the calpain system to regulate meat tenderness [6]. A previous study suggested that the expression level of CAPN3 protein and mRNA was significantly associated with tenderness [7,8]. Another member of the calpain system is calpastatin (*CAST*). Accumulated evidence suggests that different isoforms transcribed from a gene due to alternative splicing may play different biological roles. For example, distinct isoforms due to alternative splicing [9] in *CAST* have different biological functions in beef meat tenderness.

PacBio long-reads isoform sequencing (Iso-seq), a third-generation sequencing technology, can directly obtain full-length isoforms [10]. By combing with traditional RNA-sequencing (RNA-seq), Iso-seq can contribute to identify novel isoform associated with complex traits [11–13]. Although differentially expressed gene expression studies have been implemented for meat quality traits in sheep [14,15], studies focusing on the investigation of novel isoforms in sheep muscle are still scarce. Alternative splicing is the main mechanism to diversify isoforms and is regulated by numerous interacting components [16]. For example, the CCCTC-binding factor (CTCF) has been identified as a modulator of the alternative splicing process [17]. However, the role of CTCF in regulating alternative splicing to produce isoforms in sheep is largely unknown. In this context, using Iso-seq, RNA-seq and CTCF chromatin immunoprecipitation sequencing (ChIP-seq) data in sheep muscle can help to identify key isoforms associated with meat tenderness and to reveal the biological mechanisms of alternative splicing regulation to produce isoforms.

Tenderness is an important indicator of meat quality with the greatest consumer appreciation [18]. Crossbreeding is a common method for increasing sheep production. When using crossbreeding methods to increase sheep meat production, some meat quality traits, e.g., tenderness, show a significant difference across different cross-breed populations. Unveiling the genetic factors associated with tenderness is essential for guiding cross-breeding efficiently and profitably. Therefore, revealing the genetic processes that regulate meat quality between different cross-breeds is of great interest to the sheep industry.

The goals of the current study were to identify key isoforms associated with meat tenderness and to reveal the biological mechanisms of alternative splicing regulations to produce isoforms. To achieve these goals, differentially expressed isoforms (DEIs) were identified between two cross-bred populations with divergent meat tenderness by integrating Iso-seq and RNA-seq data. Then, we combined peaks detected from CTCF chromatin immunoprecipitation sequencing (ChIP-seq) data with DEIs to investigate the possible mechanism with which CTCF regulates the alternative splicing process. Our results have the potential to reveal novel isoforms associated with meat tenderness and uncover the possible regulating mechanism of alternative splicing to produce isoforms.

2. Materials and Methods

The animal experiment of this study was approved by the Experimental Animal Ethical Committee of Yangzhou University (File NO. 202103294). The overall experimental design of the current study is shown in Figure 1.

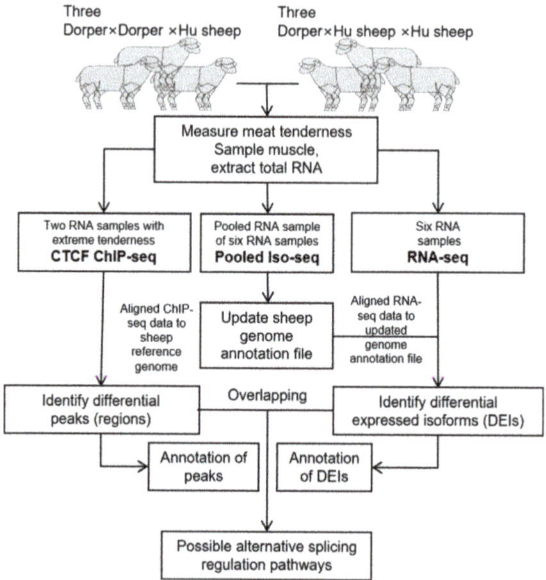

Figure 1. The overall experimental design. DEIs, differentially expressed isoforms.

2.1. Animal Samples and Meat Traits

In the current study, a total of six unrelated 6-month-old male sheep, including three Dorper × Hu × Hu (DHH) and three Dorper × Dorper × Hu (DDH), were selected from a large population for sampling and measuring meat tenderness. Six male sheep were fed in the same environment from 2 months old to 6 months old. Six sheep were selected with similar live weights (DHH = 40.80 ± 4.57 kg, DDH = 41.50 ± 0.95 kg, Table 1) to rule out the effect of live weight on meat tenderness. Live weight was measured before slaughter (24 h in advance). After slaughter, *longissimus dorsi* from between the 12th and 13th ribs from six sheep were sampled within 30 min. Each muscle sample was packed into a 1.5 mL cryotube with triplicate. All the muscle samples were frozen in liquid and stored in a −80 °C refrigerator. After slaughter, carcass weight was measured by an electronic scale. Then, 45 min after slaughter, *longissimus dorsi* were cut after the 12th rib and the samples were cooked for 35 min at 71 °C [19]. Then, six 1 cm^2 cylinders were cut and all samples were tested using a meat tenderness analyzer (CL3, NanJing, CN). The difference of meat tenderness between DDH and DHH was tested by the *t*-test function in the basic environment of R 4.1.1 [20]. A *p* value smaller than 0.05 denotes significance.

Table 1. Meat traits of Dorper × Hu × Hu (DHH) and Dorper × Dorper × Hu (DDH) sheep.

Meat Trait (Unit)	Crossing Type		Statistic (t)	*p* Value
	DDH [1] (n = 3)	DHH [2] (n = 3)		
Live weight (kg)	40.80 ± 4.57	41.50 ± 0.95	−0.2720	0.8094
Carcass weight (kg) [3]	22.00 ± 2.27	23.70 ± 0.64	−1.2717	0.3161
Shear force (N) [3]	83.20 ± 13.60	53.30 ± 8.39	3.2558	0.0406

[1] DDH denotes Dorper × Dorper × Hu. [2] DHH denotes Dorper × Hu × Hu. [3] These traits were measured on hot carcass.

2.2. RNA-Extraction, RNA-seq and Iso-seq

A Trizol reagent Kit (TaKaRa, Kusatsu, Shiga, Japan) was used to extract the total RNA from six muscle tissues. Then, Nanodrop 2000 (Thermo Fisher Scientific, Waltham, MA, USA) and a 2100 Bioanalyzer (Agilent Technologies, Waldronn, Germany) were used to evaluate the RNA Integrity Number (RIN >7 passed quality control) and 28S/18S ratio (ratio > 1.0 passed quality control). After quality control of all six RNA samples, they were used for library construction. Following library quality control, six libraries were sequenced on the MGISEQ-2000 (BIG) platform [13].

Pooled RNA samples of six muscle tissues were used for Iso-seq. Briefly, cDNA was transcribed from a pooled RNA sample using the SMARTerTM PCR cDNA synthesis kit (Takara Biotechnology, Dalian, China). Following PCR amplification, PCR product purification, size selection (>1 kb), SMRTbell library construction, and library quality control, the library was sequenced on the PacBio sequencing platform. The raw Iso-seq data were processed using the SMRT Link v8.0 pipeline. Briefly, the BAM file was processed to obtain the circular consensus sequence (CCS). Full-length CCS reads with 5′ and 3′ cDNA primers and polyA were defined as full-length non-concatemer (FLNC) reads. FLNC reads were corrected by six clean RNA-seq data using LoRDEC v0.9 software [21]. The quality control of six RNA-seq data was implemented by SOAPnuke v2.1.0 [22].

2.3. Update Sheep Reference Annotation File

The pooled Iso-seq data were used to update the sheep reference annotation file. After correction, FLNC reads were mapped to the reference genome (http://ftp.ensembl.org/pub/release-102/fasta/ovis_aries_rambouillet/dna/, accessed on 29 March 2022) using GMAP [23]. Isoforms that met at least one of the following criteria were kept: first, an isoform was identified by at least two FLNC reads; second, an isoform was identified by only one FLNC read with a percentage identity (PID) of greater than 99%; third, all splicing sites in an isoform were identified by RNA-seq data; fourth, all alternative splicing events in a identified isoform were also annotated by the reference genome annotation file. Isoforms that overlapped less than 20% of their length on the same strand were identified as distinct isoforms. A novel gene was defined as a gene locus overlapping less than 20% of their length with known genes. An isoform with a new emerged intron (exon) or a final splice site of 3′ changed ends was defined as a novel isoform. In addition, the alternative splicing events were detected by the AStalavista v3.2 software [24].

2.4. Differentially Expressed Isoform Detection and Functional Analyses

Bowtie2 was used to align short reads to novel annotation files [25]. The DESeq2 R Bioconductor package was used to identify DEIs between two groups [26]. An isoform with a fold change (FC) greater than two and a false discovery rate (FDR) smaller than 0.05 was defined as a significant DEI. All identified DEIs were blasted to GO database using Diamond [27] and blasted to the KEGG database using KOBAS [28] to extract their potential biological functions. The GOplot 1.0.2 R package was used to visualize GO enrichment results [29]. GO terms and KEGG pathways with an FDR smaller than 0.05 were regarded as having significant enrichment.

2.5. Validation of Differentially Expressed Isoforms

The isoform abundance of five DEIs were quantified for validating the result of identified DEIs by quantitative real-time PCR (qRT-PCR). Five isoform-specific paired premiers were designed by SnapGene® v2.3.2 software (from Insightful Science; available at snapgene.com) and were synthesized by Tsingke Biotechnology Co., Ltd. (Nanjing, China, Table S1). cDNA was synthesized using a FastKing gDNA Dispelling RT Super Mix (TIANGEN, Beijing, China). The qRT-PCR was performed on a CFX96 Connect™ Real-Time System (BIO-RAD, Hercules, CA, USA) using a 20 μL reaction volume, including 1 μL of cDNA in 10 μL of 2× TSINGKE Master qPCR Mix (SYBR Green I) (Tsingke, Nanjing, China), 0.8 μL (10 μm/μL) each of the forward and reverse primers, and 7.4 μL of distilled water. The abundance of the *GAPDH* was used as the control. Each biological sample was implemented in triplicate, and the $2-\Delta\Delta Ct$ method was used for calculating the relative expression level of isoforms.

2.6. CTCF ChIP-seq

We aimed to explore the possible regulating mechanism of CTCF in alternative splicing in sheep muscle. Two muscle samples with extreme meat tenderness were selected to implement CTCF ChIP-seq. CTCF ChIP-seq was implemented by a commercial sequencing provider (igenebook Technology Co., Ltd., Wuhan, China). Briefly, chromatin was crosslinked by formaldehyde, following nuclear processing, chromatin digestion, DNA-protein compound capture, decrosslinking of DNA-protein compounds, and the purification of DNA. Finally, the input and ChIP DNA samples were sequenced on the Illumina Hiseq X ten platform.

2.7. Bioinformatics Analysis of CTCF ChIP-seq

The quality of CTCF ChIP-seq data was evaluated by fastqc v0.11.5 [30]. The quality control of the raw CTCF ChIP-seq data was implemented by Trimmomatic v0.36 [31]. After quality control, clean reads were aligned to the sheep reference genome by BWA v0.7.1 [32]. The read distribution in different genomic regions was investigated using the ChIPseeker R Bioconductor package [33]. An Upset plot was plotted by the UpSetR R package [34]. Potential peaks were called by MACS v2.1.1 [35]. Differential peaks were detected by DiffBind v1.16.3 (https://bioconductor.org/packages/release/bioc/html/DiffBind.html, accessed on 29 March 2022). Peaks with an FDR < 0.05 and a Fold value >0 were defined as significant differential peaks. Motifs in significant differential peaks were predicted by HOMER v3 [36].

2.8. Overlapping between DEIs and Differential Peaks Called from ChIP-seq Data

The overlapping between genes with transcribed DEIs and genes located in differential peaks were investigated to explore the regulating role of CTCF in the alternative splicing process. The significance of overlapping was tested by the GeneOverlap R Bioconductor package (https://bioconductor.org/packages/release/bioc/html/GeneOverlap.html, accessed on 29 March 2022).

3. Results

3.1. Meat Traits

In the current study, a total of three meat traits were measured after slaughter. The mean value and standard deviation of three meat traits are documented in Table 1. Among these traits, shear force in DDH was significantly (p = 0.0406) higher than that in DHH. Other meat traits did not show a significant difference between DDH and DHH.

3.2. Update of Reference Genome Annotation File

To update the sheep genome annotation (Generic Feature Format, GFF), which is essential for accurately quantifying the abundance of isoforms, six RNA samples were pooled to implement Iso-seq. In total, 442,966 polymerase reads were produced by pooled Iso-seq (Table S2). After pre-processing, a total of 247,201 FLNC reads were kept for further analysis (Table S2). Novel genes and isoforms were identified according to the genomic position of each FLNC. In total, 18,959 gene loci (9184 known and 9775 novel) were identified (Figure S1a). Among these detected genes, 5104 (26.92%) loci had two or more transcripts (Figure S1b). In sheep muscle, 20,205 novel isoforms (57.51%) were transcribed from annotated genes (Figure S1c), followed by those transcribed from novel genes. Identified potential novel gene loci and novel isoforms as well as sheep reference annotation files (GFF file) were merged to obtain an updated genome annotation file (Table S3). Alternative splicing is the main mechanism to diversify isoforms. In this study, a total of 38,070 alternative splicing events were detected by pooled Iso-seq data (Figure S2). For example, seven novel isoforms in six pooled samples were identified in *CAST* due to alternative splicing (Figure 2).

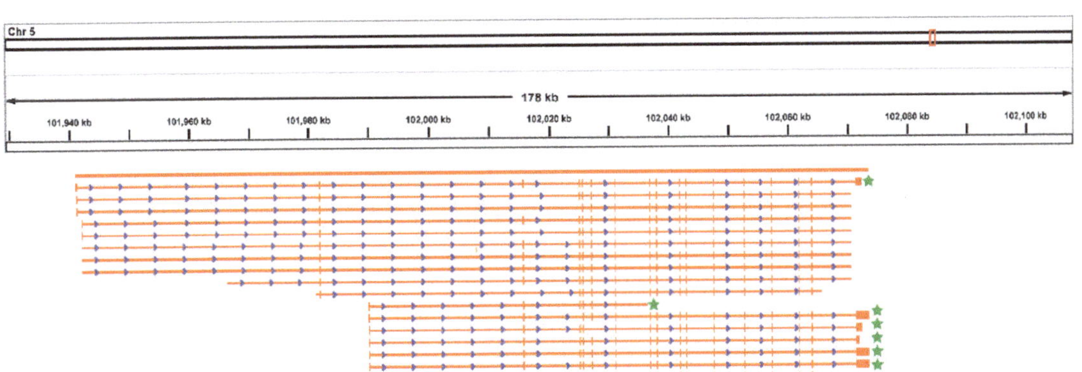

Figure 2. Identified isoforms in *CAST*. The numbers at the top of the figure represent the genome coordinates. Each yellow horizontal line represents an isoform. Each yellow vertical dash represents an exon. Stars on the right denote the novel isoforms identified in the current study. Blue arrows denote the direction of transcription. Stars in right denote novel isoforms.

3.3. Differentially Expressed Isoforms and Functional Analysis

Six RNA-seq data were aligned to the updated genome annotation file (Table S3) to identify the DEIs between DDH and DHH. As a result, a total of 624 DEIs were identified (Figure 3, Table S4). These 624 DEIs were transcribed from 492 genes, suggesting that some genes could produce more than one DEI. The most significant DEIs were X.351.3 (*FHL1*, FDR = 8.21×10^{-15}) and 11.673.91 (*MYH2*, FDR = 8.21×10^{-15}).

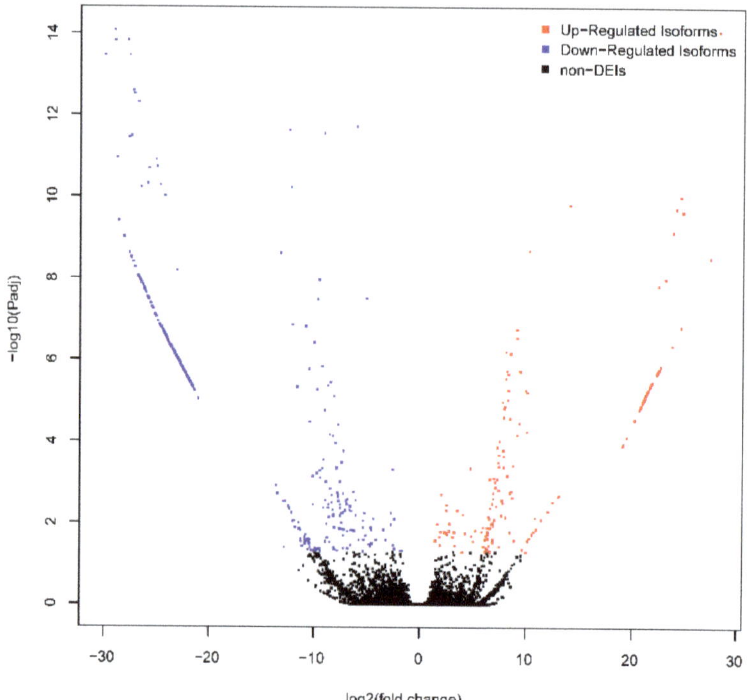

Figure 3. Volcano plot of differentially expressed isoforms (DEIs). *X*-axis denotes log2 (fold change). *Y*-axis denotes −log10 (adjusted *p* value). Red dots denote the up-regulated isoforms. Blue dots denote down-regulated isoforms.

All detected DEIs were significantly (FDR < 0.05) enriched in 280 GO terms (Figure 4a, Table S5). The most significant GO term was contractile fiber (GO:0043292, FDR = 5.38×10^{-9}, Figure 4a, Table S5). All detected DEIs were significantly (FDR < 0.05) enriched in 17 KEGG pathways (Figure 4b, Table S6). The most significant pathway was the glucagon signaling pathway (ko049222).

3.4. Validation for Target Isoforms

To validate the result of the identified DEIs, the qRT-PCR of five DEIs was implemented. Among these five isoforms, three DEIs identified from the transcriptome analysis were significantly expressed (Figure 5). The expression trends of these five isoforms, determined by qRT-PCR, were consistent with the transcriptome data.

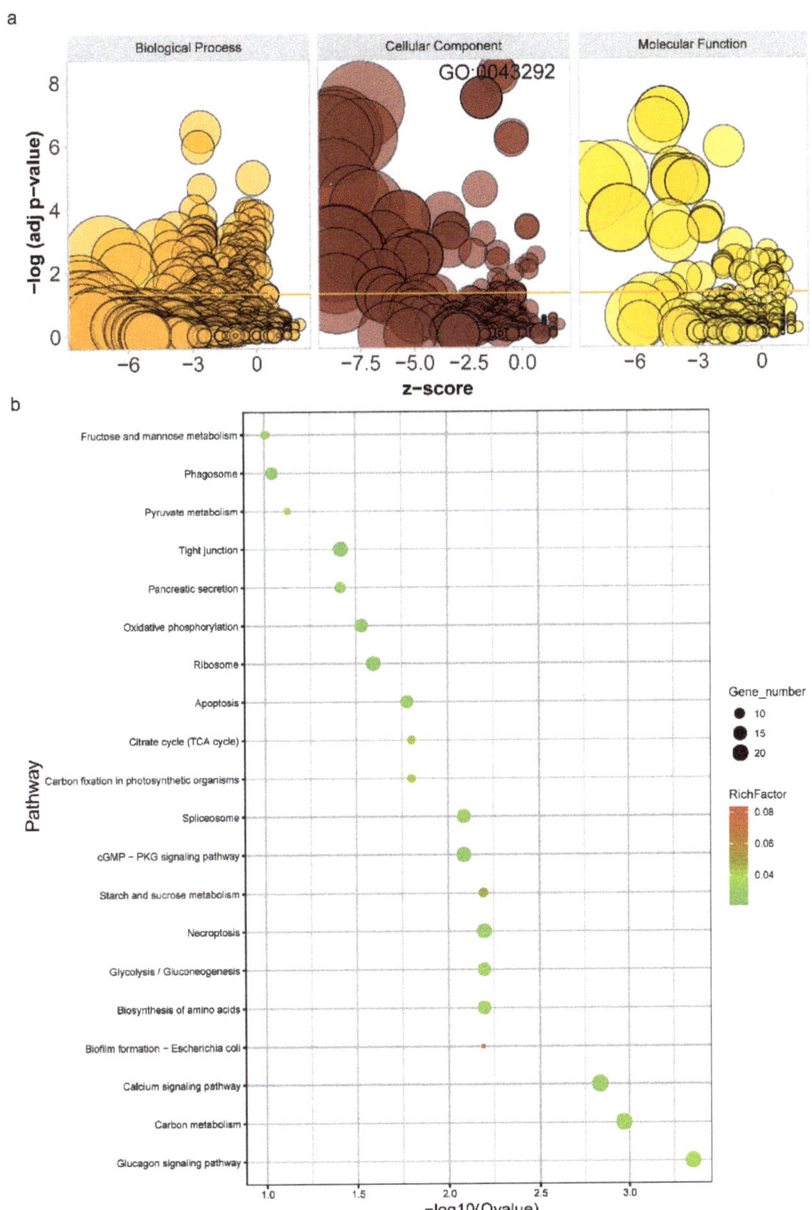

Figure 4. Functional annotation of differentially expressed isoforms (DEIs); (**a**) gene ontology (GO) enrichment analysis; *X*-axis denotes z-score; *Y*-axis denotes -log10(adjusted *p* value); bubble size denotes the enriched gene numbers; (**b**) Kyoto Encyclopedia of Genes and Genomes (KEGG) enrichment analysis; *X*-axis denotes −log10 (adjusted *p* value); *Y*-axis denotes enriched pathway; bubble size denotes the enriched gene numbers; bubble color denotes the adjusted *p* value.

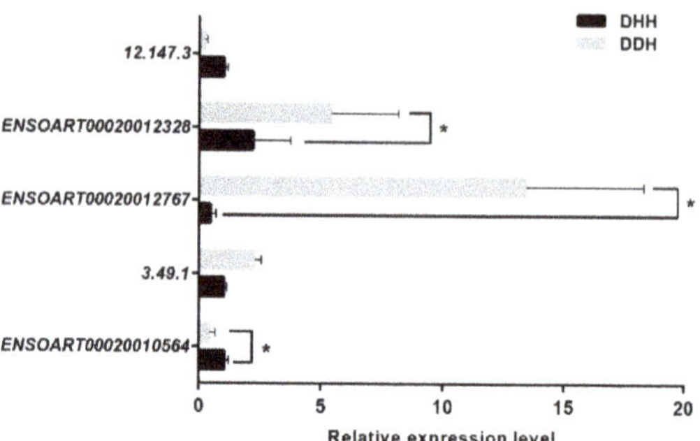

Figure 5. Validation of the differentially expressed isoforms by qRT-PCR; *Y*-axis denotes isoforms; *X*-axis denotes isoform relative expression level; DDH denotes Dorper × Dorper × Hu; DHH denotes Dorper × Hu × Hu. Results were presented as the mean ± SEM, * $p < 0.05$, significant difference.

3.5. CTCF ChIP-seq

Two muscle samples with divergent meat tenderness were selected for CTCF ChIP-seq. After quality control, more than 40 M clean reads in each sample were obtained (Table S7). All clean reads were mapped to the sheep reference genome with a mapping rate greater than 96.46% in each sample (Table S7). Above half of the mapped reads were located in the intergenic region (Figure 6), followed by an intron, promoter, exon, 3′ untranslated regions (UTR) and 5′ UTR.

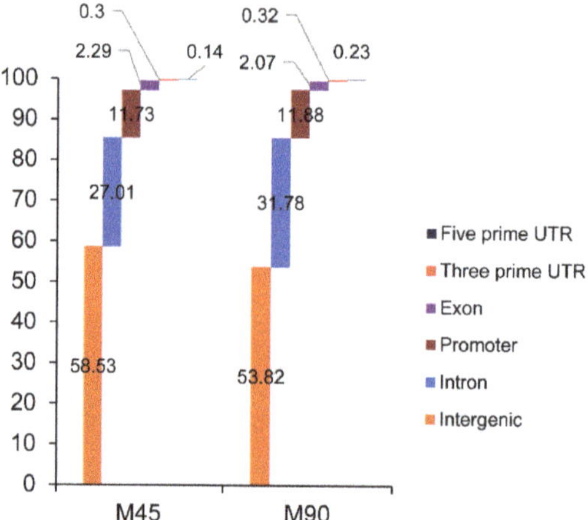

Figure 6. Read distribution of CTCF chromatin immunoprecipitation sequencing (ChIP-seq) among sheep genomes. According to the genome coordinates of reads, reads can be divided into six categories, including in intergenic region, intron region, promoter region, exon region, 3′ untranslated region (UTR) and 5′ untranslated region.

A total of 4388 differential peaks were detected between two samples with divergent meat tenderness (Table S8). The greatest number of differential peaks were distributed in the intergenic genome region, followed by the intron region, promoter region, exon region, 5′ UTR and 3′ UTR (Figure 7). The motifs in the differential peaks were predicted. A total of 66 motifs were predicted in differential peaks (Table S9).

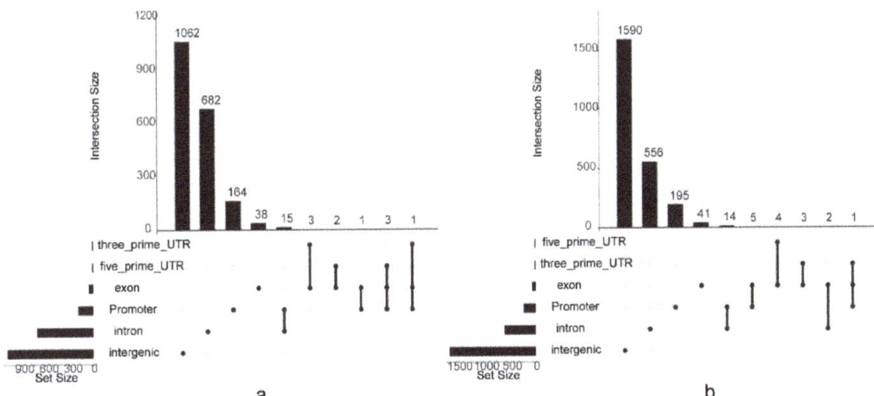

Figure 7. Distribution of differential peaks among the sheep genomes. (**a**) Up-regulated peaks; (**b**) down-regulated peaks. According to the genome coordinates of detected peaks, peaks can be divided into six categories, including in intergenic region, intron region, promoter region, exon region, 3′ untranslated region (UTR) and 5′ untranslated region. Dots in the bottom denote the peak categories. Black lines connecting the dots show peak numbers shared among several categories. The number above each bar shows the number of peaks for each category or shared category.

3.6. Overlap between DEIs and Differential Peaks

To investigate the potential role of CTCF in regulating alternative splicing in sheep muscle, the overlapping analysis between genes with transcribed DEIs and genes located in differential peaks was implemented. A total of 86 overlapped genes were found (Figure 8). The overlapping *p*-value was smaller than 0.05, indicating that the number of overlapped genes was too great to have been by chance.

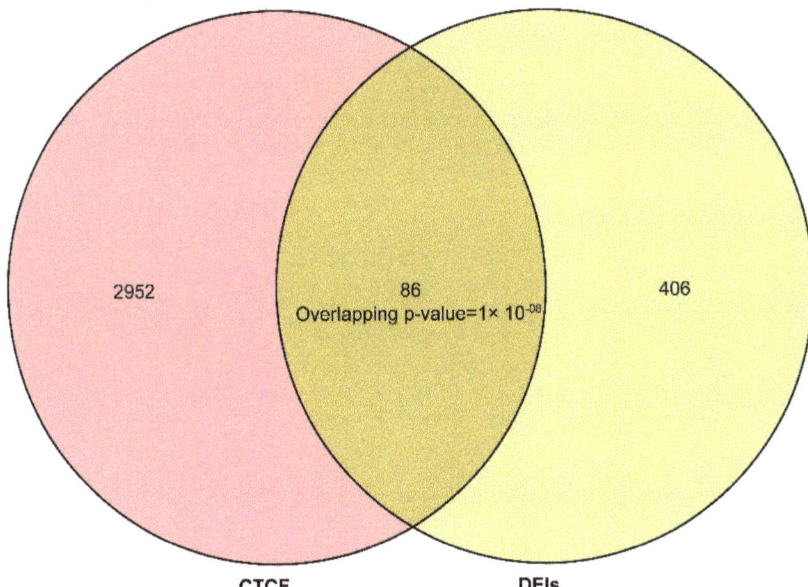

Figure 8. Veen plot of genes in differential peaks (**left**) and genes with transcribed differentially expressed isoforms (DEIs, **right**).

4. Discussion

4.1. Update Sheep Reference Genome Annotation File

It has been reported that one possible method to quantify transcripts is mapping the reads to the annotated transcriptome [37]. In this study, pooled Iso-seq data were used to update the reference genome annotation file. As a result, 9775 possible novel genes and 30,513 novel isoforms were identified (Figure S1) in sheep muscle and were added to the sheep reference genome annotation file (Table S3). In a pig study, a total of 10,465 novel genes were identified by integrating Iso-seq and RNA-seq data [38]. Similarly, in sheep tail fat, a total of 9001 novel genes and 36,667 novel isoforms were detected using pooled Iso-seq data [13]. Our results are in line with these published works, which suggests that Iso-seq is a useful method to improve genome annotation in sheep.

Alternative splicing is the main mechanism to diversify isoforms. Here, we highlight an example of an alternative splicing event that may regulate meat tenderness. *CAST* is linked with meat tenderness across many farm animals [39]. In beef, it has been reported that alternative splicing [9] in *CAST* could regulate meat tenderness. In sheep, a previous study suggested that SNPs in *CAST* may regulate exon excision events in *CAST* (Oar.3.1, Chr5:93439378-93444596), which was identified by RNA-seq [40]. In the current study, this exon excision event was further validated by Figure 2. Taken together, all of these pieces of evidence indicate that alternative splicing in *CAST* regulated by SNPs may play an important role in the meat tenderness of sheep.

4.2. Novel Isoforms Linked with Meat Tenderness

The calpain system plays an essential role in meat tenderness [2]. Calpain3 (CAPN3), a member of the calpain system, is mainly expressed in skeletal muscle and may play an important role in the calpain system to regulate meat tenderness [6]. In the current study, a novel transcript, 7.524.18, transcribed from *CAPN3*, was identified as a DEI (Table S4), suggesting its potential role in meat tenderness.

Muscle fiber type also plays an important role in meat tenderness [41,42]. Muscle fiber type is related to the abundance of myosin type [43]. The ratio of slow and fast heavy chain myosin is linked to the ratio of type I and type II fibers [44]. In the current study, 12 DEIs transcribed from four myosin heavy chain genes were identified (11.673.36, 11.673.38, 11.673.91 and 11.673.53 transcribed from *MYH2*; ENSOART00020028698 transcribed from *MYH4*; 7.469.116, 7.469.121 and 7.469.129 transcribed from *MYH7*; 14.723.9, 14.723.27, 14.723.58 and 14.723.65 transcribed from *MYBPC2*; Table S4). Previous studies have reported that *MYH2*, *MYH4*, *MYH7* and *MYBPC2* play an important role in degerming muscle fiber types [45,46]. Overall, the myosin heavy chain-related isoforms were differentially expressed between DDH and DHH, suggesting that these isoforms may regulate meat tenderness directly or indirectly.

In addition, the amount of glycogen could affect meat tenderness [47]. In the current study, seven DEIs transcribed by two glycogen-related genes were detected (14.271.7 transcribed from *GYS1*; 21.138.57, 21.138.27, 21.138.13, 21.138.58, 21.138.9 and ENSOART00020017812 transcribed from *PYGM*). A previous study suggested that *PYGM* is related to shear force in cattle [48]. These DEIs might regulate tenderness by controlling glycogen content in muscles.

GO and KEGG enrichment analyses were implemented to extract the potential function of DEIs. In pigs, differentially expressed genes between fast and slow muscle were relevant to myofibril and contractile fiber GO terms. In the current study, DEIs were most significantly enriched in contractile fiber (GO:0043292) and myofibril (GO:0030016) GO terms. Our results provide further evidence for a relationship between muscle fiber type transformation, meat tenderness and DEIs. For the KEGG pathway enrichment analysis, a previous study suggested that a gene in the calcium signaling pathway (ko04020) is related to stiffness and affects the speed of fiber degradation during the meat aging process [49]. In the current study, 23 DEIs were significantly enriched in calcium signaling pathways, indicating their potential role in the meat aging process.

4.3. CTCF Might Regulate Alternative Splicing in Sheep Muscle

In the current study, CTCF ChIP-seq was implanted to investigate its potential role in regulating alternative splicing in sheep muscle with divergent meat tenderness. The main function of alternative splicing processes are to diversify isoforms [50]. In recent years, CTCF has been identified as a modulator of the alternative splicing process [17]. In this study, 86 overlapped genes were found between genes transcribing DEIs and genes in CTCF peaks, and the number of overlapped genes was too great to be accidental (Figure 7). These results suggest that CTCF may regulate alternative splicing in sheep muscle.

The CTCF-regulated alternative splicing mechanism can be divided into co-transcriptional, genomic and epigenetic mechanisms [17]. Here, our study example suggests that CTCF may regulate alternative splicing through the CTCF-mediated DNA methylation process in sheep muscle. In the current study, two DEIs, 3.1299.2 and 26.152.10, were transcribed from *ANKRD23* (Table S4). In a cattle study, the ANKRD23-202 mRNA isoform, which is a splice form of the ANKRD, was defined as a DEI between the tough group and the tender group [49]. The result of a rabbit study suggested that *ANKRD23* methylated in promoter and gene body regions is associated with exon skipping alternative splicing in the skeletal muscle [51]. Previous results suggest that CTCF could regulate alternative splicing by controlling DNA methylation [52–54]. In the current study, a CTCF peak was observed in *ANKRD23* (Table S8). Taken together, CTCF may regulate *ANKRD23* to produce different isoforms mediated by methylation.

As the first analysis of lamb meat tenderness based on the study of novel isoforms and alternative splicing regulation pathways, our study has its limitations. In our study, the sample size was relatively small, which may reduce the power to detect DEIs. In addition, tenderness is a parameter linked to DNA, exposition, individual characteristics, and a multitude of other factors that genetics cannot fully explain. Thus, in the future, we should increase sample size and consider more factors to investigate lamb meat tenderness. Nevertheless, our results provide the first insights about lamb meat tenderness based on the study of novel isoforms and alternative splicing regulation.

5. Conclusions

In this study, a total of 624 DEIs were identified between DDH and DHH. For example, isoform 7.524.18 transcribed from *CAPN3* may be associated with meat tenderness. In addition, a total of 86 overlapped genes were found between genes with transcribed DEIs and genes in differential peaks identified by CTCF ChIP-seq. Among overlapped genes, *ANKRD23* produces different isoforms that may be regulated by CTCF mediated by methylation. As preliminary research, our results identified novel isoforms associated with meat tenderness and revealed the possible regulating mechanism of alternative splicing to produce isoforms using bioinformatic analysis. In the future, more samples should be collected and more molecular experiments, e.g., TA clone, ChIP-qPCR, etc., should be implemented to improve these preliminary results.

Supplementary Materials: The following supporting information can be downloaded at: https://www.mdpi.com/article/10.3390/foods11081068/s1: Figure S1: Identified novel genes and isoforms: (a) Venn plot of identified genes and annotated genes; (b) frequency of the number of transcripts transcribed by a gene; (c) classification of identified isoforms; Figure S2: Classification of alternative splicing events; Table S1: Designed isoform-specific primers; Table S2: Data information of pooled isoform sequencing (Iso-seq) in sheep muscle; Table S3: Updated sheep genome annotation file; Table S4: Differentially expressed isoforms; Table S5: Gene ontology (GO) enrichment analysis of differentially expressed isoforms; Table S6: Kyoto Encyclopedia of Genes and Genomes (KEGG) enrichment analysis of differentially expressed isoforms; Table S7: Summarized information of chromatin immunoprecipitation sequencing (ChIP-seq); Table S8: Differential peaks between two muscle samples with divergent tenderness; Table S9: Predicated motifs in differential peaks.

Author Contributions: Conceptualization, W.S. and Z.Y.; methodology, Z.Y.; software, Z.Y.; validation, Z.Y., L.G., W.Z. and S.W.; formal analysis, Z.Y.; data curation, L.G., W.Z., S.W., X.C., X.L. and Z.Y.; writing—original draft preparation, Z.Y.; writing—review and editing, L.G., W.Z., S.W., X.L., X.C. and W.S.; visualization, Z.Y.; supervision, W.S.; project administration, W.S.; funding acquisition, W.S. All authors have read and agreed to the published version of the manuscript.

Funding: This research was funded by the Major New Varieties of Agricultural Projects in Jiangsu Province (PZCZ201739), Natural Science Foundation of Jiangsu Province (BK20210811), National Natural Science Foundation of China (31872333, 32172689), National Natural Science Foundation of China-CGIAR (32061143036), The Projects of Domesticated Animals Platform of the Ministry of Science, Key Research and Development Plan (modern agriculture) in Jiangsu Province (BE2018354) and Jiangsu Agricultural Science and Technology Innovation Fund (CX(18)2003).

Institutional Review Board Statement: Not applicable.

Informed Consent Statement: Not applicable.

Data Availability Statement: All of the raw sequencing data will be available through NCBI SRA: PRJNA745517.

Acknowledgments: We thank Mohammed Gagaoua (Teagasc Food Research Centre) for his valuable comments on the manuscript and Ruidong Xiang (The University of Melbourne) for his work in revising English phrasing.

Conflicts of Interest: The authors declare no conflict of interest. The funders had no role in the design of the study; in the collection, analyses, or interpretation of data; in the writing of the manuscript, or in the decision to publish the results.

References

1. Raza, S.H.A.; Kaster, N.; Khan, R.; Abdelnour, S.A.; El-Hack, M.E.A.; Khafaga, A.F.; Taha, A.; Ohran, H.; Swelum, A.A.; Schreurs, N.M.; et al. The role of MicroRNAs in muscle tissue development in beef cattle. *Genes* **2020**, *11*, 295. [CrossRef] [PubMed]
2. Bhat, Z.F.; Morton, J.D.; Mason, S.L.; Bekhit, A.E.-D.A. Role of calpain system in meat tenderness: A review. *Food Sci. Hum. Wellness* **2018**, *7*, 196–204. [CrossRef]
3. Jones, S.; Parr, T.; Sensky, P.; Scothern, G.; Bardsley, R.; Buttery, P. Fibre type-specific expression of p94, a skeletal muscle-specific calpain. *J. Muscle Res. Cell Motil.* **1999**, *20*, 417–424. [CrossRef]
4. Ojima, K.; Kawabata, Y.; Nakao, H.; Nakao, K.; Doi, N.; Kitamura, F.; Ono, Y.; Hata, S.; Suzuki, H.; Kawahara, H. Dynamic distribution of muscle-specific calpain in mice has a key role in physical-stress adaptation and is impaired in muscular dystrophy. *J. Clin. Investig.* **2010**, *120*, 2672–2683. [CrossRef] [PubMed]
5. Sorimachi, H.; Imajoh-Ohmi, S.; Emori, Y.; Kawasaki, H.; Ohno, S.; Minami, Y.; Suzuki, K. Molecular cloning of a novel mammalian calcium-dependent protease distinct from both m-and μ-types: Specific expression of the mRNA in skeletal muscle. *J. Biol. Chem.* **1989**, *264*, 20106–20111. [CrossRef]
6. Basson, A.; Strydom, P.E.; van Marle-Köster, E.; Webb, E.C.; Frylinck, L. Sustained effects of muscle calpain system genotypes on tenderness pPhenotypes of South African beef Bulls during ageing up to 20 days. *Animals* **2022**, *12*, 686. [CrossRef] [PubMed]
7. Ilian, M.A.; Morton, J.D.; Bekhit, A.E.-D.; Roberts, N.; Palmer, B.; Sorimachi, H.; Bickerstaffe, R. Effect of preslaughter feed withdrawal period on longissimus tenderness and the expression of calpains in the ovine. *J. Agric. Food Chem.* **2001**, *49*, 1990–1998. [CrossRef] [PubMed]
8. Ilian, M.A.; Bekhit, A.E.-D.A.; Stevenson, B.; Morton, J.D.; Isherwood, P.; Bickerstaffe, R. Up-and down-regulation of longissimus tenderness parallels changes in the myofibril-bound calpain 3 protein. *Meat Sci.* **2004**, *67*, 433–445. [CrossRef]
9. Motter, M.M.; Corva, P.M.; Soria, L.A. Expression of calpastatin isoforms in three skeletal muscles of Angus steers and their association with fiber type composition and proteolytic potential. *Meat Sci.* **2021**, *171*, 108267. [CrossRef]
10. Rhoads, A.; Au, K.F. PacBio Sequencing and Its Applications. *Genom. Proteom. Bioinform.* **2015**, *13*, 278–289. [CrossRef]
11. Chao, Q.; Gao, Z.F.; Zhang, D.; Zhao, B.G.; Dong, F.Q.; Fu, C.X.; Liu, L.J.; Wang, B.C. The developmental dynamics of the Populus stem transcriptome. *Plant Biotechnol. J.* **2019**, *17*, 206–219. [CrossRef] [PubMed]
12. Li, Y.; Dai, C.; Hu, C.; Liu, Z.; Kang, C. Global identification of alternative splicing via comparative analysis of SMRT- and Illumina-based RNA-seq in strawberry. *Plant J.* **2017**, *90*, 164–176. [CrossRef] [PubMed]
13. Yuan, Z.; Ge, L.; Sun, J.; Zhang, W.; Wang, S.; Cao, X.; Sun, W. Integrative analysis of Iso-Seq and RNA-seq data reveals transcriptome complexity and differentially expressed transcripts in sheep tail fat. *PeerJ* **2021**, *9*, e12454. [CrossRef]
14. Cheng, S.; Wang, X.; Wang, Q.; Yang, L.; Shi, J.; Zhang, Q. Comparative analysis of Longissimus dorsi tissue from two sheep groups identifies differentially expressed genes related to growth, development and meat quality. *Genomics* **2020**, *112*, 3322–3330. [CrossRef]

15. Cheng, S.; Wang, X.; Zhang, Q.; He, Y.; Zhang, X.; Yang, L.; Shi, J. Comparative transcriptome analysis identifying the different molecular genetic markers related to production performance and meat quality in longissimus dorsi tissues of MG x STH and STH Sheep. *Genes* **2020**, *11*, 183. [CrossRef] [PubMed]

16. Wang, Y.; Liu, J.; Huang, B.O.; Xu, Y.M.; Li, J.; Huang, L.F.; Lin, J.; Zhang, J.; Min, Q.H.; Yang, W.M.; et al. Mechanism of alternative splicing and its regulation. *Biomed. Rep.* **2015**, *3*, 152–158. [CrossRef]

17. Alharbi, A.B.; Schmitz, U.; Bailey, C.G.; Rasko, J.E.J. CTCF as a regulator of alternative splicing: New tricks for an old player. *Nucleic Acids Res.* **2021**, *49*, 7825–7838. [CrossRef]

18. Hwang, I.H.; Devine, C.E.; Hopkins, D.L. The biochemical and physical effects of electrical stimulation on beef and sheep meat tenderness. *Meat Sci.* **2003**, *65*, 677–691. [CrossRef]

19. Hopkins, D.L.; Toohey, E.S.; Warner, R.D.; Kerr, M.J.; van de Ven, R. Measuring the shear force of lamb meat cooked from frozen samples: Comparison of two laboratories. *Anim. Prod. Sci.* **2010**, *50*, 382–385. [CrossRef]

20. R Core Team. *R: A Language and Environment for Statistical Computing*; R Foundation for Statistical Computing: Vienna, Austria, 2013. Available online: https://www.R-project.org (accessed on 29 March 2022).

21. Salmela, L.; Rivals, E. LoRDEC: Accurate and efficient long read error correction. *Bioinformatics* **2014**, *30*, 3506–3514. [CrossRef]

22. Chen, Y.; Chen, Y.; Shi, C.; Huang, Z.; Zhang, Y.; Li, S.; Li, Y.; Ye, J.; Yu, C.; Li, Z.; et al. SOAPnuke: A MapReduce acceleration-supported software for integrated quality control and preprocessing of high-throughput sequencing data. *Gigascience* **2018**, *7*, 1–6. [CrossRef] [PubMed]

23. Wu, T.D.; Watanabe, C.K. GMAP: A genomic mapping and alignment program for mRNA and EST sequences. *Bioinformatics* **2005**, *21*, 1859–1875. [CrossRef] [PubMed]

24. Foissac, S.; Sammeth, M. ASTALAVISTA: Dynamic and flexible analysis of alternative splicing events in custom gene datasets. *Nucleic Acids Res.* **2007**, *35*, W297–W299. [CrossRef] [PubMed]

25. Langmead, B.; Salzberg, S.L. Fast gapped-read alignment with Bowtie 2. *Nat. Methods* **2012**, *9*, 357–359. [CrossRef] [PubMed]

26. Love, M.I.; Huber, W.; Anders, S. Moderated estimation of fold change and dispersion for RNA-seq data with DESeq2. *Genome Biol.* **2014**, *15*, e550. [CrossRef] [PubMed]

27. Buchfink, B.; Xie, C.; Huson, D.H. Fast and sensitive protein alignment using DIAMOND. *Nat. Methods* **2015**, *12*, 59–60. [CrossRef]

28. Xie, C.; Mao, X.; Huang, J.; Ding, Y.; Wu, J.; Dong, S.; Kong, L.; Gao, G.; Li, C.Y.; Wei, L. KOBAS 2.0: A web server for annotation and identification of enriched pathways and diseases. *Nucleic Acids Res.* **2011**, *39*, W316–W322. [CrossRef]

29. Walter, W.; Sanchez-Cabo, F.; Ricote, M. GOplot: An R package for visually combining expression data with functional analysis. *Bioinformatics* **2015**, *31*, 2912–2914. [CrossRef]

30. Andrews, S. FastQC: A Quality Control Tool for High Throughput Sequence Data. 2010. Available online: https://www.bioinformat-ics.babraham.ac.uk/projects/fastqc (accessed on 7 January 2022).

31. Bolger, A.M.; Lohse, M.; Usadel, B. Trimmomatic: A flexible trimmer for Illumina sequence data. *Bioinformatics* **2014**, *30*, 2114–2120. [CrossRef]

32. Li, H.; Durbin, R. Fast and accurate short read alignment with Burrows-Wheeler transform. *Bioinformatics* **2009**, *25*, 1754–1760. [CrossRef]

33. Yu, G.; Wang, L.G.; He, Q.Y. ChIPseeker: An R/Bioconductor package for ChIP peak annotation, comparison and visualization. *Bioinformatics* **2015**, *31*, 2382–2383. [CrossRef] [PubMed]

34. Conway, J.R.; Lex, A.; Gehlenborg, N. UpSetR: An R package for the visualization of intersecting sets and their properties. *Bioinformatics* **2017**, *33*, 2938–2940. [CrossRef] [PubMed]

35. Zhang, Y.; Liu, T.; Meyer, C.A.; Eeckhoute, J.; Johnson, D.S.; Bernstein, B.E.; Nusbaum, C.; Myers, R.M.; Brown, M.; Li, W.; et al. Model-based analysis of ChIP-Seq (MACS). *Genome Biol.* **2008**, *9*, R137. [CrossRef] [PubMed]

36. Heinz, S.; Benner, C.; Spann, N.; Bertolino, E.; Lin, Y.C.; Laslo, P.; Cheng, J.X.; Murre, C.; Singh, H.; Glass, C.K. Simple combinations of lineage-determining transcription factors prime cis-regulatory elements required for macrophage and B cell identities. *Mol. Cell* **2010**, *38*, 576–589. [CrossRef] [PubMed]

37. Conesa, A.; Madrigal, P.; Tarazona, S.; Gomez-Cabrero, D.; Cervera, A.; McPherson, A.; Szczesniak, M.W.; Gaffney, D.J.; Elo, L.L.; Zhang, X.; et al. A survey of best practices for RNA-seq data analysis. *Genome Biol.* **2016**, *17*, e13. [CrossRef] [PubMed]

38. Beiki, H.; Liu, H.; Huang, J.; Manchanda, N.; Nonneman, D.; Smith, T.P.L.; Reecy, J.M.; Tuggle, C.K. Improved annotation of the domestic pig genome through integration of Iso-Seq and RNA-seq data. *BMC Genom.* **2019**, *20*, e344. [CrossRef] [PubMed]

39. Palma, G.A.; Carranza, P.G.; Coria, M.S. El sistema proteolítico calpaína en la tenderización de la carne: Un enfoque molecular. *Rev. MVZ Córdoba* **2018**, *23*, 6523–6536. [CrossRef]

40. Yuan, Z.; Sunduimijid, B.; Xiang, R.; Behrendt, R.; Knight, M.I.; Mason, B.A.; Reich, C.M.; Prowse-Wilkins, C.; Vander Jagt, C.J.; Chamberlain, A.J.; et al. Expression quantitative trait loci in sheep liver and muscle contribute to variations in meat traits. *Genet. Sel. Evol.* **2021**, *53*, e8. [CrossRef]

41. Şirin, E.; Aksoy, Y.; Uğurlu, M.; Çiçek, Ü.; Önenç, A.; Ulutaş, Z.; Şen, U.; Kuran, M. The relationship between muscle fiber characteristics and some meat quality parameters in Turkish native sheep breeds. *Small Rumin. Res.* **2017**, *150*, 46–51. [CrossRef]

42. Picard, B.; Gagaoua, M. Muscle Fiber Properties in Cattle and Their Relationships with Meat Qualities: An Overview. *J. Agric. Food Chem.* **2020**, *68*, 6021–6039. [CrossRef]

43. Duris, M.-P.; Picard, B.; Geay, Y. Specificity of different anti-myosin heavy chain antibodies in bovine muscle. *Meat Sci.* **2000**, *55*, 67–78. [CrossRef]

44. Sazili, A.Q.; Parr, T.; Sensky, P.L.; Jones, S.W.; Bardsley, R.G.; Buttery, P.J. The relationship between slow and fast myosin heavy chain content, calpastatin and meat tenderness in different ovine skeletal muscles. *Meat Sci.* **2005**, *69*, 17–25. [CrossRef] [PubMed]
45. Yu, H.; Waddell, J.N.; Kuang, S.; Tellam, R.L.; Cockett, N.E.; Bidwell, C.A. Identification of genes directly responding to DLK1 signaling in Callipyge sheep. *BMC Genom.* **2018**, *19*, 283. [CrossRef] [PubMed]
46. Vuocolo, T.; Byrne, K.; White, J.; McWilliam, S.; Reverter, A.; Cockett, N.E.; Tellam, R.L. Identification of a gene network contributing to hypertrophy in callipyge skeletal muscle. *Physiol. Genom.* **2007**, *28*, 253–272. [CrossRef]
47. Onopiuk, A.; Półtorak, A.; Wierzbicka, A. Influence of post-mortem muscle glycogen content on the quality of beef during aging. *J. Vet. Res.* **2016**, *60*, 301–307. [CrossRef]
48. Gagaoua, M.; Terlouw, E.M.C.; Mullen, A.M.; Franco, D.; Warner, R.D.; Lorenzo, J.M.; Purslow, P.P.; Gerrard, D.; Hopkins, D.L.; Troy, D.; et al. Molecular signatures of beef tenderness: Underlying mechanisms based on integromics of protein biomarkers from multi-platform proteomics studies. *Meat Sci.* **2021**, *172*, 108311. [CrossRef]
49. Muniz, M.M.M.; Fonseca, L.F.S.; Dos Santos Silva, D.B.; de Oliveira, H.R.; Baldi, F.; Chardulo, A.L.; Ferro, J.A.; Canovas, A.; de Albuquerque, L.G. Identification of novel mRNA isoforms associated with meat tenderness using RNA sequencing data in beef cattle. *Meat Sci.* **2021**, *173*, 108378. [CrossRef] [PubMed]
50. Keren, H.; Lev-Maor, G.; Ast, G. Alternative splicing and evolution: Diversification, exon definition and function. *Nat. Rev. Genet.* **2010**, *11*, 345–355. [CrossRef] [PubMed]
51. Li, Y.; Wang, J.; Elzo, M.A.; Fan, H.; Du, K.; Xia, S.; Shao, J.; Lai, T.; Hu, S.; Jia, X.; et al. Molecular profiling of DNA methylation and alternative splicing of genes in skeletal muscle of obese rabbits. *Curr. Issues Mol. Biol.* **2021**, *43*, 1558–1575. [CrossRef] [PubMed]
52. Guastafierro, T.; Cecchinelli, B.; Zampieri, M.; Reale, A.; Riggio, G.; Sthandier, O.; Zupi, G.; Calabrese, L.; Caiafa, P. CCCTC-binding factor activates PARP-1 affecting DNA methylation machinery. *J. Biol. Chem.* **2008**, *283*, 21873–21880. [CrossRef] [PubMed]
53. Zampieri, M.; Guastafierro, T.; Calabrese, R.; Ciccarone, F.; Bacalini, M.G.; Reale, A.; Perilli, M.; Passananti, C.; Caiafa, P. ADP-ribose polymers localized on Ctcf-Parp1-Dnmt1 complex prevent methylation of Ctcf target sites. *Biochem. J.* **2012**, *441*, 645–652. [CrossRef] [PubMed]
54. Dubois-Chevalier, J.; Oger, F.; Dehondt, H.; Firmin, F.F.; Gheeraert, C.; Staels, B.; Lefebvre, P.; Eeckhoute, J. A dynamic CTCF chromatin binding landscape promotes DNA hydroxymethylation and transcriptional induction of adipocyte differentiation. *Nucleic Acids Res.* **2014**, *42*, 10943–10959. [CrossRef] [PubMed]

Article

A Simple and Reliable Single Tube Septuple PCR Assay for Simultaneous Identification of Seven Meat Species

Zhendong Cai [1,2], Song Zhou [1,2], Qianqian Liu [3,*], Hui Ma [1,2], Xinyi Yuan [1,2], Jiaqi Gao [1,2], Jinxuan Cao [1,2] and Daodong Pan [1,2,*]

1 State Key Laboratory for Managing Biotic and Chemical Threats to the Quality and Safety of Agro-Products, Ningbo University, Ningbo 315211, China; zhendongcai@hotmail.com (Z.C.); zhousongluchen@163.com (S.Z.); m17805843103@163.com (H.M.); yxy13065810608@163.com (X.Y.); caidan1389@126.com (J.G.); caojinxuan@nbu.edu.cn (J.C.)
2 Key Laboratory of Animal Protein Deep Processing Technology of Zhejiang Province, College of Food and Pharmaceutical Sciences, Ningbo University, Ningbo 315800, China
3 Institute of Environmental Research at Greater Bay Area, Guangzhou University, Guangzhou 510006, China
* Correspondence: csuliuqian@foxmail.com (Q.L.); daodongpan@163.com (D.P.)

Abstract: Multiplex PCR methods have been frequently used for authentication of meat product adulteration. Through screening of new species-specific primers designed based on the mitochondrial DNA sequences, a septuple PCR method is ultimately developed and optimized to simultaneously detect seven species including turkey (110 bp), goose (194 bp), pig (254 bp), sheep (329 bp), beef (473 bp), chicken (612 bp) and duck (718 bp) in one reaction. The proposed method has been validated to be specific, sensitive, robust and inexpensive. Taken together, the developed septuple PCR assay is reliable and efficient, not only to authenticate animal species in commercial meat products, but also easily feasible in a general laboratory without special infrastructures.

Keywords: septuple PCR; adulteration; meat species; mitochondrial genes; multiplex PCR

Citation: Cai, Z.; Zhou, S.; Liu, Q.; Ma, H.; Yuan, X.; Gao, J.; Cao, J.; Pan, D. A Simple and Reliable Single Tube Septuple PCR Assay for Simultaneous Identification of Seven Meat Species. *Foods* **2021**, *10*, 1083. https://doi.org/10.3390/foods10051083

Academic Editors: Mohammed Gagaoua and Brigitte Picard

Received: 14 March 2021
Accepted: 10 May 2021
Published: 13 May 2021

Publisher's Note: MDPI stays neutral with regard to jurisdictional claims in published maps and institutional affiliations.

1. Introduction

Meat authentication is an important concern to protect consumers from illegal and unwanted ingredients [1–4]. However, meat adulteration such as unlisted, mislabeled or fraudulent ingredients has frequently been reported around the world and has become a severe global issue [4,5]. Although some laws have been enacted for ensuring the quality and safety of meat products, adulteration is still widespread due to the purpose of economic pursuit [1,6]. Poultry meat (chicken, duck and goose), especially, is frequently adulterated to red meat due to their low cost of production in Chinese markets [1,7]. As is known, soy allergy has become an arising public health concern regarding food allergies, as a small amount of soybean may elicit allergic reactions in both children and adults [8]. Notably, there is increasing evidence supporting that meat may trigger allergic reactions, especially for sensitized patients, which may cause a severe health risk of infectious diseases, metabolic disorders and allergies [9]. In addition, meat adulteration could also violate religious concerns; as is known, meat products containing pork ingredients are not permitted by Kosher and Halal food laws [10,11]. Therefore, it remains a pressing need for identifying meat species with 100% accuracy in real-world foodstuffs.

In recent years, the techniques for authenticating meat species have been continuously evolving. Many analytical methods of biological, immunological, physical, chemical, anatomical and histological analyses have been developed [2–4,7,12]. Among them, protein-based methods are widely used to identify meat species in composite mixtures [5,13]. Proteins are likely to be degraded, denatured or damaged in processed meat products, limiting the accuracy of the identification of meat species in thermally treated foods [13,14]. In comparison, DNA-based analytical methods coupled with polymerase chain reaction

(PCR) present a reliable alternative to protein-based methods in the discrimination and mislabeling detection of meat species, as DNA molecules possess high stability and are present in every type of cell [4,5]. Both multiplex and real-time PCR techniques are highly specific and efficient in the identification of meat adulteration [15]. Real-time PCR techniques are widely used to quantify the amount of a target sequence in a reaction system. The levels of PCR amplification are monitored in real time, once per cycle, by measuring specific fluorescence signals, whose intensities reflect the amount of PCR product [16]. With the progress of molecular biology in recent years, multiplex real-time PCR techniques depending on melting curve analysis have been developed and widely adopted in the identification of meat species, and they have characteristics such as time saving, high specificity and high sensitivity [17–20]. Collectively, real-time PCR analysis shows more detailed information with regard to the identification and quantification of meat species. However, accurate quantification can only be achieved with a proper reference material because the matrix may interfere with the amplification process [21], indicating that quantification of meat fractions in real-world foodstuffs is difficult, and the results may not truly correlate to the recipe of the meat products. In comparison, multiplex PCR assay can be easily implemented with minimum effort, but much gain, to verify the identification of meat species. Recently, many studies have also constructed multiplex PCR with electrophoresis analysis to authenticate meat species with satisfying results [1]. Although much is known about multiplex PCR as duplex, triplex, tetraplex, pentaplex (quintuple) and hexaplex (sextuple) PCRs, little information is available on a multiplex PCR authenticating more than six animal species simultaneously.

Mitochondrial DNA possesses high copy numbers per cell and strong stability, which ensures a low limit of detection and its availability in both raw and cooked meat products, and it has been broadly adopted for PCR protocols [22,23]. For example, cytochrome *b* gene, D-loop, 12S and 16S rRNA genes, ATPase subunits 8 and 6 genes, and NADH dehydrogenase genes are common targets for identifying meat species [21,24,25]. All data provide reliable evidence for the roles of mitochondrial DNA sequences in animal species identification. Using mitochondrial DNA sequences obtained from turkey, goose, pig, sheep, beef, chicken and duck, we designed a set of primer pairs that specifically amplified for seven species with differential lengths through PCR assays. We next performed the specific, sensitive and cost-effective detection of the indicated primers through simplex and multiplex PCR assays. Through screening, this study develops a septuple PCR assay for identifying seven ingredients of turkey, goose, pig, sheep, beef, chicken and duck simultaneously in commercial foodstuffs.

2. Materials and Methods

2.1. Preparation of Meat Samples and DNA Extraction

Fresh meat samples of turkey, goose, pig, sheep, beef, chicken and duck were purchased from local retailers and markets and transported on ice to the laboratory for immediate processing. Samples were stored at $-20\,^{\circ}\mathrm{C}$ to prevent DNA degradation. Total DNA was extracted from various meat samples using the EasyPure® Genomic DNA Kit (Beijing Trans Gen Biotech Co., Ltd., Beijing, China) following the manufacturer's instructions. DNA concentration was determined by a NanoDrop 2000 UV–Vis spectrophotometer (Thermo Scientifc, Wilmington, DE, USA).

2.2. Design of Species-Specific Primers

Primers were designed by targeting mitochondrial DNA sequences based on both high divergence and conservation within the species. As shown in Table 1, sequences of 16S rRNA gene of turkey (GenBank Accession No. EF153719.1), 16S rRNA gene of goose (KJ124555.1), cytochrome *c* oxidase subunit I gene of pig (KJ746666.1), cytochrome *c* oxidase subunit I gene of sheep (KP702285.1), 16S rRNA gene of cattle (MN714195.1), cytochrome *b* gene of chicken (MK163565.1) and 12S rRNA gene of duck (MK770342.1) were obtained from the National Centre of Biotechnology Information (NCBI) database. Next, the MEGA6 alignment tool was employed for identifying the conservative and variable regions. Using Oligo 7.0 and BLAST programs (www.ncbi.nlm.nih.gov/blast/ accessed on 1 April 2021), primers were newly designed according to their physical parameters, such as melting temperature, secondary structures, self-complementarity and cross-reactivity. The primer pairs were synthesized by Shanghai Sangon Biological Engineering Technology & Services Co., Ltd. (Shanghai, China). To determine the mismatch between the target and nontarget species, each set of primers was in silico screened with 13 land animals: turkey (*Meleagris gallopavo*), goose (*Anser cygnoides*), pig (*Sus scrofa*), cattle (*Bos taurus*), sheep (*Ovis aries*), chicken (*Gallus gallus*), duck (*Anas platyrhynchos*), horse (*Equus caballus*), camel (*Camelus bactrianus*), ostrich (*Struthio camelus*), dog (*Canis lupus*), rabbit (*Oryctolagus cuniculus*), cat (*Felis catus*) and 3 aquatic species, namely, small yellow croaker (*Larimichthys polyactis*), tuna (*Thunnus orientalis*) and black carp (*Mylopharyngodon piceus*), using ClustalW software. The final specificity of each primer pair was examined through PCR assays against templates of all species mentioned above.

Table 1. Oligonucleotide primers for meat species used in this study.

Primers	Genes	Sequence (5′–3′ Direction)	Amplicons (bp)	Reference or Source
Turkey	16S rRNA	CTCTAGCCCAACCACCCAT GCGCCTAAGGTCTTTTCTATCAC	110	this study
Goose	16S rRNA	TTAGACGCGATAGAGACCCCA GTTCGCTCTCTTTAACTGCTTG	194	this study
Pig	cytochrome c oxidase subunit I	CAGCCCGGAACCCTACTTG GTTCATCCAGTACCCGCTCC	254	this study
Sheep	cytochrome c oxidase subunit I	AGATATCGGCACCCTTTACCTTC CTGCTCCGGCCTCAACCAT	329	this study
Beef	16S rRNA	GTGCCTGATAATACTCTGACCAC CACCCCAACCGAAACTACCAA	473	this study
Chicken	cytochrome b	TTTCGGCTCCCTATTAGCAGTC AGTATGAGAGTTAAGCCCAGA	612	this study
Duck	12S rRNA	TGCCCTCAATAGCCTTCACC CATACTTCTTTCCGTGTTGCC	718	this study
Eukaryotes	12S rRNA	CAACTGGGATTAGATACCCCACTAT GAGGGTGACGGGCGGTGTGT	456	[26]
Eukaryotes	16S rRNA	AAGACGAGAAGACCCTATGGA GATTGCGCTGTTATCCCTAGGGTA	240	[27]
Eukaryotes	18S rRNA	AGGATCCATTGGAGGGCAAGT TCCAACTACGAGCTTTTTAACTGCA	99	[28]

2.3. Simplex and Multiplex PCR

To develop a multiplex PCR method, simplex PCR was firstly carried out for each of the target species with their own primers to ensure that each target was amplified. PCR amplification (in one reaction of 25 µL, including 2.5 µL of 10 × EasyTaq® Buffer, 2 µL of 2.5 mM dNTPs, 0.5 µL of EasyTaq DNA Polymerase, 0.5 µL of 10 µM each primer, and 0.01–10 ng genomic DNA of each species) was achieved using EasyTaq® DNA Polymerase kit (TransGen Biotech Co., Ltd., Beijing, China). The reaction was initiated by a 5 min denaturation at 94 °C, followed by 34 cycles of denaturation at 94 °C for 30 s, annealing at 63 °C for 30 s, elongation at 72 °C for 45 s and a final elongation at 72 °C for 5 min. For universal primer pairs, the annealing was set at 56 °C. After simplex PCR assays for each species was achieved, a septuple PCR assay was developed. The 4% agarose gels were visualized and subsequently photographed in a Bio-rad GelDoc 1000 gel documentation system.

2.4. Test of Primers' Specificity, Sensitivity and Reproducibility

The specificity of species-specific primers was corroborated by using template DNA isolated from all species (turkey, goose, pig, cattle, sheep, chicken, duck, horse, camel, ostrich, dog, rabbit, cat, small yellow croaker, tuna, black carp). In the preliminary phase of this experiment, simplex and septuple PCRs were respectively performed by using the DNA extracted from raw animal species. The PCR product was run on agarose gel and then visualized for the proper amplification.

The sensitivity of septuple PCR assay was confirmed by serial dilutions of the premixed DNA templates of all target species indicated, starting with 10 ng and progressing downward in one reaction. Seven concentrations (10 ng to 0.01 ng) of the target templates were used for PCR amplification determining the minimal concentration detected. PCR fragments were run on 4% agarose gel to confirm the limit of detection.

To assess the reproducibility, template DNA isolated from raw, boiled (97–99 °C, 30 min), autoclaved (121 °C, 150 kPa for 15 min) and microwave-cooked (750 W, 10 min, 119–121 °C) meat samples were individually analyzed by using the simplex PCR. The PCR product was run on agarose gel.

2.5. Commercial Samples

Using the present multiplex PCR method, 60 raw or thermally processed meat products including meat balls (15), meat slices (13), kebab (10), sausages (5), cutlets (10), jerky (2) and breast (5) were purchased from retail markets and supermarkets in Ningbo City, PR China, as well as online supermarket platforms, which were used for assessing the authentication of meat species. Details of the samples are listed in Table 2. All samples of meat balls, slices, kebabs, sausages, cutlets and breasts were raw materials with mechanical processing but not heat treatment, while two jerky samples, within turkey, were subjected to heat processing.

Table 2. Results of multiplex PCR assay performed on commercial meat products.

Products	No	Labelled	Detected Species							Adulteration
			Turkey	Goose	Pig	Sheep	Beef	Chicken	Duck	
Beef	15									5 (33.3%)
meat balls	5	beef	1/5 [b]	—	1/5 [a], 1/5 [b]	—	5/5	1/5 [a]	—	
meat slices	5	beef	—	—	1/5	—	5/5	—	—	
kebab	5	beef	—	—	2/5	—	5/5	—	—	
Mutton	15									6 (40.0%)
meat balls	5	mutton	—	—	1/5 [a], 1/5 [b]	5/5	—	1/5 [a]	1/5 [b]	
meat slices	5	mutton	—	—	2/5	5/5	—	—	—	
kebab	5	mutton	—	—	2/5	5/5	—	—	—	
Pork	15									4 (26.7%)
meat balls	5	pig	—	1/5 [b]	5/5	—	—	1/5 [a], 1/5 [b]	1/5 [a]	
sausages	5	pig	—	—	5/5	—	—	1/5 [a]	1/5 [b]	
cutlets	5	pig	—	—	5/5	—	—	—	—	
Turkey	15									1 (6.7%)
cutlets	5	turkey	5/5	—	—	—	—	1/5	—	
meat slices	3	turkey	3/3	—	—	—	—	—	—	
breast	5	turkey	5/5	—	—	—	—	—	—	
jerky	2	turkey	2/2	—	—	—	—	—	—	

A horizontal line (—) denotes no PCR product detected. In each row, the meat samples labeled with same letter ([a] or [b]) represent the identical meat samples, while different letters indicate a difference in meat samples.

3. Results

3.1. Specificity Assays of Simplex PCR

To determine the species-specific primers, we designed many pairs of primers for each species as candidates by using Oligo 7.0 and BLAST programs. Each set of primers was compared against 16 species (turkey, goose, pig, cattle, sheep, chicken, duck, horse, camel, ostrich, dog, rabbit, cat, small yellow croaker, tuna and black carp) by simplex PCR assays (data not shown). Through gel electrophoresis, we ultimately selected the primer pairs for each species in Table 1. PCR fragments showed distinguishable bands with the predicted size of 110, 194, 254, 329, 473, 612 and 718 bp for turkey, goose, pig, sheep, beef, chicken and duck species, respectively (Figure 1A). Three pairs of universal eukaryotic primers, which target 12S rRNA, 16S rRNA and 18S rRNA genes with individual 456, 240 and 99 bp PCR fragments in all meat species, were used as positive controls for ensuring the quality of template DNAs in one PCR reaction (Figure 1B). To further test the efficiency and specificity of primers, simplex PCRs were carried out using a DNA mixture of all seven meat species. In these experiments, each set of species-specific primers yielded the expected PCR fragment by using only the template DNA mixture of seven meat species, but not with nontarget species (Figure 1C), further confirming that the new primers were highly specific for the target species.

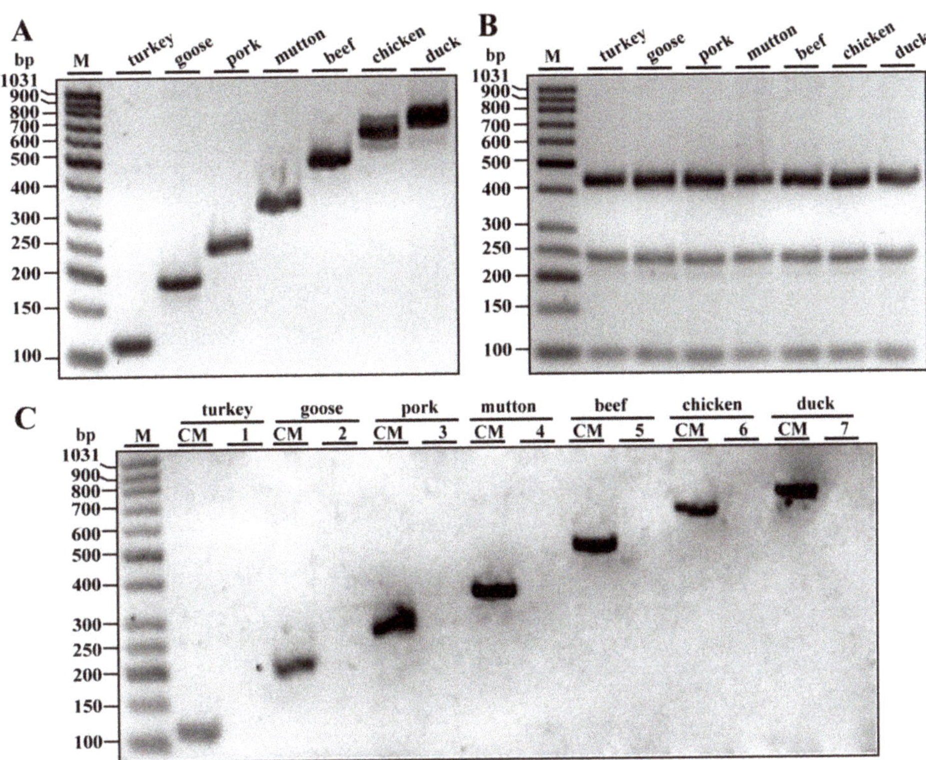

Figure 1. Specificity assays of simplex PCR. (**A**) Gel image of the products generated by PCR amplification with species-specific primers for turkey, goose, pig, cattle, sheep, chicken and duck using corresponding genomic DNA as a template, respectively. (**B**) As positive controls, gel image of the PCR products generated after amplification with premixed universal eukaryotic primer pairs of 12S rRNA, 16S rRNA and 18S rRNA genes for all meat species. (**C**) Gel image of the products through simplex PCR amplification using species-specific primers for turkey, goose, pig, cattle, sheep, chicken and duck. CM, a complete mixture of turkey, goose, pig, cattle, sheep, chicken and duck; 1–7, a complete DNA mixture except target species DNA. Lane M is ladder DNA.

3.2. Sensitivity Assays of Septuple PCR

A septuple PCR system was constructed by using seven sets of species-specific primers. To validate the sensitivity of the multiplex PCR assay, extracted DNA of each target species was serially diluted (10, 5, 1, 0.5, 0.1, 0.05 and 0.01 ng). PCR products were subsequently run on an agarose gel to assess the sensitivity. As can be seen from Figure 2A, the expected bands of seven meat species were obtained by multiplex PCR under the conditions of all tested concentrations (10–0.01 ng). In accordance with that of gel-view, each electropherogram clearly represented seven peaks corresponding to the seven different bands displayed in the gel-view (Figure 2B). Both intensities of bands and peaks were dramatically decreased in a concentration-dependent manner. However, even at the concentration of 0.01 ng per reaction, some PCR products of meat species can be clearly recognized in Figure 2A,B. Thus, the limit of detection of the developed septuple PCR assay was concluded to be 0.01–0.05 ng DNA.

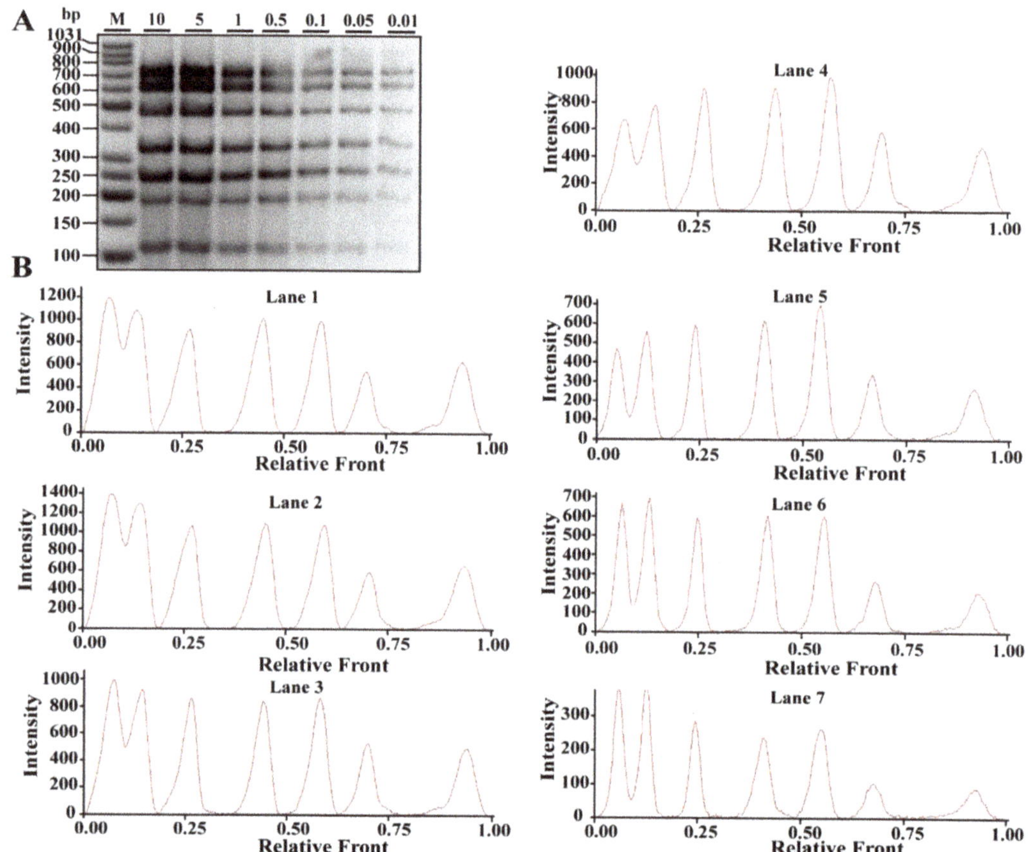

Figure 2. Sensitivity of the developed septuple PCR assay. (**A**) Gel image of the products generated after multiplex PCR amplifications of premixed DNA templates of all species (10, 5, 1, 0.5, 0.1, 0.05, 0.01 ng) with primers of seven meat species mixtures including turkey, goose, pig, cattle, sheep, chicken and duck. (**B**) The corresponding electropherograms of gel image (**A**). Lanes 1–7 are presented with labels (10, 5, 1, 0.5, 0.1, 0.05, 0.01) in (**A**). Lane M is ladder DNA.

3.3. Validation of Reproducibility Assay in Thermally Processed Meat

To assess the efficiency of designed primers in detecting thermally processed meat, three different heat treatment processes were selected to treat raw meat samples as described in Section 2.4. The quality of template DNA extracted from processed meat samples was examined by simplex PCR assays. As shown in Figure 3A–D, using DNA extracted from raw, boiled, autoclaved and microwave-cooked meat samples, PCR amplification of turkey, goose, pig, sheep, beef, chicken and duck species generated the expected PCR products with 100% accuracy in meat authentication, indicating that our designed primers can be successfully employed for authenticating animal species in processed meat products.

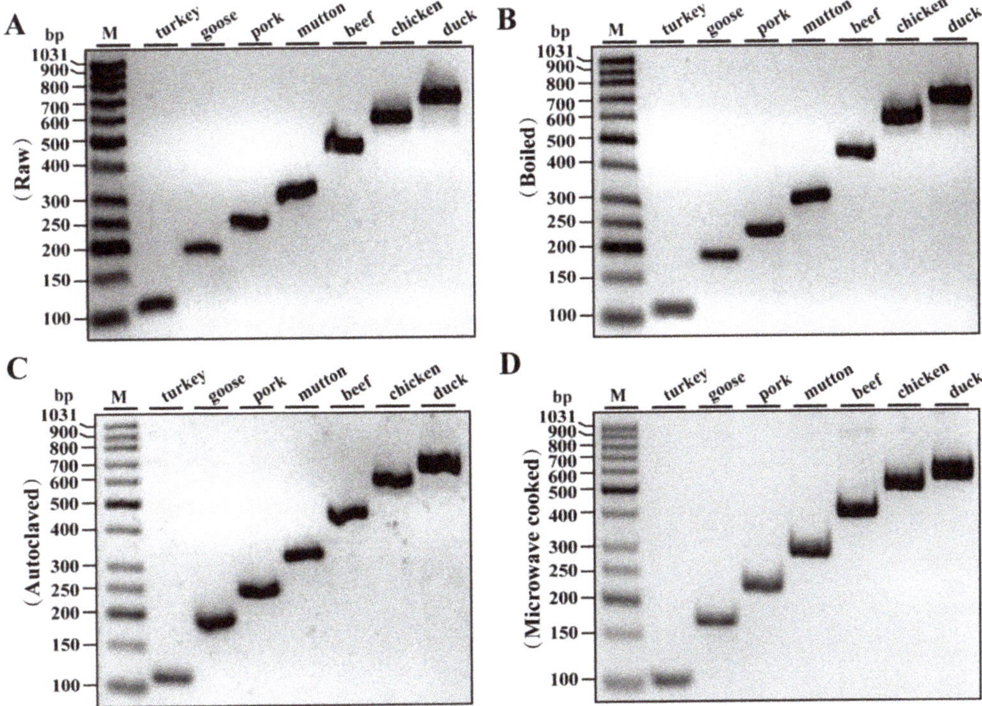

Figure 3. Gel image of the PCR products generated by simplex PCR amplifications with turkey, goose, pig, cattle, sheep, chicken and duck DNA extracted from raw (**A**), boiled (**B**), autoclaved (**C**) and microwave-cooked meat samples (**D**) using each species-specific primer pair. Lane M is ladder DNA.

3.4. Application of Multiplex PCR Assay on Commercially Processed Meat Products

The real-world food products were examined using the developed septuple PCR. The survey was conducted with 60 commercial samples of beef, mutton, pork and turkey (15 samples each). As shown in Figure 4 and summarized in Table 2, most of the samples had the same ingredients as labeled, without contamination. However, 5 of 15 (33.3%) beef samples, 6 of 15 (40.0%) mutton samples, 4 of 15 (26.7%) pork samples and 1 of 15 (6.7%) turkey samples contained some unlisted meat species. The survey revealed that inexpensive chicken, duck and pork meats were frequently adulterated products. The results further corroborated the efficiency of the developed septuple PCR assay in identification of commonly consumed meats.

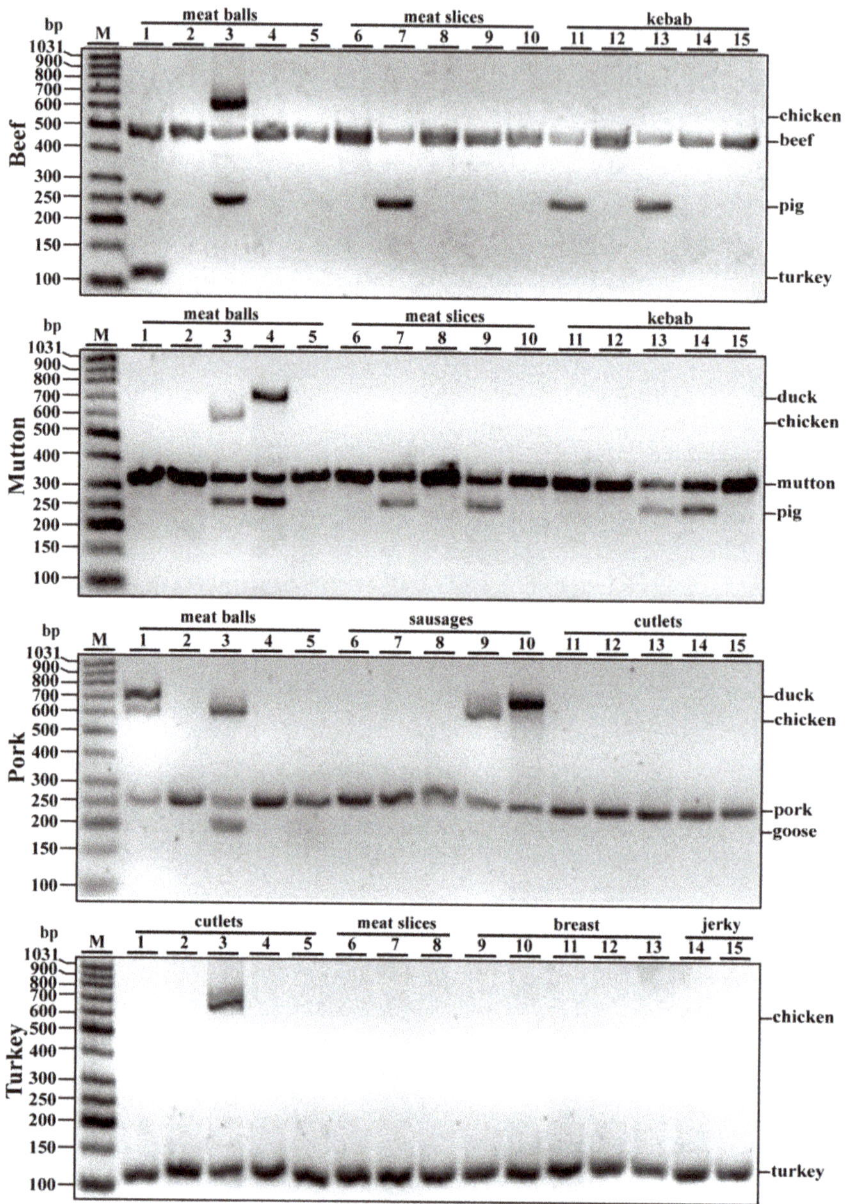

Figure 4. Analysis of commercial foodstuffs using the developed septuple assay. Gel image of the fragments generated by multiplex PCR amplifications using DNA obtained from commercial meat products with premixed primers for seven meat species including turkey, goose, pig, cattle, sheep, chicken and duck.

4. Discussion

Frequent meat frauds have aroused significant social attention because adulteration risks food safety, breaches market rules and even threatens public health [8]. In recent years, adulteration practice has been ingeniously applied to treated meat products showing similar morphological and physical appearance to pure meat. Nowadays, PCR-based techniques are the effective methods for species authentication. Real-time PCR techniques and microchip electrophoresis-dependent multiplex PCR methods require special infrastructures [11,29,30]; multiplex PCR assays through simple agarose gel analysis minimizes the cost drastically on a large scale and can be easily carried out to verify the identity of ingredients in foodstuffs [6,31,32]. As seen in Table 3, much is known about multiplex PCR assays that simultaneously verify two to six meat ingredients in one reaction. Relatively little information is available on multiplex PCR methods for authenticating more meat species simultaneously. Although one study authenticated 10 animal species (beef, sheep, pork, chicken, turkey, cat, dog, mouse, rat, human), it was achieved by two tube multiplex assays, where every five animal species were verified by a pentaplex PCR assay in one reaction [33]. Similarly, using two tube independent pentaplex PCR assays with ten pairs of primers, 14 animal species including cattle, donkey, *Canidae* (dog, fox, raccoon-dog), deer and horse, pig, *Ovis* (sheep, goat), poultry (chicken, duck), cat and mouse were detected through chip electrophoresis; however, the multiplex PCR failed to accurately distinguish sheep and goat within Ovis, dog, fox and raccoon-dog within Canidae, and chicken and duck within poultry [11]. Therefore, there is still a lack of more efficient detection methods with low cost for supervising more meat content. The goal of the present study was to develop a multiplex PCR method for reliable and efficient identification of ruminant, poultry and pork materials.

The choice of animal species was considered based on actual adulteration cases, with a higher practicability in Chinese markets. We found that multiplex PCR with increased species-specific primers in one reaction led to more opportunities of cross-reactivity with each other, or generated unexpected bands, which may limit the availability of multiplex PCR for verifying more animal species. To provide a multiplex PCR method that detects more animal species in a single assay platform, we designed many sets of primers throughout target mitochondrial DNA sequences such as cytochrome b gene, D-loop, 12S and 16S rRNA genes, ATPase subunits 8 and 6 genes and NADH dehydrogenase genes using Oligo 7.0 and Primer-BLAST programs. Through screening species-specific primer pairs, a species-specific septuple PCR method was ultimately developed and optimized to simultaneously detect turkey (110 bp), goose (194 bp), pig (254 bp), sheep (329 bp), beef (473 bp), chicken (612 bp) and duck (718 bp) in one reaction. To ensure the quality of template genomic DNA in one PCR reaction, a universal eukaryotic primer set that amplifies a bigger PCR fragment than that of all meat species tested should be chosen as the preferred positive control. However, to our knowledge, little information is available on a universal eukaryotic primer set amplifying the fragment with more than 700 bp length. As alternatives, we chose three pairs of universal eukaryotic primers, which target different mitochondrial DNA sequences, including 12S rRNA, 16S rRNA, 18S rRNA genes, and amplifies distinguishable 456, 240 and 99 bp PCR fragments in all meat species, respectively [26–28]. In addition, the control primer set should be inserted in the multiplex assay; these PCR fragments were too close to that of turkey (110 bp), pig (254 bp) and beef (473 bp) to discriminate each other. Accordingly, these universal primer pairs were used in a single PCR in this study. Figure 1B shows the expected bands of each primer set in all meat species, implying the high quality of genomic DNA used in this study.

Table 3. Comparative analysis of recently published multiplex PCR assays for the identification of meat species.

Multiplex PCR Type	Sp. No [a]	Detection Items	Detection Limit	Detection Method [b]	Reference or Source
Septuple	7	turkey, goose, pig, sheep, beef, chicken, duck	0.01–0.05 ng DNA	Gel	This study
Multiplex	4	ruminant, poultry, pork, and donkey	0.01–0.1 ng/µL DNA	Gel	[25]
Hexaplex	6	chicken, cow/buffalo, sheep/goat, horse/donkey, pork, dog	0.03–0.05 ng DNA	Gel	[31]
Multiplex	5	sheep/goat, bovine, chicken, duck, pig	0.5 ng DNA	Gel	[6]
Multiplex	2	cattle, buffalo	2.23–2.31 ng/µL DNA	Gel	[34]
Quadruple	4	fox, mink, or raccoon in beef and mutton	1% for each species	Gel	[35]
Pentaplex	5	dog, duck, buffalo, goat, sheep	0.1–0.32 ng DNA	Gel	[21]
Multiplex (two-tube)	14	cattle, donkey, Canidae (dog, fox, raccoon-dog), deer and horse, pig, Ovis (sheep, goat), poultry (chicken, duck), cat, mouse	0.02–0.2 ng DNA	Chip	[11]
Quadruplex	4	chicken, mutton, beef, pork	16 pg DNA, 0.01% of each species	Gel	[36]
Multiplex (two-tube)	10	beef, sheep, pork, chicken, turkey, cat, dog, mouse, rat, human	30 pg DNA	Gel	[33]
Tetraplex	3	pig, cattle, fish, eukaryotic18S rRNA	0.001–0.1 ng DNA	Gel	[37]
Hexaplex	6	horse, soybean, sheep, poultry, pork, cow	0.01% for each species	Gel	[38]
Octuplex	8	dog, chicken, cattle, pig, horse, donkey, fox, and rabbit	0.05 ng/µL DNA	Gel	[27]
Multiplex	3	chicken, duck and goose	0.05 ng DNA, 1% for each species	Gel	[39]
Multiplex	5	cat, dog, pig, monkey, rat	0.01–0.02 ng DNA	Chip	[24]
Quadruple	4	beef, pork, mutton, duck	0.1 ng DNA	Gel	[40]

[a] Species number; [b] Chip, microchip electrophoresis; Gel, agarose gel electrophoresis.

Through the specificity test, we validated that the primers were highly specific to each of particular species and had no cross-reactivity, at least with the 15 animal species tested. The detection limit of this particular assay on reference meat samples was 0.01–0.05 ng, indicating that this method is highly sensitive and reliable. Using the DNA isolated from raw, boiled, autoclaved and microwave-cooked samples of seven meat species, simplex PCR assays generated all expected PCR products, suggesting that PCR assay with our primers had a high reproducibility in processed meat samples. Most importantly, multiplex PCR assay on commercially available processed meat products revealed that inexpensive chicken, duck and pork meats were adulterated products (Table 2 and Figure 4). Consistent with previous reports, inexpensive poultry meat readily evades visual detection and is frequently adulterated into other meat products, in particular, processed beef, mutton and pork [41–43]. Perhaps, economically driven thoughts of manufacturers or peddlers are a critical factor for the replacement of expensive, high-quality meat with inferior and low-cost ones. Collectively, the developed septuple PCR assay is not only reliable and efficient but is also a sensitive detection method for the identification of meat species in actual adulteration events. However, vegetable proteins such as soybean are found to a substitute ingredient for muscle proteins, due to their low cost of production [8,38,44]. In addition, some surveys demonstrate that inexpensive fish species are adulterated into meat products [44,45]. Therefore, we still cannot exclude the possibility that vegetable proteins and fish sources may be present in commercial meat products. Considering the fact that multiplex PCR with increased species-specific primers in one reaction may cause more opportunities of cross-reactivity with each other, or generated unexpected bands, a more efficient method for identification of meat adulteration should be developed in future study.

5. Conclusions

The aim of this study is to provide reliable adulteration detection, by means of septuple PCR, which can simultaneously authenticate seven animal species of turkey, goose, pig, sheep, beef, chicken and duck. The assay is also quite sensitive to enable detection of 0.01–0.05 ng DNA templates for each species per reaction, thus making it qualified for authenticating meat species in commercial, real-world foodstuffs. By simple agarose gel analysis, without expensive equipment or a high level of technical skill, this septuple PCR method could be more broadly used for detecting sources of meat species in foodstuffs in which adulteration is suspected.

Author Contributions: Z.C., Q.L. and D.P. contributed to the conceptualization, methodology, supervision, investigation and data curation; Z.C. and D.P. contributed to the funding acquisition; S.Z. and H.M. contributed to the software and data curation; X.Y., J.G. and J.C. contributed to the project administration and methodology; Z.C. and Q.L. contributed to the writing—original draft. All authors have read and agreed to the published version of the manuscript.

Funding: This work was financially supported by the National Natural Science Foundation of China (NSFC) (31901668 and 31972048), Scientific Research Fund of Zhejiang Provincial Education Department (Y201940932), the Natural Science Foundation of Ningbo (2019A610436), School Research Project in Ningbo University (XYL19011) and the Open Project Program of Beijing Key Laboratory of Flavor Chemistry, Beijing Technology and Business University (BTBU) (SPFW2020YB13).

Institutional Review Board Statement: Not applicable.

Informed Consent Statement: Not applicable.

Data Availability Statement: Not applicable.

Conflicts of Interest: The authors have declared no conflict of interest. This paper is our original unpublished work; it has not been published previously and is not under consideration for publication elsewhere, in whole or in part. All authors of this paper have read and approved the final, submitted version.

References

1. Li, Y.; Liu, S.; Meng, F.; Liu, D.; Zhang, Y.; Wang, W.; Zhang, J. Comparative review and the recent progress in detection technologies of meat product adulteration. *Compr. Rev. Food Sci. Food Saf.* **2020**, *19*, 2256–2296. [CrossRef] [PubMed]
2. Chung, S.M.; Hellberg, R.S. Effects of poor sanitation procedures on cross-contamination of animal species in ground meat products. *Food Control.* **2020**, *109*, 106927. [CrossRef]
3. Da Costa Filho, P.A.; Cobuccio, L.; Mainali, D.; Rault, M.; Cavin, C. Rapid analysis of food raw materials adulteration using laser direct infrared spectroscopy and imaging. *Food Control.* **2020**, *113*, 107114. [CrossRef]
4. Mansouri, M.; Khalilzadeh, B.; Barzegari, A.; Shoeibi, S.; Isildak, S.; Bargahi, N.; Omidi, Y.; Dastmalchi, S.; Rashidi, M.-R. Design a highly specific sequence for electrochemical evaluation of meat adulteration in cooked sausages. *Biosens. Bioelectron.* **2020**, *150*, 111916. [CrossRef] [PubMed]
5. Mokhtar, N.F.K.; El Sheikha, A.F.; Azmi, N.I.; Mustafa, S. Potential authentication of various meat-based products using simple and efficient DNA extraction method. *J. Sci. Food Agric.* **2019**, *100*, 1687–1693. [CrossRef] [PubMed]
6. Wang, W.; Wang, X.; Zhang, Q.; Liu, Z.; Zhou, X.; Liu, B. A multiplex PCR method for detection of five animal species in processed meat products using novel species-specific nuclear DNA sequences. *Eur. Food Res. Technol.* **2020**, *246*, 1351–1360. [CrossRef]
7. Barbin, D.F.; Badaró, A.T.; Honorato, D.C.; Ida, E.Y.; Shimokomaki, M. Identification of turkey meat and processed products using near infrared spectroscopy. *Food Control.* **2020**, *107*, 106816. [CrossRef]
8. Safdar, M.; Junejo, Y.; Arman, K.; Abasıyanık, M. A highly sensitive and specific tetraplex PCR assay for soybean, poultry, horse and pork species identification in sausages: Development and validation. *Meat Sci.* **2014**, *98*, 296–300. [CrossRef]
9. Tanabe, S.; Miyauchi, E.; Muneshige, A.; Mio, K.; Sato, C.; Sato, M. PCR method of detecting pork in foods for verifying allergen labeling and for identifying hidden pork ingredients in processed foods. *Biosci. Biotechnol. Biochem.* **2007**, *71*, 1663–1667. [CrossRef]

10. Li, T.T.; Jalbani, Y.M.; Zhang, G.L.; Zhao, Z.Y.; Wang, Z.Y.; Zhao, X.Y.; Chen, A.L. Detection of goat meat adulteration by real-time PCR based on a reference primer. *Food Chem.* **2019**, *277*, 554–557. [CrossRef]
11. Li, J.; Li, J.; Xu, S.; Xiong, S.; Yang, J.; Chen, X.; Wang, S.; Qiao, X.; Zhou, T. A rapid and reliable multiplex PCR assay for simultaneous detection of fourteen animal species in two tubes. *Food Chem.* **2019**, *295*, 395–402. [CrossRef]
12. Leng, T.; Li, F.; Xiong, L.; Xiong, Q.; Zhu, M.; Chen, Y. Quantitative detection of binary and ternary adulteration of minced beef meat with pork and duck meat by NIR combined with chemometrics. *Food Control.* **2020**, *113*, 107203. [CrossRef]
13. Asensio, L.; González, I.; García, T.; Martín, R. Determination of food authenticity by enzyme-linked immunosorbent assay (ELISA). *Food Control.* **2008**, *19*, 1–8. [CrossRef]
14. Sentandreu, M.Á.; Sentandreu, E. Authenticity of meat products: Tools against fraud. *Food Res. Int.* **2014**, *60*, 19–29. [CrossRef]
15. Liu, G.; Luo, J.; Xu, W.; Li, C.; Guo, Y.; Guo, L. Improved triplex real-time PCR with endogenous control for synchronous identification of DNA from chicken, duck, and goose meat. *Food Sci. Nutr.* **2021**. [CrossRef]
16. Li, J.; Li, J.; Liu, R.; Wei, Y.; Wang, S. Identification of eleven meat species in foodstuff by a hexaplex real-time PCR with melting curve analysis. *Food Control.* **2021**, *121*, 107599. [CrossRef]
17. Dolch, K.; Andrée, S.; Schwägele, F. Comparison of real-time PCR quantification methods in the identification of poultry species in meat products. *Foods* **2020**, *9*, 1049. [CrossRef] [PubMed]
18. Li, J.; Wei, Y.; Li, J.; Liu, R.; Xu, S.; Xiong, S.; Guo, Y.; Qiao, X.; Wang, S. A novel duplex SYBR green real-time PCR with melting curve analysis method for beef adulteration detection. *Food Chem.* **2021**, *338*, 127932. [CrossRef]
19. Zhu, T.; Zhou, X.; Zhang, W.; Wu, Y.; Yang, J.; Xu, L.; Chen, M.; Dong, W.; Xu, H. Multiplex and real-time PCR for qualitative and quantitative donkey meat adulteration. *J. Food Meas. Charact.* **2021**, *15*, 1161–1168. [CrossRef]
20. Li, T.; Wang, J.; Wang, Z.; Qiao, L.; Liu, R.; Li, S.; Chen, A. Quantitative determination of mutton adulteration with single-copy nuclear genes by real-time PCR. *Food Chem.* **2020**, *344*, 128622. [CrossRef]
21. Thanakiatkrai, P.; Dechnakarin, J.; Ngasaman, R.; Kitpipit, T. Direct pentaplex PCR assay: An adjunct panel for meat species identification in Asian food products. *Food Chem.* **2019**, *271*, 767–772. [CrossRef]
22. Fajardo, V.; González, I.; Rojas, M.; García, T.; Martín, R. A review of current PCR-based methodologies for the authentication of meats from game animal species. *Trends Food Sci. Technol.* **2010**, *21*, 408–421. [CrossRef]
23. Kumar, A.; Kumar, R.R.; Sharma, B.D.; Gokulakrishnan, P.; Mendiratta, S.K.; Sharma, D. Identification of species origin of meat and meat products on the dna basis: A review. *Crit. Rev. Food Sci. Nutr.* **2013**, *55*, 1340–1351. [CrossRef]
24. Ali, M.E.; Razzak, M.A.; Hamid, S.B.A.; Rahman, M.M.; Amin, M.A.; Rashid, N.R.A. Multiplex PCR assay for the detection of five meat species forbidden in Islamic foods. *Food Chem.* **2015**, *177*, 214–224. [CrossRef] [PubMed]
25. Galal-Khallaf, A. Multiplex PCR and 12S rRNA gene sequencing for detection of meat adulteration: A case study in the Egyptian markets. *Gene* **2021**, *764*, 145062. [CrossRef]
26. Vaithiyanathan, S.; Vishnuraj, M.R.; Reddy, G.N.; Kulkarni, V.V. Application of DNA technology to check misrepresentation of animal species in illegally sold meat. *Biocatal. Agric. Biotechnol.* **2018**, *16*, 564–568. [CrossRef]
27. Liu, W.; Tao, J.; Xue, M.; Ji, J.; Zhang, Y.; Zhang, L.; Sun, W. A multiplex PCR method mediated by universal primers for the identification of eight meat ingredients in food products. *Eur. Food Res. Technol.* **2019**, *245*, 2385–2392. [CrossRef]
28. Martín, I.; García, T.; Fajardo, V.; Rojas, M.; Pegels, N.; Hernández, P.E.; González, I.; Martín, R. SYBR-green real-time PCR approach for the detection and quantification of pig DNA in feedstuffs. *Meat Sci.* **2009**, *82*, 252–259. [CrossRef]
29. Pegels, N.; González, I.; Garcia, T.; Martín, R. Avian-specific real-time PCR assay for authenticity control in farm animal feeds and pet foods. *Food Chem.* **2014**, *142*, 39–47. [CrossRef]
30. Cheng, X.; He, W.; Huang, F.; Huang, M.; Zhou, G. Multiplex real-time PCR for the identification and quantification of DNA from duck, pig and chicken in Chinese blood curds. *Food Res. Int.* **2014**, *60*, 30–37. [CrossRef]
31. Iqbal, M.; Saleem, M.S.; Imran, M.; Khan, W.A.; Ashraf, K.; Zahoor, M.Y.; Rashid, I.; Rehman, H.-U.; Nadeem, A.; Ali, S.; et al. Single tube multiplex PCR assay for the identification of banned meat species. *Food Addit. Contam. Part B* **2020**, *13*, 284–291. [CrossRef]
32. Mafra, I.; Ferreira, I.M.P.L.V.O.; Oliveira, M.B.P.P. Food authentication by PCR-based methods. *Eur. Food Res. Technol.* **2008**, *227*, 649–665. [CrossRef]
33. Prusakova, O.V.; Glukhova, X.A.; Afanas'Eva, G.V.; Trizna, Y.A.; Nazarova, L.F.; Beletsky, I.P. A simple and sensitive two-tube multiplex PCR assay for simultaneous detection of ten meat species. *Meat Sci.* **2018**, *137*, 34–40. [CrossRef] [PubMed]
34. Dantas, V.V.; Cardoso, G.V.F.; Araújo, W.S.C.; De Oliveira, A.C.D.S.; Da Silva, A.S.; Da Silva, J.B.; Pedroso, S.C.D.S.; Roos, T.B.; De Moraes, C.M.; Lourenço, L.D.F.H. Application of a multiplex polymerase chain reaction (mPCR) assay to detect fraud by substitution of bovine meat cuts with water buffalo meat in Northern Brazil. *CyTA J. Food* **2019**, *17*, 790–795. [CrossRef]
35. Li, X.; Guan, Y. Specific identification of the adulterated components in beef or mutton meats using multiplex PCR. *J. AOAC Int.* **2019**, *102*, 1181–1185. [CrossRef]
36. Balakrishna, K.; Sreerohini, S.; Parida, M. Ready-to-use single tube quadruplex PCR for differential identification of mutton, chicken, pork and beef in processed meat samples. *Food Addit. Contam. Part A* **2019**, *36*, 1435–1444. [CrossRef] [PubMed]
37. Sultana, S.; Hossain, M.; Zaidul, I.; Ali, E. Multiplex PCR to discriminate bovine, porcine, and fish DNA in gelatin and confectionery products. *LWT* **2018**, *92*, 169–176. [CrossRef]
38. Safdar, M.; Junejo, Y. The development of a hexaplex-conventional PCR for identification of six animal and plant species in foodstuffs. *Food Chem.* **2016**, *192*, 745–749. [CrossRef]

39. Qin, P.; Hong, Y.; Kim, H.-Y. Multiplex-PCR assay for simultaneous identification of lamb, beef and duck in raw and heat-treated meat mixtures. *J. Food Saf.* **2015**, *36*, 367–374. [CrossRef]

40. He, H.; Hong, X.; Feng, Y.; Wang, Y.; Ying, J.; Liu, Q.; Qian, Y.; Zhou, X.; Wang, D. Application of quadruple multiplex PCR detection for beef, duck, mutton and pork in mixed meat. *J. Food Nutr. Res.* **2015**, *3*, 392–398. [CrossRef]

41. Mane, B.; Mendiratta, S.; Tiwari, A. Polymerase chain reaction assay for identification of chicken in meat and meat products. *Food Chem.* **2009**, *116*, 806–810. [CrossRef]

42. Hou, B.; Meng, X.; Zhang, L.; Guo, J.; Li, S.; Jin, H. Development of a sensitive and specific multiplex PCR method for the simultaneous detection of chicken, duck and goose DNA in meat products. *Meat Sci.* **2015**, *101*, 90–94. [CrossRef] [PubMed]

43. Yacoub, H.A.; Sadek, M.A. Identification of fraud (with pig stuffs) in chicken-processed meat through information of mitochondrial cytochrome b. *Mitochondrial DNA Part A* **2016**, *28*, 855–859. [CrossRef] [PubMed]

44. Kim, M.-J.; Kim, H.-Y. Species identification of commercial jerky products in food and feed using direct pentaplex PCR assay. *Food Control.* **2017**, *78*, 1–6. [CrossRef]

45. Nejad, F.P.; Tafvizi, F.; Ebrahimi, M.T.; Hosseni, S.E. Optimization of multiplex PCR for the identification of animal species using mitochondrial genes in sausages. *Eur. Food Res. Technol.* **2014**, *239*, 533–541. [CrossRef]

Article

Interlaboratory Validation of a DNA Metabarcoding Assay for Mammalian and Poultry Species to Detect Food Adulteration

Stefanie Dobrovolny [1], Steffen Uhlig [2], Kirstin Frost [2], Anja Schlierf [2], Kapil Nichani [2], Kirsten Simon [2], Margit Cichna-Markl [3],* and Rupert Hochegger [1],*

[1]   Austrian Agency for Health and Food Safety (AGES), Department for Molecular Biology and Microbiology, Institute for Food Safety Vienna, Spargelfeldstrasse 191, 1220 Vienna, Austria; stefanie.dobrovolny@ages.at
[2]   QuoData GmbH, Prellerstrasse 14, 01309 Dresden, Germany; steffen.uhlig@quodata.de (S.U.); kirstin.frost@quodata.de (K.F.); anja.schlierf@quodata.de (A.S.); kapil.nichani@quodata.de (K.N.); kirsten.simon@quodata.de (K.S.)
[3]   Department of Analytical Chemistry, Faculty of Chemistry, University of Vienna, Währinger Strasse 38, 1090 Vienna, Austria
*   Correspondence: margit.cichna@univie.ac.at (M.C.-M.); rupert.hochegger@ages.at (R.H.)

Abstract: Meat species authentication in food is most commonly based on the detection of genetic variations. Official food control laboratories frequently apply single and multiplex real-time polymerase chain reaction (PCR) assays and/or DNA arrays. However, in the near future, DNA metabarcoding, the generation of PCR products for DNA barcodes, followed by massively parallel sequencing by next generation sequencing (NGS) technologies, could be an attractive alternative. DNA metabarcoding is superior to well-established methodologies since it allows simultaneous identification of a wide variety of species not only in individual foodstuffs but even in complex mixtures. We have recently published a DNA metabarcoding assay for the identification and differentiation of 15 mammalian species and six poultry species. With the aim to harmonize analytical methods for food authentication across EU Member States, the DNA metabarcoding assay has been tested in an interlaboratory ring trial including 15 laboratories. Each laboratory analyzed 16 anonymously labelled samples (eight samples, two subsamples each), comprising six DNA extract mixtures, one DNA extract from a model sausage, and one DNA extract from maize (negative control). Evaluation of data on repeatability, reproducibility, robustness, and measurement uncertainty indicated that the DNA metabarcoding method is applicable for meat species authentication in routine analysis.

Keywords: DNA metabarcoding; animal species; species identification; NGS; food adulteration; validation; interlaboratory ring trial

Citation: Dobrovolny, S.; Uhlig, S.; Frost, K.; Schlierf, A.; Nichani, K.; Simon, K.; Cichna-Markl, M.; Hochegger, R. Interlaboratory Validation of a DNA Metabarcoding Assay for Mammalian and Poultry Species to Detect Food Adulteration. *Foods* **2022**, *11*, 1108. https://doi.org/10.3390/foods11081108

Academic Editors: Mohammed Gagaoua and Brigitte Picard

Received: 5 March 2022
Accepted: 8 April 2022
Published: 12 April 2022

1. Introduction

Food authentication is known to be a challenging task. The methodology applied depends on several factors, including sample type, type of adulteration, and the information required. Meat products are most commonly adulterated by the replacement of high-priced animal species by lower-quality or cheaper ones. Since DNA-based methodologies are highly suitable to detect genetic variations such as single nucleotide polymorphisms (SNPs), insertions, and deletions, they play a crucial role in the identification and differentiation of animal species in meat products [1–3]. DNA-based methodologies target either species-specific sequences in nuclear DNA or conserved regions in mitochondrial DNA [4]. Currently, authentication of meat products in official food laboratories is mainly based on real-time polymerase chain reaction (PCR) assays and/or DNA arrays. Multiplex real-time PCR assays, allowing the identification and quantification of multiple species in one and the same well, are particularly applicable for routine analysis because they allow saving time and costs. Multiplex real-time PCR assays have not only been developed for domesticated species, e.g., beef, pig, chicken, and turkey [5], and beef, pig, horse, and sheep [6], but also for game species, e.g., roe deer, red deer, fallow deer, and sika deer [7]. However, the low number of optical channels of real-time PCR instruments limits the number of species that can be targeted simultaneously.

Next generation sequencing (NGS) technologies, in particular massively parallel sequencing of PCR products based on the analysis of species-specific differences in DNA sequences (DNA barcoding), is being considered a promising alternative [8–10]. So-called DNA metabarcoding offers the possibility to identify a wide variety of species not only in individual foodstuffs but even in complex mixtures. Moreover, in contrast to real-time PCR assays, it is an untargeted approach, allowing the detection of species one has not been looking for.

We have recently developed a DNA metabarcoding method for 15 mammalian species and six poultry species, which are quite frequently contained in European foodstuff [11]. In order to detect both mammalian and poultry species, a primer pair for mammals and a primer pair for poultry species was combined in a duplex PCR assay. A ~120 bp fragment of the mitochondrial 16S ribosomal DNA gene serves as barcode region. The DNA metabarcoding method has been validated with regard to specificity, repeatability, robustness, accuracy, and limit of detection (LOD) [11]. In-house validation data showed that the DNA metabarcoding method can be used for routine applications. Meat species can be identified down to a concentration of 0.1%. Very recently, the applicability of the DNA metabarcoding method for routine analysis was further investigated by the analysis of a total of 104 samples (25 reference samples, 56 food products, and 23 pet food products). Results obtained by DNA metabarcoding were in line with those obtained by real-time PCR and/or a commercial DNA array [12]. However, interlaboratory evaluation of novel methods is a prerequisite for standardization and harmonization.

In this study, we summarized the results of an interlaboratory ring trial for the DNA metabarcoding method, initiated by the §64 German Food and Feed Code (LFGB) working group "NGS Species Identification", chaired by the Federal Office of Consumer Protection and Food Safety (BVL) in Germany. One goal of the working group is the validation and standardization of (screening) methods for the identification and differentiation of animal species based on next generation amplicon sequencing for food authentication. The interlaboratory ring trial was coordinated by the Austrian Agency for Health and Food Safety (AGES) in 2020 and involved 15 participating laboratories. The aim was to evaluate the performance (e.g., repeatability, reproducibility, accuracy) of the DNA metabarcoding method in detail.

2. Materials and Methods

2.1. Participating Laboratories

The interlaboratory ring trial was organized by the AGES on behalf of the BVL. The following laboratories participated in the ring trial (in alphabetical order): Bavarian State Office for Food Safety and Health (LGL), Oberschleißheim, Germany; Chemical and Veterinary Analytical Institute Muensterland-Emscher-Lippe (CVUA-MEL), Muenster, Germany; Chemical and Veterinary Investigation Office Freiburg (CVUA-FR), Freiburg, Germany; Chemical and Veterinary Investigation Office Karlsruhe (CVUA-KA), Karlsruhe, Germany; Eurofins Genomics Europe Applied Genomics GmbH, Ebersberg, Germany; StarSEQ GmbH, Mainz, Germany; Labor Kneissler GmbH and Co. KG, Burglengenfeld, Germany; Saxony-Anhalt State Office for Consumer Protection (LAV S-A), Halle, Germany; State Office Laboratory Hessen (LHL), Kassel, Germany; Max Rubner-Institut (MRI)/National Reference Centre for Authentic Food (NRZ-Authent), Kulmbach, Germany; Max Planck Institute for Plant Breeding Research (MPIPZ), Köln, Germany; Lower Saxony State Office for Consumer Protection and Food Safety (LAVES), Niedersachsen, Germany; AGES, Vienna, Austria; PLANTON Laboratory for Analysis and Biotechnology GmbH, Kiel, Germany; SGS Institute Fresenius GmbH, Freiburg, Germany.

2.2. Samples

In the course of the interlaboratory ring trial, eight samples had to be analyzed: six DNA extract mixtures (samples 1–6), one DNA extract from a model sausage (sample 7), and one DNA extract from maize (sample 8), serving as a negative control (Table 1).

Table 1. Sample composition. Samples 1–6: DNA extract mixtures; percentage refers to DNA (*v/v*). Sample 7: extract from a model sausage; percentage refers to meat content (*w/w*). Sample 8: DNA extract from maize (negative control). - indicates that the species is not in the DNA extract mixture/sample.

Sample	Chicken	Horse	Turkey	Beef	Sheep	Pig	Goat
				Percentage (%)			
1	1	1	1	1	1	94	1
2	-	1.9	0.5	65.7	1.9	30	-
3	1.9	66.1	1.9	-	0.5	-	30
4	-	0.5	-	30	67.5	0.1	1.9
5	67.5	-	-	1.9	30	0.5	0.1
6	0.1	30	67.5	0.5	-	1.9	-
7	5	-	5	50	-	40	-
8	-	-	-	-	-	-	-

In total, seven animal species, including five mammalian species (*Sus scrofa domesticus* (pig), *Bos taurus* (cattle), *Equus caballus* (horse), *Ovis gmelini aries* (sheep), and *Capra aegagrus hircus* (goat)), and two poultry species (*Gallus gallus domesticus* (chicken) and *Meleagris gallopavo* (turkey)) were covered by the samples. All samples originated from muscle meat and were purchased from local meat suppliers.

Sample 1 contained DNA from seven animal species: DNA from pig as major component (94%, *v/v*), and DNA from six animal species (cattle, horse, sheep, goat, chicken, turkey; 1% (*v/v*) each). Samples 2–6 consisted of DNA from five animal species in varying proportions, ranging from 0.1% (*v/v*) to 67.5% (*v/v*). All DNA extract mixtures were prepared at the AGES.

The model sausage was produced according to the Codex Alimentarius Austriacus by the Higher Technical College for Food Technology Hollabrunn (Hollabrunn, Austria). The model sausage consisted of 50% (*w/w*) beef, 40% (*w/w*) pig, 5% (*w/w*) chicken, and 5% (*w/w*) turkey.

2.3. Genomic DNA Extraction

Extraction of genomic DNA from muscle meat of the seven animal species (pig, beef, horse, sheep, goat, chicken, turkey) was carried out at the CVUA-FR by applying the official method L 00.00–119 [13]. Identity of the animal species was verified by subjecting DNA extracts to Sanger sequencing and matching a ~464 base pair (bp) fragment of the mitochondrial cytochrome b gene against public databases provided by the National Center for Biotechnology Information (NCBI, Bethesda, MD, USA) [14,15]. After verification by Sanger sequencing and isolating genomic DNA fourfold, individual DNA extracts were combined. Total DNA of the (combined) DNA extracts was quantified by spectroscopy employing a UV/VIS spectrophotometer, adjusted to a DNA concentration of 20 ng/μL and sent to AGES. Isolation of DNA from the homogenized model sausage was performed at AGES [13].

The copy number of the mitochondrial 16S ribosomal DNA gene in the extracts from the respective animal species was determined by droplet digital PCR (ddPCR, QX200 Droplet Generator, QX200 Droplet Reader (Bio-Rad, Hercules, CA, USA)) using the Eva-Green Supermix. DNA extract mixtures were prepared at the AGES, by taking the copy numbers (pig (870 copies/μL), beef (1069 copies/μL), horse (1795 copies/μL), sheep (520 copies/μL), goat (790 copies/μL), chicken (620 copies/μL), turkey (673 copies/μL)), into account. The percentages of samples 1 to 6 given in Table 1 were calculated by relating the DNA copy number of the respective animal species to the total number of copies of animal species in the sample.

2.4. Study Design

The interlaboratory ring trial for validation of the DNA metabarcoding method for mammalian and poultry species [11] was conducted in the framework of the §64 LFGB working group "NGS Species Identification" under the coordination of AGES. Statistical data analysis was performed by QuoData GmbH (Dresden, Germany).

For sequencing, three benchtop NGS instruments from two companies were used. Benchtop instruments from Illumina (Illumina, San Diego, CA, USA) were employed by eleven laboratories, whereof eight used the MiSeq instrument, three the iSeq 100 instrument, and one participant used both the MiSeq and the iSeq 100 instrument. Four laboratories applied the Ion GeneStudio S5 instrument from Thermo Fisher Scientific (Thermo Fisher Scientific, Waltham, MA, USA).

Each participant obtained 16 anonymously labelled samples, comprising two sub-samples of each of the eight samples (Table 1). Sixteen samples were chosen to allow the iSeq 100 platform to be included in the ring trial. This also enabled the use of the most cost-effective MiSeq Reagent Micro Kit v2 for a small number of samples on the Illumina platforms. Participants directly used all individual DNA extracts for DNA library preparation and subsequently for amplicon sequencing on a next-generation sequencing instrument. Together with the "ready-to-use" DNA extracts, the participants obtained reagents for creating DNA libraries, a sequencing kit, and a step-by-step instruction.

In order to be able to perform sequencing on the Ion GeneStudio S5 instrument (Thermo Fisher Scientific, Waltham, MA, USA), the protocol for preparation of DNA libraries and sequencing published previously by Dobrovolny et al. (2019) [11] had to be adapted as follows. Each of the two forward primers were elongated by a 3 bp barcode adapter, one of 16 different 10 bp barcodes (index sequence) and the overhang adapter sequence (A adapter). Each of the two reverse primers was linked to an overhang adapter sequence (trP1 adapter). The PCR setup did not include additional magnesium chloride solution. In the magnetic bead cleaning step, a total of 37.5 µL Agencourt® AMPure® XP beads (Beckman Coulter, Brea, CA, USA) was used and the DNA was eluted with 50 µL Tris-EDTA (TE) buffer. The average library size was 190 bp, and all DNA libraries were adjusted to 100 pM and were mixed together in a single 1.5 mL tube. A 25 pM DNA pool was used for sequencing. In general, any deviations from the protocols had to be reported by the participants.

Paired-end sequencing on an Illumina instrument was performed using either the MiSeq Reagent Micro Kit v2 (300-cycles) or the iSeq 100 i1 Reagent v2 (300-cycles), which included a 5% PhiX spike-in. The Ion Chef instrument was used with the Ion 510^{TM} and Ion 520^{TM} and Ion 530^{TM} Kit-Chef and the Ion 520^{TM} Chip Kit to perform template preparation, enrichment and chip loading. Finally, the sequencing reaction was started on the Ion GeneStudio S5 instrument.

To obtain information about the presence of the animal species in the samples, the sequencing output in FastQ format was processed with a multi-step analysis pipeline by using Galaxy (version 19.01) as described previously [12]. Before the resulting FastQ files were used as input for the data analysis, the raw binary base call (bcl) files generated by Illumina devices were converted to text files using the conversion software bcl2fastq2-v2.19.0.316 (Illumina, San Diego, CA, USA). The default demultiplexing option of one allowed mismatch in the barcode recognition of the Illumina software (-- barcode mismatches) was thereby set to zero (value: 0) and the step was also integrated into the pipeline. Preliminary tests had shown that this can increase the quality of index recognition. The Thermo Fisher instrument software uses this setting by default. The analysis pipeline for sequencing data of the Thermo Fisher Scientific (Waltham, MA, USA) platforms was modified because paired-end FastQ files do not exist in this case. Consequently, the primer sequences were adapted according to the requirements of the analysis tool Cutadapt (Galaxy version 1.16.6 [16]) and the tool fastq-join (Galaxy version 1.1.2-806.1) [17] was removed. Dereplicated reads were directly matched against a customized database (AGES database) including 51 mitochondrial genomes from animals (Supplementary Table S1) and the public databases provided by NCBI using BLASTn [18]. The AGES database contains verified sequences from the NCBI database exclusively from food-relevant animal species. This reduces the time needed for alignment and is intended to avoid nonsense results. For each of the samples, results were listed automatically in a table according to taxonomy and abundance and a formula calculated the proportions of animal species by relating the number of reads for the respective species to the number of total reads (after pipeline) across all animal species obtained for the subsample. For further statistical analysis, all Excel spreadsheets were sent to QuoData GmbH.

2.5. Statistical Data Analysis

Statistical analyses were performed by QuoData GmbH. Even though the DNA metabarcoding method used in this interlaboratory comparison is commonly regarded as a qualitative method, the underlying decision process is based on the comparison of a quantitative value, namely the proportion of a single species, with a specific threshold. The performance of such a method can be assessed both on the basis of the qualitative result (yes/no) and on the basis of the underlying quantitative data. Because the information content of the quantitative data can be far greater than the corresponding qualitative data, the quantitative data were used in addition to the qualitative data to describe the performance of the DNA metabarcoding method.

In addition, the study of quantitative data also aimed to verify the extent to which this method can also be used for quantitative determinations.

2.5.1. Quantitative Statistical Analyses

Proportions of animal species ranged between 0.1% and 94%. To avoid asymmetric distributions for proportions near 0% and 100%, and to ensure equality of variances for the individual combinations of samples/animal species, the proportions were subjected to a logit transformation:

$$\mathrm{logit(proportion)} = \ln \left(\frac{\mathrm{proportion}}{1 - \mathrm{proportion}} \right)$$

The logit-transformed proportions can be retransformed as follows:

$$\mathrm{proportion} = \frac{e^{\mathrm{logit(proportion)}}}{1 + e^{\mathrm{logit(proportion)}}}$$

The logit-transformed proportions were then subjected to several outlier tests. Data were checked for systematic errors across samples and/or animal species affecting the mean values (Mandel h statistics) and/or variances (Mandels k statistics). In addition, the occurrence of sample- and animal species-specific outliers regarding the laboratory mean values and variances was tested for by means of the Grubbs and the Cochran tests (significance level 1%), respectively. Proportions identified as outliers were excluded from further statistical analyses.

Logit-transformed and outlier-cleaned data were checked for normal distribution using the Shapiro–Wilk test. Then, repeatability, reproducibility, and accuracy of the proportions of animal species were determined according to the criteria of QuoData certified with the aid of the software solution for method comparison studies and interlaboratory comparison studies PROLab Plus, version 2021.7.22.0 [19] (QuoData, Dresden, Germany), using the statistical methods according to DIN ISO 5725-2 and according to the specifications in the Official Collection of Test Methods ASU §64 LFGB for the statistical evaluation of ring trials for method validation [20]. Taking into account the obtained repeatability and reproducibility standard deviations for samples 1–7, variance functions describing the functional relationship between standard deviations and overall mean for the individual combinations of samples/animal species were modelled.

For each of the combinations, the bias (difference) between this overall mean and the proportion of the animal species added to the sample was also determined. Furthermore, the standard deviation of this bias was calculated.

Prediction profiles and measurement uncertainty profiles were constructed, both not considering the bias (based on reproducibility standard deviations), and considering the bias (based on reproducibility standard deviations as well as on the standard deviation of the bias).

In addition, the z scores for each combination of lab/sample/animal species were determined, providing a measure for the standardized deviations of laboratory mean values from the respective overall mean value.

2.5.2. Qualitative Statistical Analyses

A sample was classified as false positive if for at least one animal species that had not been added, a proportion above a defined threshold was obtained. By contrast, a sample was considered false negative for a specific animal species if the proportion was below a defined threshold for this species.

The probability of detection for an arbitrary animal at a defined threshold for (1) a laboratory with average performance, (2) a laboratory with positive bias, and (3) a laboratory with negative bias was determined based on the variance functions by the quantitative statistical analysis.

3. Results and Discussion

Fourteen of the fifteen laboratories submitted their sequencing results in time. Ten laboratories applied the Illumina platform, with seven laboratories (01, 02, 03, 04, 06, 08, 14) using the MiSeq, two laboratories (07, 13) the iSeq 100, and one laboratory utilizing both the MiSeq and the iSeq 100 (referred to as "laboratory 15" and "laboratory 20", respectively). The remaining four laboratories (09, 10, 11, 12) applied the Ion GeneStudio S5 system from Thermo Fisher Scientific.

Each of the laboratories submitted 16 sequencing results in total (eight samples, two subsamples each), with the exception of laboratories 07, 12, and 13. Laboratory 07 did not provide the result for subsample 8B (negative control), whereas datasets submitted by laboratories 12 and 13 were lacking results for both subsamples of sample 8. With the exception of laboratory 03, sequencing was done by using the test kit provided by the AGES. The suitability of the MiSeq Reagent Kit v2 applied by laboratory 03 had been demonstrated in preliminary experiments.

FastQ data provided by the participating laboratories was evaluated by the AGES by using the analysis pipeline in Galaxy. For identification of animal species, the DNA sequences (reads) were aligned, once with the customized AGES database and once with the NCBI database. Supplementary Table S2 summarizes the total number of reads for each laboratory, taking into account the results obtained for each of the fourteen subsamples containing animal species (samples 1–7).

Total numbers of raw reads that passed the analysis pipeline were quite different between laboratories. Very low total numbers of raw reads can, for example, be caused by errors during wet-lab activities, e.g., pipetting errors or error rate of DNA polymerase. In addition, problems with adapter- and index-recognition are known to have an impact. Loss of reads or their elimination by pipeline tools can also be caused by errors in PCR amplification of the library (e.g., index hopping), sequencing errors (e.g., inserts, substitutions, or deletions) or insufficient cluster resolution [21]. All these errors may affect the quality of raw data (FastQ file) and thus the number of DNA sequences (reads) after analysis pipeline.

Total numbers of reads obtained with the Ion GeneStudio S5 were significantly higher than those obtained with the MiSeq ($p < 0.001$) and the iSeq 100 ($p < 0.001$). Differences observed between the Illumina and the Thermo Fisher technology are caused by differences in data filtering. The instrument-specific software of the Ion GeneStudio S5 removes datasets of lower quality by filtering before starting the analysis pipeline. Thus, considerably more sequences remain after primary data analysis compared to the instruments from Illumina.

Differences in recoveries, by relating the total number of reads to the number of raw reads before analysis pipeline, between laboratories using the same instrument type (MiSeq, iSeq 100, or Ion GeneStudio S5, Table S3) hint at differences in the quality of the sequencing run and unintended loss of reads. Significant differences ($p < 0.001$) in recoveries between laboratories using Illumina instruments and those applying the Ion GeneStudio S5 were, however, expected. These differences are caused by the fact that the pipeline of the Ion GeneStudio S5 neither included paired-end sequencing nor a "joining step" as was the case with the Illumina platforms.

3.1. Quantitative Evaluation of Ring Trial Data

The aim of quantitative evaluation of ring trial data was to determine average proportions of the animal species that had been added to samples 1–7, and to identify resulting error components within and in between laboratories.

3.1.1. Proportions of Animal Species in Samples 1–7

Proportions of animal species were calculated by relating the number of reads for the respective species to the number of total reads (after pipeline) across all animal species obtained for the sample. Table 2 gives the proportions of animal species determined for sample 1 containing seven animal species (Table 2a), sample 2 (as a representative of samples consisting of five animal species; Table 2b) and sample 7, a model food sample (Table 2c). Results for samples 3, 4, 5, and 6 are shown as stacked bar plots (Figure 1).

Preliminary evaluation of the results indicated that the proportions of animal species determined considerably depended on the sequencing platform/technology applied. Due to the low number of laboratories using the sequencing technology from Thermo Fisher Scientific, only results obtained by laboratories applying Illumina platforms were included into statistical evaluation. Results obtained by laboratories 09–12 using the Ion GeneStudio S5 are only shown for comparison.

3.1.2. Logit Transformation

Proportions of animal species in samples 1–7 were quite different, ranging from 0.1% to 94%. However, a prerequisite for the evaluation of ring trial data according to ASU §64 LFGB is that the proportions of animal species follow normal or at least symmetric distribution. In order to allow assumption of normal distribution and ensure equality of variances of the individual combinations of samples/animal species (after elimination of outliers), proportions of animal species were subjected to a logit transformation. The logit is the logarithm of the proportion of the animal species divided by 1 minus the proportion of the animal species. Proportions of, e.g., 0.1%, 1%, 10%, 30%, 50%, and 70%, resulted in logit values of -6.91, -4.60, -2.20, -0.85, 0, and 0.85, respectively. Since the logit for 0% and 100% is not defined, it was set to surrogate values of -15 and $+15$, respectively.

3.1.3. Outlier Tests

In the course of evaluating ring trial data according to ASU §64 LFGB, logit-transformed proportions of animal species were subjected to several outlier tests (see also Section 2.5.1). Table 3 summarizes the outliers and reasons for their elimination.

Table 2. Proportion of animal species determined for samples 1 (**a**), 2 (**b**), and 7 (**c**). A and B refer to subsamples A and B, respectively.

(a). Sample 1

Species	Spiking Level (%)	Laboratory	AGES Database		NCBI Database		Laboratory	AGES Database		NCBI Database	
			A	B	A	B		A	B	A	B
Pig	94		91.25	91.77	91.21	91.74		95.58	94.83	95.56	94.81
Chicken	1		1.42	1.37	1.42	1.37		0.73	1.08	0.73	1.08
Horse	1		1.42	1.34	1.42	1.33		0.60	0.67	0.60	0.66
Turkey	1	01	1.17	1.03	1.16	1.02	10	0.49	0.66	0.41	0.56
Beef	1		1.55	1.51	1.55	1.51		0.57	0.58	0.57	0.58
Sheep	1		1.65	1.57	1.62	1.56		1.05	1.15	1.05	1.15
Goat	1		1.54	1.42	1.56	1.44		0.97	1.03	0.98	1.04
Pig	94		90.31	90.33	90.29	90.30		94.62	94.90	94.59	94.87
Chicken	1		1.50	1.57	1.50	1.57		0.47	0.30	0.47	0.30
Horse	1		1.64	1.59	1.64	1.59		1.01	1.09	1.00	1.09
Turkey	1	02	1.32	1.41	1.31	1.41	11	0.54	0.43	0.51	0.40
Beef	1		1.77	1.67	1.77	1.67		1.02	1.03	1.02	1.03
Sheep	1		1.81	1.73	1.79	1.72		1.19	1.16	1.19	1.16
Goat	1		1.66	1.69	1.68	1.71		1.15	1.10	1.16	1.11
Pig	94		92.67	92.48	92.65	92.46		94.30	94.68	94.27	94.65
Chicken	1		1.04	1.11	1.04	1.11		0.56	0.36	0.57	0.36
Horse	1		1.04	1.14	1.04	1.14		1.10	1.13	1.10	1.13
Turkey	1	03	0.90	0.90	0.89	0.89	12	0.66	0.48	0.64	0.47
Beef	1		1.60	1.55	1.60	1.55		0.98	1.03	0.98	1.04
Sheep	1		1.44	1.48	1.43	1.46		1.24	1.19	1.24	1.19
Goat	1		1.31	1.34	1.32	1.35		1.16	1.13	1.16	1.14
Pig	94		92.08	92.28	92.06	92.25		92.01	91.80	91.98	91.77
Chicken	1		1.16	1.21	1.16	1.21		1.27	1.33	1.27	1.33
Horse	1		1.22	1.14	1.22	1.14		1.35	1.38	1.35	1.38
Turkey	1	04	1.09	1.00	1.09	0.99	13	1.22	1.17	1.22	1.16
Beef	1		1.56	1.56	1.56	1.56		1.45	1.52	1.45	1.52
Sheep	1		1.45	1.49	1.45	1.48		1.38	1.37	1.38	1.37
Goat	1		1.43	1.33	1.44	1.34		1.32	1.43	1.34	1.44

Table 2. *Cont.*

(a). Sample 1

Species	Spiking Level (%)	Laboratory	AGES Database		NCBI Database		Laboratory	AGES Database		NCBI Database	
			A	B	A	B		A	B	A	B
Pig	94	06	90.36	90.41	90.33	90.38	14	90.44	93.48	90.41	93.46
Chicken	1		1.51	1.53	1.51	1.53		1.57	0.97	1.57	0.97
Horse	1		1.52	1.56	1.51	1.56		1.32	0.92	1.32	0.92
Turkey	1		1.36	1.33	1.36	1.32		1.14	0.82	1.13	0.82
Beef	1		1.87	1.84	1.87	1.84		2.17	1.50	2.17	1.50
Sheep	1		1.72	1.67	1.71	1.66		1.81	1.24	1.80	1.24
Goat	1		1.66	1.66	1.68	1.67		1.56	1.06	1.58	1.07
Pig	94	07	92.48	92.19	92.47	92.17	15	92.22	92.23	92.18	92.20
Chicken	1		1.08	1.17	1.08	1.17		1.19	1.18	1.19	1.18
Horse	1		1.21	1.27	1.21	1.27		1.14	1.21	1.14	1.21
Turkey	1		1.03	1.10	1.02	1.10		0.98	0.96	0.98	0.96
Beef	1		1.51	1.50	1.51	1.50		1.60	1.55	1.60	1.55
Sheep	1		1.36	1.38	1.37	1.38		1.48	1.45	1.47	1.44
Goat	1		1.32	1.39	1.32	1.40		1.39	1.41	1.41	1.42
Pig	94	08	91.31	91.28	91.28	91.25	20	92.59	92.63	92.58	92.61
Chicken	1		1.34	1.40	1.34	1.40		1.03	1.09	1.03	1.09
Horse	1		1.40	1.32	1.40	1.32		1.24	1.28	1.24	1.28
Turkey	1		1.28	1.23	1.27	1.23		1.11	1.03	1.11	1.02
Beef	1		1.68	1.62	1.68	1.62		1.44	1.39	1.44	1.39
Sheep	1		1.53	1.59	1.52	1.57		1.31	1.30	1.31	1.29
Goat	1		1.46	1.56	1.48	1.58		1.27	1.28	1.28	1.29
Pig	94	09	94.62	93.61	94.60	93.59					
Chicken	1		0.64	1.07	0.64	1.08					
Horse	1		1.06	1.19	1.06	1.19					
Turkey	1		0.63	1.10	0.61	1.07					
Beef	1		0.92	0.90	0.92	0.90					
Sheep	1		1.10	1.05	1.10	1.05					
Goat	1		1.04	1.07	1.05	1.08					

Table 2. *Cont.*

			(b). Sample 2									
Species	Spiking Level (%)	Laboratory	AGES Database		NCBI Database		Laboratory	AGES Database		NCBI Database		
			A	B	A	B		A	B	A	B
Beef	65.7		64.59	65.91	64.42	65.73		53.26	55.43	53.14	55.32
Pig	30		30.04	29.36	29.93	29.24		41.44	40.09	41.25	39.91
Horse	1.9	01	2.24	1.97	2.23	1.96	10	2.21	1.63	2.19	1.61
Sheep	1.9		2.61	2.33	2.58	2.30		2.89	2.71	2.88	2.70
Turkey	0.5		0.46	0.40	0.45	0.39		0.19	0.13	0.16	0.11
Beef	65.7		62.99	63.18	62.83	63.01		66.01	65.78	65.89	65.68
Pig	30		31.30	31.24	31.20	31.12		29.52	29.75	29.40	29.63
Horse	1.9	02	2.49	2.44	2.48	2.43	11	2.03	2.00	2.02	1.98
Sheep	1.9		2.60	2.56	2.57	2.53		2.16	2.27	2.15	2.26
Turkey	0.5		0.61	0.54	0.60	0.53		0.28	0.20	0.26	0.18
Beef	65.7		71.67	70.55	71.48	70.36		66.29	64.97	66.22	64.88
Pig	30		24.75	25.50	24.68	25.43		29.13	30.25	28.98	30.08
Horse	1.9	03	1.37	1.44	1.36	1.44	12	2.08	2.23	2.06	2.21
Sheep	1.9		1.90	2.16	1.88	2.13		2.23	2.31	2.22	2.30
Turkey	0.5		0.30	0.33	0.29	0.33		0.27	0.24	0.25	0.23
Beef	65.7		67.22	67.55	67.05	67.39		62.21	61.73	62.00	61.52
Pig	30		28.26	28.10	28.17	28.01		32.37	32.75	32.26	32.64
Horse	1.9	04	1.90	1.76	1.89	1.75	13	2.52	2.55	2.51	2.54
Sheep	1.9		2.18	2.16	2.17	2.14		2.31	2.31	2.30	2.30
Turkey	0.5		0.40	0.37	0.39	0.37		0.56	0.62	0.55	0.62
Beef	65.7		66.46	65.45	66.28	65.24		70.20	68.65	69.99	68.47
Pig	30		28.63	29.35	28.54	29.24		26.07	27.18	25.99	27.11
Horse	1.9	06	2.09	2.10	2.08	2.09	14	1.44	1.63	1.43	1.62
Sheep	1.9		2.32	2.54	2.31	2.52		1.94	2.15	1.92	2.13
Turkey	0.5		0.47	0.52	0.47	0.52		0.32	0.36	0.32	0.36
Beef	65.7		65.64	66.96	65.46	66.77		69.77	67.63	69.61	67.48
Pig	30		29.78	28.89	29.68	28.80		26.43	27.98	26.36	27.90
Horse	1.9	07	1.93	1.73	1.92	1.72	15	1.49	1.77	1.48	1.76
Sheep	1.9		2.17	2.04	2.17	2.03		1.96	2.20	1.93	2.16
Turkey	0.5		0.47	0.37	0.47	0.37		0.31	0.37	0.31	0.37

Table 4. Statistical parameters for relative proportions of animal species according to ASU §64 LFGB. AGES: AGES database, NCBI: NCBI database.

Species	Parameter [1]	Sample 1 AGES	1 NCBI	2 AGES	2 NCBI	3 AGES	3 NCBI	4 AGES	4 NCBI	5 AGES	5 NCBI	6 AGES	6 NCBI	7 AGES	7 NCBI
Chicken	Number of labs	11	11			11	11			11	11	11	11	11	11
	Number of labs after outlier elimination	10	10			11	11			9	9	10	10	11	11
	Proportion (v/m)	1%				1.9%				67.5%		0.1%		5%	
	Mean value	1.27%	1.26%			2.12%	2.11%			55.49%	55.34%	0.15%	0.15%	6.89%	6.88%
	s_R	0.18%	0.18%			0.23%	0.23%			0.84%	0.84%	0.03%	0.03%	0.72%	0.71%
	s_r	0.04%	0.04%			0.08%	0.08%			0.57%	0.57%	0.01%	0.01%	0.24%	0.24%
	Logit proportion	−4.60	−4.60			−3.94	−3.94			0.73	0.73	−6.91	−6.91	−2.94	−2.94
	Logit mean value	−4.36	−4.36			−3.83	−3.83			0.22	0.21	−6.53	−6.53	−2.60	−2.61
	Logit s_R	0.14	0.14			0.11	0.11			0.03	0.03	0.21	0.21	0.11	0.11
	Logit s_r	0.03	0.03			0.04	0.04			0.02	0.02	0.05	0.05	0.04	0.04
Horse	Number of labs	11	11	11	11	11	11	11	11			11	11		
	Number of labs after outlier elimination	10	10	11	11	11	11	11	11			11	11		
	Proportion (v/m)	1%		1.9%		66.1%		0.5%				30%			
	Mean value	1.31%	1.31%	1.89%	1.89%	60.3%	60.0%	0.47%	0.47%			42.0%	41.9%		
	s_R	0.17%	0.17%	0.37%	0.37%	1.1%	1.1%	0.08%	0.08%			1.9%	1.9%		
	s_r	0.05%	0.05%	0.14%	0.14%	0.3%	0.3%	0.04%	0.04%			0.7%	0.7%		
	Logit proportion	−4.60	−4.60	−3.94	−3.94	0.67	0.67	−5.29	−5.29			−0.85	−0.85		
	Logit mean value	−4.32	−4.32	−3.95	−3.95	0.42	0.40	−5.35	−5.36			−0.32	−0.33		
	Logit s_R	0.13	0.13	0.20	0.20	0.05	0.05	0.18	0.17			0.08	0.08		
	Logit s_r	0.04	0.04	0.08	0.08	0.01	0.01	0.08	0.08			0.03	0.03		
Turkey	Number of labs	11	11	11	11	11	11					11	11	11	11
	Number of labs after outlier elimination	10	10	11	11	11	11					11	11	11	11
	Proportion (v/m)	1%		0.5%		1.9%						67.5%		5%	
	Mean value	1.12%	1.12%	0.42%	0.42%	1.86%	1.85%					54.0%	53.8%	4.03%	4.02%
	s_R	0.16%	0.16%	0.09%	0.09%	0.19%	0.19%					2.1%	2.0%	0.46%	0.46%
	s_r	0.05%	0.05%	0.04%	0.04%	0.07%	0.07%					0.9%	0.9%	0.14%	0.14%
	Logit proportion	−4.60	−4.60	−5.29	−5.29	−3.94	−3.94					0.73	0.73	−2.94	−2.94
	Logit mean value	−4.48	−4.48	−5.47	−5.48	−3.97	−3.97					0.16	0.15	−3.17	−3.17
	Logit s_R	0.14	0.14	0.22	0.22	0.10	0.10					0.08	0.08	0.12	0.12
	Logit s_r	0.05	0.05	0.10	0.10	0.04	0.04					0.04	0.04	0.04	0.04

Table 4. *Cont.*

Species	Parameter [1]	Sample														
		1		2		3		4		5		6		7	
		AGES	NCBI	AGES	NCBI	AGES	NCBI	AGES	NCBI	AGES	NCBI	AGES	NCBI	AGES	NCBI
Beef	Number of labs	11	11	11	11			11	11	11	11	11	11	11	11
	Number of labs after outlier elimination	10	10	11	11			11	11	11	11	11	11	11	11
	Proportion (v/m)	1%		65.7%				30%		1.9%		0.5%		50%	
	Mean value	1.58%	1.58%	66.9%	66.7%			33.02%	32.87%	2.79%	2.79%	0.91%	0.91%	48.7%	48.7%
	s_R	0.13%	0.13%	2.8%	2.8%			0.55%	0.56%	0.20%	0.20%	0.09%	0.09%	2.6%	2.6%
	s_r	0.04%	0.04%	1.0%	1.0%			0.31%	0.31%	0.10%	0.10%	0.05%	0.05%	0.4%	0.4%
	Logit proportion	−4.60	−4.60	0.65	0.65			−0.85	−0.85	−3.94	−3.94	−5.29	−5.29	0	0
	Logit mean value	−4.13	−4.13	0.70	0.69			−0.71	−0.71	−3.55	−3.55	−4.69	−4.69	−0.05	−0.05
	Logit s_r	0.08	0.08	0.13	0.13			0.02	0.03	0.07	0.07	0.10	0.10	0.11	0.10
	Logit s_r	0.02	0.02	0.04	0.04			0.01	0.01	0.04	0.04	0.06	0.06	0.02	0.02
Sheep	Number of labs	11	11	11	11	11	11	11	11	11	11				
	Number of labs after outlier elimination	10	10	11	11	9	9	11	11	9	9				
	Proportion (v/m)	1%		1.9%		0.5%		67.5%		30%					
	Mean value	1.50%	1.49%	2.21%	2.19%	0.14%	0.14%	64.41%	63.67%	40.92%	40.62%				
	s_R	0.15%	0.14%	0.22%	0.22%	0.01%	0.01%	0.65%	0.60%	0.62%	0.65%				
	s_r	0.03%	0.03%	0.14%	0.13%	0.01%	0.01%	0.27%	0.27%	0.49%	0.46%				
	Logit proportion	−4.60	−4.60	−3.94	−3.94	−5.29	−5.29	0.73	0.73	−0.85	−0.85				
	Logit mean value	−4.18	−4.19	−3.79	−3.80	−6.58	−6.60	0.59	0.56	−0.37	−0.38				
	Logit s_r	0.10	0.10	0.10	0.10	0.10	0.09	0.03	0.03	0.03	0.03				
	Logit s_r	0.02	0.02	0.06	0.06	0.08	0.08	0.01	0.01	0.02	0.02				
Pig	Number of labs	11	11	11	11			11	11	11	11	11	11	11	11
	Number of labs after outlier elimination	10	10	11	11			10	10	11	11	11	11	11	11
	Proportion (v/m)	94%		30%				0.1%		0.5%		1.9%		40%	
	Mean value	91.77%	91.74%	28.5%	28.4%			0.04%	0.04%	0.61%	0.61%	2.90%	2.90%	40.2%	40.1%
	s_R	0.83%	0.83%	2.2%	2.2%			0.02%	0.02%	0.12%	0.12%	0.26%	0.26%	2.3%	2.3%
	s_r	0.16%	0.16%	0.7%	0.7%			0%	0%	0.04%	0.04%	0.13%	0.13%	0.4%	0.4%
	Logit proportion	2.75	2.75	−0.85	−0.85			−6.91	−6.91	−5.29	−5.29	−3.94	−3.94	−0.41	−0.41
	Logit mean value	2.41	2.41	−0.92	−0.92			−7.73	−7.74	−5.09	−5.10	−3.51	−3.51	−0.40	−0.40
	Logit s_r	0.11	0.11	0.11	0.11			0.41	0.41	0.19	0.19	0.09	0.09	0.10	0.10
	Logit s_r	0.02	0.02	0.03	0.03			0.10	0.10	0.07	0.07	0.05	0.05	0.02	0.02

Table 4. *Cont.*

Species	Parameter [1]	Sample													
		1		2		3		4		5		6		7	
		AGES	NCBI	AGES	NCBI	AGES	NCBI	AGES	NCBI	AGES	NCBI	AGES	NCBI	AGES	NCBI
Goat	Number of labs	11	11			11	11	11	11	11	11				
	Number of labs after outlier elimination	10	10			11	11	11	11	11	11				
	Proportion (v/m)	1%				30%		1.9%		0.1%					
	Mean value	1.44%	1.45%			35.4%	35.6%	2.00%	2.01%	0.09%	0.09%				
	s_R	0.14%	0.14%			1.0%	1.0%	022%	0.23%	0.02%	0.02%				
	s_r	0.05%	0.05%			0.3%	0.3%	0.10%	0.10%	0.02%	0.02%				
	Logit proportion	−4.60	−4.60			0.85	−0.85	−3.94	−3.94	−6.91	−6.91				
	Logit mean value	−4.23	−4.22			−0.60	−0.59	−3.89	−3.89	−7.05	−6.99				
	Logit s_r	0.10	0.10			0.04	0.05	0.11	0.11	0.19	0.19				
	Logit s_r	0.04	0.04			0.01	0.01	0.05	0.05	0.19	0.17				

1 s_R: reproducibility standard deviation; s_r: repeatability standard deviation; v: volume; m: mass.

As expected, the reproducibility standard deviation and repeatability standard deviation were found to be higher for lower proportions of animal species than for proportions of about 50% (logit = 0), independent of the database selected for alignment. It was found that pig tends to have higher standard deviations in reproducibility, but not in repeatability, compared to other animal species (Figure 2A). A tendency towards higher reproducibility standard deviations was also observed in case beef was the predominant animal species in the sample. This also held true for the repeatability standard deviation, although to a lower extent.

The bias between the proportion of the animal species added to the sample and the overall mean determined in the ring trial was in the range from −0.6 to 0.6 logits, with just two exceptions. Neither the animal species nor the proportion of the animal species was found to have a systematic effect on the bias. The animal species and the proportion of the animal species did not have a systematic impact on the reproducibility standard deviation either.

Thus, the standard deviation induced by the bias ("bias standard deviation"), absolute reproducibility standard deviation, and absolute repeatability standard deviation could be modeled across animal species and samples, for both the AGES and NCBI databases. The modeled variance function was similar for both databases. The lowest bias standard deviation, reproducibility standard deviation, and repeatability standard deviation were found for a proportion of 50% (logit = 0). The closer the proportion to 0% or 100%, the higher the standard deviations. Table 5 summarizes the modeled and re-transformed standard deviations, which were found to be independent of the database used for alignment.

Table 5. Bias standard deviation, reproducibility standard deviation, and repeatability standard deviation (absolute, i.e., retransformed to proportions of animal species) depending on the proportion of animal species.

Standard Deviation	Proportion of Animal Species	
	5%/95%	50%
Absolute bias standard deviation	1.8%	7.2%
Absolute reproducibility standard deviation	0.5%	1.8%
Absolute repeatability standard deviation	0.2%	0.6%

3.1.7. Variability across Animal Species and Measuring Uncertainty

To allow predictions for further analyses, the measuring uncertainty was evaluated. Since the database (AGES or NCBI) was not found to have an impact on bias standard deviation, reproducibility standard deviation, or repeatability standard deviation, the measuring uncertainty was only evaluated for the AGES database, representative for both databases.

Based on the reproducibility standard deviation, a prediction profile was established in terms of the 95% confidence interval of the results of all laboratories across all animal species. In addition, the 95% confidence interval was established by considering both the reproducibility standard deviation and the bias standard deviation.

The upper part of Figure 3 shows the 95% confidence interval of the (outlier-cleaned) results depending on the respective overall mean value of the proportion of an animal species (without considering the bias). The left side of the figure shows the entire range, and the right side an enlarged view of proportions from 0 to 10%. The figure indicates that for a "true proportion" (assuming that the overall mean across laboratories applying Illumina platforms reflects the "true proportion") of, e.g., 5%, the 95% confidence interval is 4.1–6.2%. In total, 4.7% of the individual values are outside the 95% confidence interval.

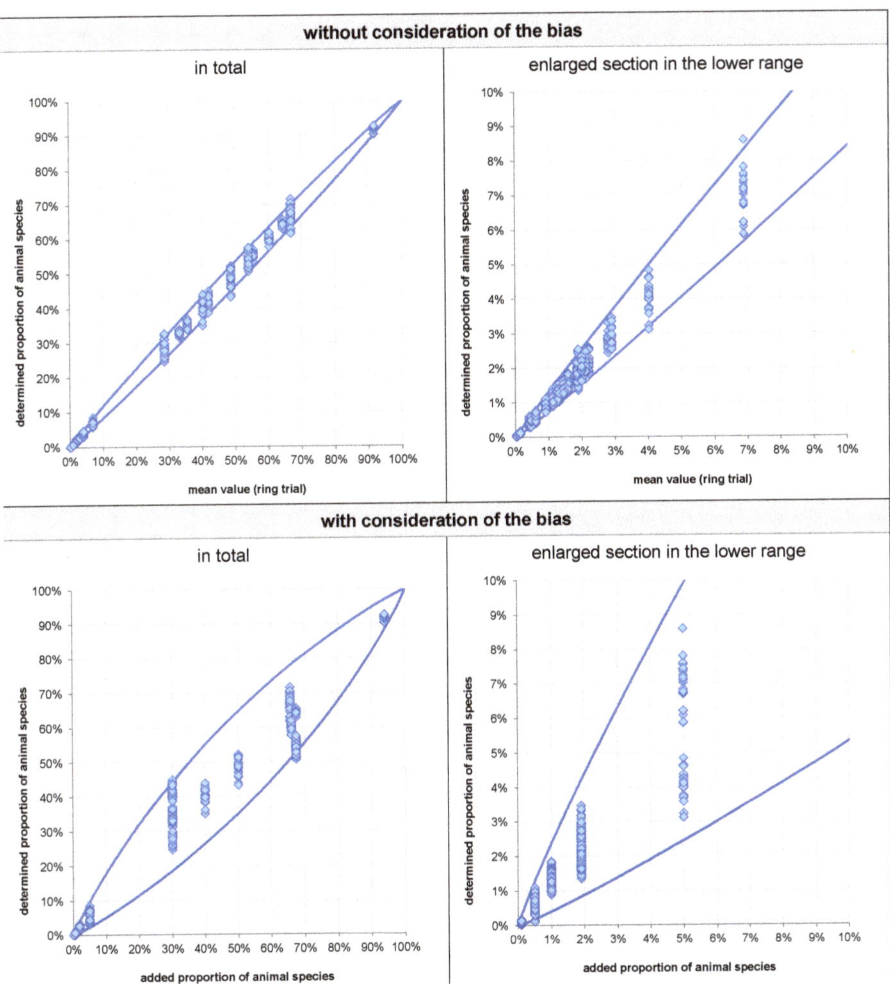

Figure 3. 95% confidence interval for the mean (**top**) and added (**bottom**) proportion of animal species based on a single measurement, independent of the animal species.

The 95% confidence interval of the (outlier-cleaned) results of all laboratories, depending on the respective proportion added (by considering the bias) is shown in the lower part of Figure 3. The left and right sides show the entire range and an enlarged view of proportions from 0 to 10%, respectively. For example, for an added proportion of 5%, the 95% confidence interval is 2.5–9.9%. In total, 3.7% of the individual values are outside the 95% confidence interval.

From the prediction profiles, a measurement uncertainty profile was established, indicating how far the proportion determined may deviate from the "true" proportion of the animal species. Figure 4 shows the 95% measurement uncertainty intervals depending on the proportion of the animal species determined, on the left side without consideration of the bias (assuming that the "true" proportion equals the overall mean across laboratories applying Illumina platforms), on the right side under consideration of the bias (assuming that the "true" proportion equals the proportion of animal species added to the sample).

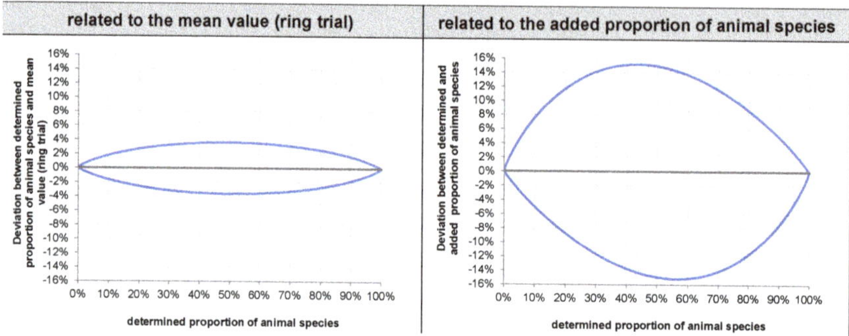

Figure 4. 95% measuring uncertainty interval for the proportion of the animal species determined, based on a single measurement, independent of the animal species.

For a 50% proportion of the animal species, the "true value" can be 3.6% (percentage points) lower or higher, if the bias is not taken into account, or even 15.2%, if the bias is taken into consideration.

3.1.8. z Scores

z scores were calculated for interlaboratory evaluation across samples and animal species (Figure 5). z scores measure standardized deviations of laboratory mean values from the overall mean value. An absolute z score > 2 hints at a statistically significant deviation of the respective laboratory.

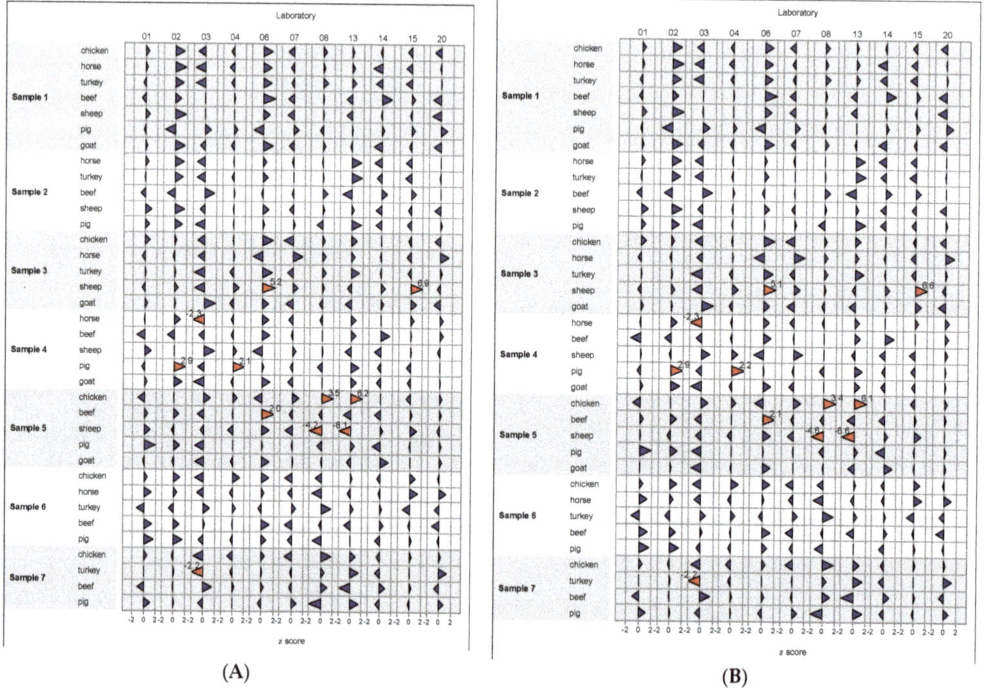

Figure 5. z scores for determination of proportions of animal species. (**A**) AGES database, (**B**) NCBI database. Absolute z scores < 2 are shown in blue, absolute z scores > 2 in red.

LASSO regression was applied to check whether absolute z scores depended on the instrument (MiSeq, iSeq 100), NGS experience of the respective laboratory, and/or activation of the function "adapter trimming" before starting the sequencing run. Analyses were performed excluding data previously identified as outliers. NGS experience of the laboratory was found to significantly affect the z score. For laboratories more experienced in NGS (laboratories 01, 04, 08, 15, 20), lower absolute z scores were determined compared to those with lower NGS experience. By contrast, neither the instrument (MiSeq, iSeq 100) nor activation of the function "adapter trimming" before starting the sequencing run was found to significantly affect the z score.

3.2. Qualitative Evaluation of Ring Trial Data

3.2.1. False Positive Rate

Next, it was investigated whether animal species were identified that had not been added to the samples, and whether the false positive rate depended on the database selected for alignment. A sample was classified as positive if for at least one animal species that had not been added, a proportion above a defined threshold was obtained. The threshold was set to 0.1%, 0.5%, 1.0%, or 2.0% proportion of the animal species.

Table 6 indicates that alignment against the AGES database (Table 6A) resulted in less false positive reads compared to alignment against the NCBI database (Table 6B).

Table 6. False positive reads obtained for samples 01–07. (**A**): AGES database, (**B**): NCBI database. Laboratories 01–06, 08, 14, 15: MiSeq; laboratories 07, 13, 20: iSeq 100.

(A)

Sample	Subsample	Laboratory										
		01	02	03	04	06	07	08	13	14	15	20
1	A	6	2	1	-	2	-	2	-	2	2	-
	B	2	5	-	3	3	-		-	1	4	-
2	A	99	41	21	95	78	6	76	41	44	87	5
	B	43	89	31	121	73	8	74	43	34	121	
3	A	49	325	14	1234	177	91	47	359	91	88	64
	B	51	369	40	1356	196	85	44	355	79	52	35
4	A	55	5	12	38	106	69	57	3	90	57	10
	B	38	34	20	64	117	37	80	10	100	67	10
5	A	6	51	2	286	173	121	56	124	271	79	65
	B	9	57	3	286	212	62	66	137	396	39	32
6	A	8	37	2	111	274	80	110	33	161	41	64
	B	5	50	7	102	314	46	106	50	169	71	27
7	A	45	188	42	608	99	62	69	250	61	62	39
	B	80	717	42	596	140	83	86	294	49	28	25

(B)

Sample	Subsample	Laboratory										
		01	02	03	04	06	07	08	13	14	15	20
1	A	104	75	122	49	72	40	65	20	48	72	19
	B	120	91	164	59	81	24	58	35	29	54	25
2	A	630	715	612	743	804	242	891	465	526	637	388
	B	489	783	1047	753	898	388	752	540	432	682	345
3	A	387	766	456	1639	497	394	209	583	355	297	242
	B	241	798	633	1792	606	383	229	621	365	177	127
4	A	2133	400	2738	1482	1908	1095	2214	670	1164	2428	585
	B	1584	2043	2896	1717	1941	1477	2182	837	1390	2585	668

Table 6. *Cont.*

		(B)										
		Laboratory										
Sample	Subsample	01	02	03	04	06	07	08	13	14	15	20
5	A	987	846	1963	1127	1259	1177	988	531	863	924	487
	B	1711	1179	2261	1063	1368	468	1202	628	1165	547	297
6	A	501	590	2354	634	796	584	556	437	525	371	261
	B	627	580	2619	584	853	394	518	440	612	543	200
7	A	250	452	496	788	322	210	426	459	244	229	175
	B	260	950	422	819	627	366	388	519	245	192	132

Alignment against the AGES database only yielded false positive samples at threshold values of 0.05% or 0.1% (Figure 6).

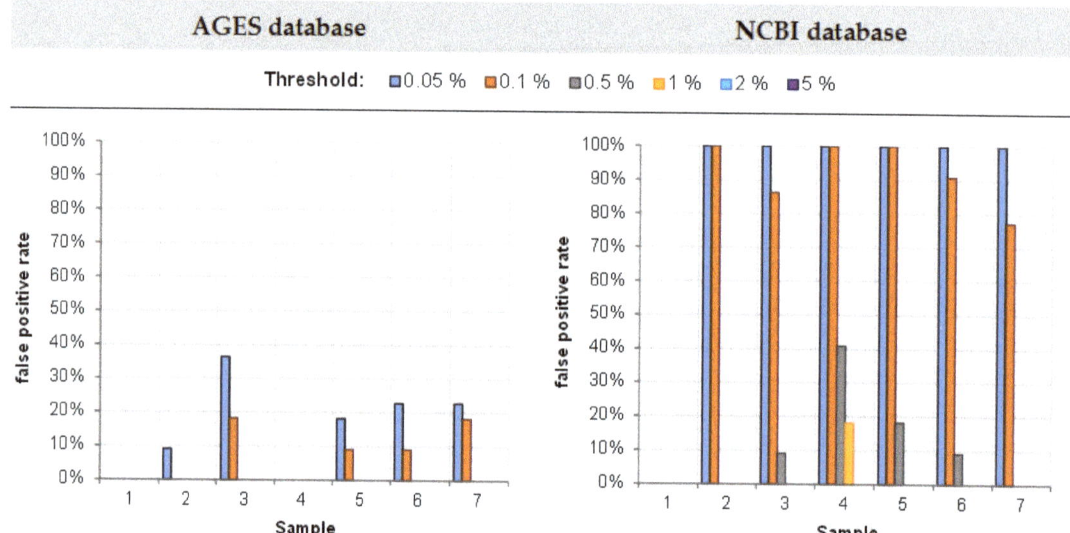

Figure 6. False positive rates by sample, depending on the defined threshold for the AGES (**left**) and the NCBI (**right**) database.

Higher false positive rates for the NCBI database were inevitably caused by the higher number of entries in the NCBI database compared to the AGES database. Most species resulting in false positive reads when using the NCBI database were not contained in the AGES database and thus could not be identified with the latter.

From the ring trial data it can be concluded that by using an appropriate customized database and by setting the threshold to 0.5%, false positive rates < 1% will be obtained.

3.2.2. False Negative Rate

Next, the false negative rate was evaluated at threshold values of 0.05%, 0.1%, 0.5%, and 1% for both the AGES and the NCBI databases (Table 7).

Table 7. Proportion of results below a thresholds of 0.05%, 0.1%, 0.5%, and 1%.

Sample	Species	Spiking Level (%)	Proportion of Results below a Threshold of							
			0.05%	0.1%	0.5%	1%	0.05%	0.1%	0.5%	1%
			AGES Database				NCBI Database			
1	Pig	94	-	-	-	-	-	-	-	-
	Chicken	1	-	-	-	1/22 (5%)	-	-	-	1/22 (5%)
	Horse	1	-	-	-	1/22 (5%)	-	-	-	1/22 (5%)
	Turkey	1	-	-	-	5/22 (23%)	-	-	-	6/22 (27%)
	Beef	1	-	-	-	-	-	-	-	-
	Sheep	1	-	-	-	-	-	-	-	-
	Goat	1	-	-	-	-	-	-	-	-
2	Beef	65.7	-	-	-	-	-	-	-	-
	Pig	30.0	-	-	-	-	-	-	-	-
	Horse	1.9	-	-	-	-	-	-	-	-
	Sheep	1.9	-	-	-	-	-	-	-	-
	Turkey	0.5	-	-	17/22 (77%)	22/22 (100%)	-	-	17/22 (77%)	22/22 (100%)
3	Horse	66.1	-	-	-	-	-	-	-	-
	Goat	30.0	-	-	-	-	-	-	-	-
	Chicken	1.9	-	-	-	-	-	-	-	-
	Turkey	1.9	-	-	-	-	-	-	-	-
	Sheep	0.5	-	-	22%22 (100%)	22/22 (100%)	-	-	22/22 (100%)	22/22 (100%)
4	Sheep	67.5	-	-	-	-	-	-	-	-
	Beef	30.0	-	-	-	-	-	-	-	-
	Goat	1.9	-	-	-	-	-	-	-	-
	Horse	0.5	-	-	13/22 (59%)	22/22 (100%)	-	-	14/22 (64%)	22/22 (100%)
	Pig	0.1	14/22 (64%)	20/22 (91%)	22/22 (100%)	22/22 (100%)	14/22 (64%)	20/22 (91%)	22/22 (100%)	22/22 (100%)
5	Chicken	67.5	-	-	-	-	-	-	-	-
	Sheep	30.0	-	-	-	-	-	-	-	-
	Beef	1.9	-	-	-	-	-	-	-	-
	Pig	0.5	-	-	4/22 (18%)	22/22 (100%)	-	-	4/22 (18%)	22/22 (100%)
	Goat	0.1	-	15/22 (68%)	22/22 (100%)	22/22 (100%)	-	12/22 (55%)	22/22 (100%)	22/22 (100%)

Table 7. *Cont.*

Sample	Species	Spiking Level (%)	Proportion of Results below a Threshold of							
			0.05%	0.1%	0.5%	1%	0.05%	0.1%	0.5%	1%
			AGES Database				NCBI Database			
6	Turkey	67.5	-	-	-	-	-	-	-	-
	Horse	30.0	-	-	-	-	-	-	-	-
	Pig	1.9	-	-	-	-	-	-	-	-
	Beef	0.5	-	-	-	19/22 (86%)	-	-	-	19/22 (86%)
	Chicken	0.1	-	-	22/22 (100%)	22/22 (100%)	-	-	22/22 (100%)	22/22 (100%)
7	Beef	50.0	-	-	-	-	-	-	-	-
	Pig	40.0	-	-	-	-	-	-	-	-
	Chicken	5.0	-	-	-	-	-	-	-	-
	Turkey	5.0	-	-	-	-	-	-	-	-

There was no considerable difference between the AGES and the NCBI databases regarding the proportion of false negative results obtained for samples 1–7. At a threshold of 0.05%, false negative results were only obtained for pig in sample 4. At a threshold of 0.1%, none of the combinations of sample–animal species led to false negative results, with the exception of proportions being close to the threshold (0.1%). The data indicate that at a threshold of ≥0.5%, the probability of obtaining false negative results is very low.

3.2.3. Probability of Detection (Qualitative Evaluation)

Figure 7 shows the probability of detection for three thresholds (0.1%, 0.5%, and 1%) and three scenarios, namely a laboratory with average performance, a laboratory with positive bias, and a laboratory with negative bias.

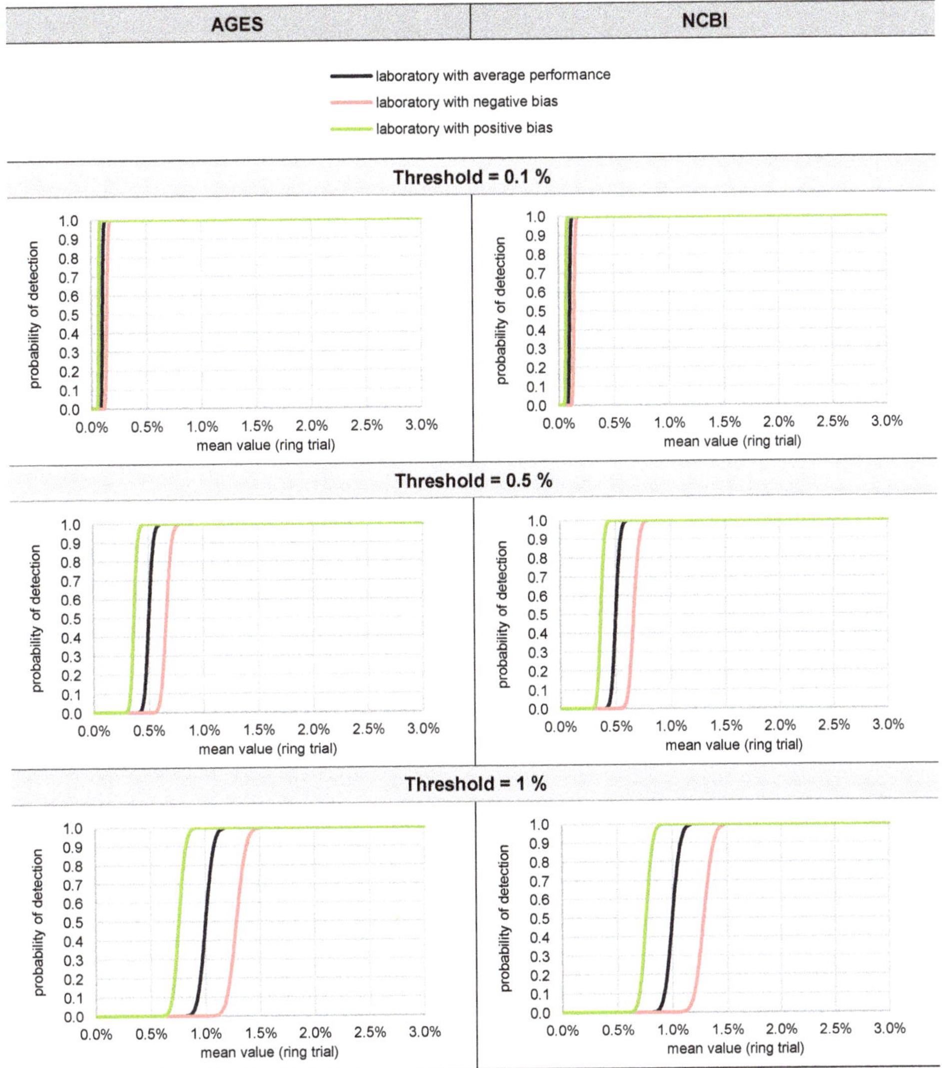

Figure 7. Probability of detection for a laboratory with average performance, a laboratory with positive bias, and a laboratory with negative bias, depending on the threshold and the database.

Figure 7 indicates that a threshold value of 1% seems to be a good compromise, provided that the variance function determined in the ring trial is equal to the actual variance function. A threshold value of 1% guarantees that a laboratory with a positive bias (overestimating the actual proportion) does not complain if the proportion of a certain animal species is 0.5%, whereas even a laboratory with a negative bias (underestimating the actual proportion) will be able to identify proportions >1.5% reliably.

3.3. Negative Control

Sample 8 was a DNA extract from maize, serving as a negative control. Since the marker system designed for mammals and poultry species does not detect maize, all reads that were obtained for sample 8 had to be regarded as false positive.

Table 8 lists the number of total reads (after pipeline) per laboratory and species. Laboratories 12 and 13 did not submit results for sample 8, laboratory 07 only provided results for one of the two subsamples.

In total, eight animal species were identified in the negative control by alignment against the AGES database. Fourteen further animal species, including the species *Homo sapiens* were identified, when the NCBI database was used. Per laboratory, up to five animal species were only identified with the NCBI database, with the exception of laboratory 06, which even detected eight additional animal species in subsample A.

In most cases, within a laboratory, the number of reads per animal species was similar for both subsamples. When the AGES database was used for alignment, most reads were assigned to beef and pig. Sample multiplexing in general, together with an inappropriate index layout, carries the risk of index misassignment. This is obviously the reason for the over-represented number of reads for pig and beef in the negative controls, as these animal species represent the main quantities in the samples. Although the index kit was used according to the manufacturer's instructions, the number of reads of these animal species could be reduced to the expected level in a supplementary experiment with an alternative index layout. Alignment against the NCBI database also resulted in considerably high numbers of reads (laboratories 06, 07, and 14 > 300, laboratory 07 even > 1000) for *Homo sapiens*, which was not contained in the AGES database.

As mentioned above, the marker system applied does not detect maize. Thus, the high number of reads for maize obtained by laboratory 06 for one subsample seems to be caused by a random error.

In general, the total number of false reads obtained for both subsamples was similar within a laboratory. Larger differences between the total reads of subsamples was observed for laboratories 03 and 14 (AGES and NCBI databases) and laboratory 06 (only NCBI database).

Table 8. Reads obtained for sample 8 (negative control). (**A**): AGES database, (**B**): NCBI database.

(A)

Species	Laboratory 01 A	01 B	02 A	02 B	03 A	03 B	04 A	04 B	06 A	06 B	07 A	07 B	08 A	08 B	14 A	14 B	15 A	15 B	20 A	20 B
Chicken	-	-	13	28	-	-	42	30	4	98	-	-	2	-	10	177	2	10	8	5
Horse	-	-	7	6	-	2	57	51	-	2	-	-	-	-	-	1	-	-	7	5
Turkey	-	-	5	11	-	6	51	52	3	7	1	-	-	3	232	6	6	15	13	9
Beef	10	9	25	22	2	171	17	65	101	97	222	-	43	51	185	148	57	99	110	55
Pig	-	1	1	3	23	87	11	3	216	114	563	-	21	21	91	56	38	54	50	26
Sheep	5	14	51	79	2	2	82	74	3	9	19	-	3	1	11	3	-	-	-	10
Goat	1	1	-	1	-	-	1	1	1	1	2	-	-	-	-	1	-	-	-	1
Bison bonasus	-	-	-	-	-	-	-	-	-	-	-	-	-	1	-	2	-	-	-	-
Total reads	16	25	102	150	27	268	261	276	328	328	807	-	69	77	529	394	103	178	197	111

(B)

Species	Laboratory 01 A	01 B	02 A	02 B	03 A	03 B	04 A	04 B	06 A	06 B	07 A	07 B	08 A	08 B	14 A	14 B	15 A	15 B	20 A	20 B
Chicken	-	-	13	28	-	-	42	30	4	98	-	-	2	-	10	177	2	10	8	5
Horse	-	-	7	6	-	2	57	51	-	2	-	-	-	-	-	1	-	-	7	5
Turkey	-	-	5	11	-	6	51	50	3	7	1	-	-	3	231	6	6	15	13	9
Beef	10	9	25	22	2	171	17	65	101	97	222	-	43	52	187	149	57	99	111	55
Pig		1	1	3	23	99	11	3	226	114	564	-	21	21	91	57	38	54	50	27
Sheep	6	14	50	76	2	2	80	74	3	9	19	-	3	1	7	3	-	-	9	10
Goat	1	1	-	1	-	-	1	1	1	1	2	-	-	-	-	1	-	-	-	1
Bison bonasus	-	-	-	-	-	-	-	-	-	-	-	-	-	-	-	-	-	1	-	-
Bos mutus	-	-	-	-	-	-	-	-	-	1	-	-	-	1	-	-	1	-	-	-
Brachypodium sylvaticum	-	-	-	-	-	-	-	-	1	-	-	-	-	-	-	-	-	-	-	-
Coregonus migratorius	-	-	-	-	-	-	-	-	2	-	-	-	-	-	-	-	-	-	-	-
Equus zebra	-	-	-	-	-	-	-	-	-	-	-	-	-	-	-	-	-	-	-	-
Eukaryotic synthetic	-	-	-	15	-	-	-	-	-	-	-	-	-	-	-	-	-	-	-	-

Table 8. *Cont.*

(B)

Species	Laboratory																			
	01		02		03		04		06		07		08		14		15		20	
	A	B	A	B	A	B	A	B	A	B	A	B	A	B	A	B	A	B	A	B
Homo sapiens	-	-	58	46	30	188	82	215	302	784	1086	-	5	124	687	451	-	15	-	49
Meleagris ocellata	-	-	-	-	-	-	-	2	-	-	-	-	-	-	1	-	-	-	-	-
Oncorhynchus environmental	-	-	-	-	-	-	-	-	1	1	-	-	-	-	-	-	-	-	-	-
Oncorhynchus mykiss	-	-	-	-	-	-	-	-	37	91	-	-	-	-	-	-	-	-	-	-
Ovis ammon	-	1	1	5	-	-	2	1	1	-	-	-	-	-	4	-	-	-	-	-
Ovis vignei	-	-	-	-	-	-	1	1	-	-	-	-	-	-	-	-	-	-	-	-
Phascolosoma esculenta	-	-	-	-	-	1	-	-	-	-	-	-	-	-	-	-	-	-	-	-
Synthetic construct	-	-	-	-	-	-	-	-	20	-	-	-	-	-	-	-	-	-	-	-
Zea mays	-	-	-	2	-	-	-	3	851	3	-	-	-	-	-	-	-	-	2	-
total reads	17	26	160	215	57	469	344	496	1553	1208	1894	-	74	202	1218	846	104	193	200	161

4. Conclusions

In summary, evaluation of data from the interlaboratory ring trial indicates that the DNA metabarcoding method performed on an Illumina platform is applicable for determining the proportion of the seven animal species with the given precision. Furthermore, the applicability of the method for testing foodstuff was demonstrated by the correct identification of the ingredients of a model sausage, which also supports the results in our study published recently.

Based on the data of the ring trial, a threshold of 0.5% is suitable to reliably assess whether a certain animal species is contained in a sample. The DNA metabarcoding method turned out to be rather robust and is therefore suitable to be implemented in routine analysis in official food control laboratories. Even laboratories that did not have much experience in NGS were able to provide reliable results. We suggest strictly following the given protocol. The results of the interlaboratory ring trial indicate that even alternative test kits or various sequencing platforms might be applied. However, the impact of any deviations from the experimental conditions has obviously to be tested before implementation in routine analysis.

Correct index recognition is of particular importance for pooled DNA libraries. We recommend frequently changing the index kits or the use of longer index sequences to avoid false positive and/or false negative results.

For taxonomic assignment, we suggest applying a customized database, as the pipeline is completed significantly faster and no nonsense results from erroneous database entries occur. However, if unexpected read losses and non-identifiable reads occur, the additional use of the entire NCBI database or any other appropriate sequence database is recommended.

In order to increase interlaboratory comparability of results obtained by DNA metabarcoding methods, it would be necessary to establish a reference database with verified sequence entries of relevant species. Access to adequate reference material would also facilitate harmonization of the methods used.

In general, determination of the meat content (w/w) from the number of NGS reads or the determined target DNA concentration is a well-known difficulty, especially in the quantification of meat species in processed foods. The result is also influenced by the degree of processing of the sample present and by the type of animal ingredients used. Data from testing reference samples out of proficiency testing schemes confirm the limitations known for DNA quantification in meat products [12]. Quantitative results should therefore serve only as rough estimates for weight ratios of different species in food.

Supplementary Materials: The following supporting information can be downloaded at: https://www.mdpi.com/article/10.3390/foods11081108/s1, Supplementary Table S1: Sequences included into the reference database, Supplementary Table S2: Total number of reads after pipeline ($n = 14$, samples 1–7, two subsamples each). min: minimal value, max: maximal value, mean: arithmetic mean, RSD: relative standard deviation, Supplementary Table S3: Recovery (%) (total number of reads after pipeline related to the number of raw reads before analysis pipeline). ($n = 14$, samples 1–7, two subsamples each). min: minimal value, max: maximal value, mean: arithmetic mean, RSD: relative standard deviation.

Author Contributions: Conceptualization, R.H.; methodology, S.D. and R.H.; validation, S.U., K.F., A.S., K.N. and K.S.; writing—original draft preparation, M.C.-M., S.D. and R.H.; writing—review and editing, R.H., M.C.-M., S.D., S.U., K.F., A.S.; visualization, M.C.-M., K.F. and A.S.; supervision, R.H. All authors have read and agreed to the published version of the manuscript.

Funding: The ring trial was supported by the Federal Office of Consumer Protection and Food Safety, Germany (BVL). BVL funded all reagents for performing the tests and the statistical evaluation by QuoData GmbH.

Institutional Review Board Statement: Ethical review and approval was waived for this study as no live animals were used or slaughtered to achieve the aims of the study.

Data Availability Statement: The datasets generated during the current study are available from the corresponding authors upon reasonable request.

Acknowledgments: We thank the members of the working group "NGS Species identification" within the scope of the official method collection according to §64 of the German Food and Feed Code (LFGB) for their participation in the study.

Conflicts of Interest: The authors declare no conflict of interest.

References

1. Ballin, N.Z.; Vogensen, F.K.; Karlsson, A.H. Species determination—Can we detect and quantify meat adulteration? *Meat Sci.* **2009**, *83*, 165–174. [CrossRef] [PubMed]
2. Ballin, N.Z. Authentication of meat and meat products. *Meat Sci.* **2010**, *86*, 577–587. [CrossRef] [PubMed]
3. Sajali, N.; Wong, S.C.; Abu Bakar, S.; Khairil Mokhtar, N.F.; Manaf, Y.N.; Yuswan, M.H.; Mohd Desa, M.N. Analytical approaches of meat authentication in food. *Int. J. Food Sci. Technol.* **2021**, *56*, 1535–1543. [CrossRef]
4. Kumar, A.; Kumar, R.R.; Sharma, B.D.; Gokulakrishnan, P.; Mendiratta, S.K.; Sharma, D. Identification of species origin of meat and meat products on the DNA basis: A review. *Crit. Rev. Food Sci. Nutr.* **2015**, *55*, 1340–1351. [CrossRef] [PubMed]
5. Köppel, R.; Ruf, J.; Zimmerli, F.; Breitenmoser, A. Multiplex real-time PCR for the detection and quantification of DNA from beef, pork, chicken and turkey. *Eur. Food Res. Technol.* **2008**, *227*, 1199–1203. [CrossRef]
6. Köppel, R.; Ruf, J.; Rentsch, J. Multiplex real-time PCR for the detection and quantification of DNA from beef, pork, horse and sheep. *Eur. Food Res. Technol.* **2011**, *232*, 151–155. [CrossRef]
7. Kaltenbrunner, M.; Hochegger, R.; Cichna-Markl, M. Tetraplex real-time PCR assay for the simultaneous identification and quantification of roe deer, red deer, fallow deer and sika deer for deer meat authentication. *Food Chem.* **2018**, *269*, 486–494. [CrossRef] [PubMed]
8. Staats, M.; Arulandhu, A.J.; Gravendeel, B.; Holst-Jensen, A.; Scholtens, I.; Peelen, T.; Prins, T.W.; Kok, E. Advances in DNA metabarcoding for food and wildlife forensic species identification. *Anal. Bioanal. Chem.* **2016**, *408*, 4615–4630. [CrossRef] [PubMed]
9. Nehal, N.; Choudhary, B.; Nagpure, A.; Gupta, R.K. DNA barcoding: A modern age tool for detection of adulteration in food. *Crit. Rev. Biotechnol.* **2021**, *41*, 767–791. [CrossRef] [PubMed]
10. Galimberti, A.; Casiraghi, M.; Bruni, I.; Guzzetti, L.; Cortis, P.; Berterame, N.M.; Labra, M. From DNA barcoding to personalized nutrition: The evolution of food traceability. *Curr. Opin. Food Sci.* **2019**, *28*, 41–48. [CrossRef]
11. Dobrovolny, S.; Blaschitz, M.; Weinmaier, T.; Pechatschek, J.; Cichna-Markl, M.; Indra, A.; Hufnagl, P.; Hochegger, R. Development of a DNA metabarcoding method for the identification of fifteen mammalian and six poultry species in food. *Food Chem.* **2019**, *272*, 354–361. [CrossRef] [PubMed]
12. Preckel, L.; Brünen-Nieweler, C.; Denay, G.; Petersen, H.; Cichna-Markl, M.; Dobrovolny, S.; Hochegger, R. Identification of mammalian and poultry species in food and pet food samples using 16s rdna metabarcoding. *Foods* **2021**, *10*, 2875. [CrossRef] [PubMed]
13. *ISO 21571 I*; Foodstuffs—Methods of Analysis for the Detection of Genetically Modified Organisms and Derived Products— Nucleic Acid Extraction; ISO: Geneva, Switzerland, 2005.
14. Meyer, R.; Höfelein, C.; Lüthy, J.; Candrian, U. Polymerase chain reaction-restriction fragment length polymorphism analysis: A simple method for species identification in food. *J. AOAC Int.* **1995**, *78*, 1542–1551. [CrossRef] [PubMed]
15. Palumbi, S.R.; Martin, A.; Romano, S.; McMillan, W.O.; Stice, L.; Grabowski, G. *The Simple Fool's Guide to PCR*; Deptartment of Zoology and Kewalo Marine Laboratory, University of Hawaii: Honolulu, HI, USA, 2002.
16. Martin, M. Cutadapt removes adapter sequences from high-throughput sequencing reads. *EMBnet J.* **2011**, *17*, 10. [CrossRef]
17. Aronesty, E. Comparison of sequencing utility programs. *Open Bioinform. J.* **2013**, *7*, 1–8. [CrossRef]
18. Edgar, R.C. Search and clustering orders of magnitude faster than BLAST. *Bioinformatics* **2010**, *26*, 2460–2461. [CrossRef] [PubMed]
19. Quodata. PROLab Plus Software for Planning, Organizing, Performing and Analyzing Interlaboratory Studies. Available online: www.quodata.de (accessed on 10 December 2021).
20. *Official Collection of Test Methods A. Statistik—Planung und Statistische Auswertung von Ringversuchen zur Methodenvalidierung*; Federal Office of Consumer Protection and Food Safety (BVL), Beuth Verlag: Berlin, Germany, 2006.
21. MacConaill, L.E.; Burns, R.T.; Nag, A.; Coleman, H.A.; Slevin, M.K.; Giorda, K.; Light, M.; Lai, K.; Jarosz, M.; McNeill, M.S.; et al. Unique, dual-indexed sequencing adapters with UMIs effectively eliminate index cross-talk and significantly improve sensitivity of massively parallel sequencing. *BMC Genom.* **2018**, *19*, 30. [CrossRef] [PubMed]

Article

Development of a Duck Genomic Reference Material by Digital PCR Platforms for the Detection of Meat Adulteration

Xiaoyun Chen [1], Yi Ji [1], Kai Li [2], Xiaofu Wang [1], Cheng Peng [1], Xiaoli Xu [1], Xinwu Pei [2], Junfeng Xu [1,*] and Liang Li [2,*]

[1] State Key Laboratory for Managing Biotic and Chemical Threats to the Quality and Safety of Agro-Products, Zhejiang Academy of Agricultural Sciences, Hangzhou 310021, China; xiaoyunchen_2016@163.com (X.C.); jymemory12138@163.com (Y.J.); yywxf1981@163.com (X.W.); pc_phm@163.com (C.P.); xuxiaoli@zju.edu.cn (X.X.)

[2] Biotechnology Research Institute, Chinese Academy of Agricultural Sciences, Beijing 100081, China; likaij@163.com (K.L.); peixinwu@caas.cn (X.P.)

[*] Correspondence: njjfxu@163.com (J.X.); liliang@caas.cn (L.L.)

Abstract: Low-cost meat, such as duck, is frequently used to adulterate more expensive foods like lamb or beef in many countries. However, the lack of DNA-based reference materials has limited the quality control and detection of adulterants. Here, we report the development and validation of duck genomic DNA certified reference materials (CRMs) through the detection of the duck *interleukin 2* (*IL2*) gene by digital PCR (dPCR) for the identification of duck meat in food products. The certified value of *IL2* in CRMs was $5.78 \pm 0.51 \times 10^3$ copies/μL with extended uncertainty (coverage factor $k = 2$) based on *IL2* quantification by eight independent collaborating laboratories. Quantification of the mitochondrial gene *cytb* revealed a concentration of 2.0×10^6 copies/μL, as an information value. The CRMs were also used to determine the limit of detection (LOD) for six commercial testing kits, which confirmed that these kits meet or exceed their claimed sensitivity and are reliable for duck detection.

Keywords: certified reference material; digital PCR; duck *interleukin 2*; mitochondrial gene; meat adulteration

Citation: Chen, X.; Ji, Y.; Li, K.; Wang, X.; Peng, C.; Xu, X.; Pei, X.; Xu, J.; Li, L. Development of a Duck Genomic Reference Material by Digital PCR Platforms for the Detection of Meat Adulteration. *Foods* **2021**, *10*, 1890. https://doi.org/10.3390/foods10081890

Academic Editors: Mohammed Gagaoua and Brigitte Picard

Received: 13 July 2021
Accepted: 13 August 2021
Published: 15 August 2021

Publisher's Note: MDPI stays neutral with regard to jurisdictional claims in published maps and institutional affiliations.

1. Introduction

Meat adulteration has become a major global issue. Fraud by substitution or adulteration with inexpensive raw materials poses an attractive shortcut for unscrupulous food producers, although it is fraught with potential public health risks [1]. In order to definitely show that a product has been adulterated or fraudulently labeled, the product composition must be first determined for comparison of its authenticity with the description provided on its label. This process requires quantitative analysis of characteristic compounds or analytes specific to the ingredient in question, or other evidence that it is present in concentrations at or above levels required by regulatory agencies [2]. Therefore, the development of an accurate and reliable method for determining the content of specific meat components in food products has major economic implications and far-reaching social significance.

Traditional methods for species identification have historically relied on anatomy, histology, sensory judgment, chemistry, electrophoresis, chromatography, and immunology [3]. However, these approaches are each accompanied by limitations in their accuracy and/or sensitivity. To address this issue, recent studies have developed polymerase chain reaction (PCR)-based assays to accommodate the discriminatory capability necessary for identification of specific ingredients or contaminants. The most widely used methods are DNA-based screens [4], including gel-based PCR [5], real-time qPCR [6], multiplex PCR [7], digital PCR [8], isothermal nucleic acid amplification [9–11], and other PCR techniques [12,13]. The recent and extremely rapid expansion in detection methods and the standardization of detection methods for animal-based food products has inadvertently circumvented some steps that are essential to ensure the rigor and quality of data, which is compounded by a lack of reference materials. These factors together can incur a bottle-neck in data quality assurance. To rectify this issue in data quality assurance in food production, many reference materials are urgently needed for the ongoing evaluation of the methods used for quality control.

A reference material (RM) is sufficiently homogeneous and stable with respect to one or more specified properties that have been established to be fit for its intended use in a measurement process. Certified reference materials (CRMs) are reference materials (RMs) characterized using a metrologically valid procedure for one or more specified properties, accompanied by an RM certificate [14], which can be used for calibration of a measurement system, assessment of a measurement procedure, for assigning values to other materials, and quality control. Several DNA-based CRMs have been developed, such as for genetically modified organisms [15–17], cancer diagnosis [18–20], foodborne pathogens [21], and forensic science [22,23].

Due to the increasingly high frequency of adulteration of meat products with inexpensive duck, this work aimed to develop a novel DNA reference material for duck meat through the targeted detection and quantification of the duck *IL2* gene. Here, we describe the preparation, homogeneity, and short- and long-term stability of the gDNA materials. The mean *IL2* copy number values of these CRMs for duck were validated through digital PCR (dPCR) by eight independent collaborating diagnostic laboratories, and the materials were certified by China's State Administration for Market Regulation. In addition, the reference materials were used to evaluate the limit of detection for six commercial testing kits. These CRMs are intended for qualitative and quantitative screening for duck meat in meat products through detection of the *IL2* gene. These CRMs can also be used for validation of other quantitative DNA-based methods and laboratory quality control.

2. Materials and Methods

2.1. Extraction and Evaluation of Duck Genomic DNA

Duck leg meat was provided by the Institute of Animal Husbandry and Veterinary Science, Zhejiang Academy of Agricultural Sciences. Duck leg meat was used, and a sterile scalpel was used to remove the skin and cut it into small pieces, after which it was blended in Philips Mixture HR2027 (Hongkong, China), and then transferred to a Nalgene® 3118-0050 Oak Ridge Centrifuge Tube. Genomic DNA extraction was performed using a Simgen Animal Tissue DNA Midi Kit (Hangzhou, China). The integrity of the DNA critically affects the success of the characterization, which was analyzed by gel electrophoresis using a DYY-6C gel electrophoresis system (Liuyi Biotechnology, Beijing, China). The quality and purity of gDNA was evaluated at 230 nm, 260 nm, and 280 nm by the UV absorbance method (Nanodrop™ 2000, Thermo Fisher, Wilmington, DE, USA). The extracted gDNA was quantified using the Quant-iT™ dsDNA PicoGreen® Kits (Invitrogen, Shanghai, China) according to the manufacturer's instructions using a lambda DNA standard solution.

2.2. Digital PCR Assay

2.2.1. Certified Value

The target gene is a single-copy nuclear gene in the genome. The primers and probe (Figure 1) used for specific quantification were as described previously [24]. The primers and probe were synthesized by Sangon (Shanghai, China). The amplicon size was 212 bp. The primer and probe sequences are shown as follows:

5′-GGAGCACCTCTATCAGAGAAAGACA-3′;
5′-GTGTGTAGAGCTCAAGATCAATCCC-3′;
5′-FAM-TGGGAACAAGCATGAATGTAAGTGGATGGT-BHQ1-3′.

>AY821656.1 Anas platyrhynchos interleukin-2 precursor (IL-2) gene
TTCTATGGGGGAGTGGGACAGCGCATGAGGGAAAAAGTGTTTCATGCAGAAAGCACAAAGAGACT
GCCAAAAGTGAGTGTGGGCTTCTCTTTCCCCAATGAATGTAGGTAAAATCCCTCTTGCTTTAAAA
AATTCCAAAGTGTCATCAGGGGAGGAAAAACAAAAGTAATGATTCTTGCCATACAGGTAAAGCA
TATGAAAAAATGTGTAATAAAACCTCTTTTACATGACGCCCCCATCTTTTTCCCTCCAGAAAGAA
GAGTATAAATACACTAAACAGCCTAATGACAACATATCAGCTCTCATTACATATCACAACTGAAA
TACTAGCACAGAGACAACCAGAACACTGACAAGATGTGCAAAGTACTCATCTTCAGCTGCCTTTC
AGTACTAATGCTTATGACTACAGCTTAT<u>GGAGCACCTCTATCAGAGAAAGACA</u>ACACTCTTAAAA
CTTTAATAAAAGATTTAGAAAACC<u>TGGGAACAAGCATGAATGTAAGTGGATGGT</u>TTTCCTGTATT
AATTCTTTACTATTTGTCTTTTCTACTTATTTTTGTCTATTAACTTTGTTCACAAAAATTAACTT
TTTCCCCCTCTTCTCTACAG<u>GGGATTGATCTTGAGCTCTACACAC</u>CAAATGACACAAAGGTAAGT
TCAAGTTTTCTGTCTTCAGGACTGCTAAGATACGGCTATTTTGAAGTTATGTGTTATTGAATTAG
ACCAAAGTAATTCACTTACCAGAAGGCCCGATACAAAAGACAAATCTGAGCTGTCTGAACTTTGT
GTGAGGATAGCATAAGCTAACCTAGAAGTACTGGGCACAAGGATAGCAGGCTAACCTAGAAGACA
TTTCTGCTTTTGATTTGTAT

Figure 1. Sequence of the *interleukin* 2 precursor gene (AY821656.1). The primers (red) and probe (green) for the digital PCR are underlined. The amplicon size was 212 bp.

Optimization of the PCR condition is critical for developing a PCR-based method for the specific detection of duck. This study fully optimized the key parameters of real-time qPCR and digital PCR, including specificity, oligonucleotide concentration, annealing temperature, dynamic range, the limit of detection, and the limit of quantification (data not shown). The optimized PCR reaction contains 10 μL of 2X ddPCR Supermix for probes (no dUTP) (Bio-Rad, Hercules, CA, USA), 1 μL of each primer (0.5 μM), 0.5 μL of probe (0.25 μM), 2 μL of DNA template, and 5.5 μL of ddH2O. After that, the 20 μL of the reaction mixture was then loaded on eight-channel disposable droplet generator cartridges (Bio-Rad). Droplets were generated with 70 μL of droplet generation oil (Bio-Rad) in the droplet generator of the QX200 system (Bio-Rad). The generated droplets were transferred to a 96-well PCR plate (Bio-Rad). The amplification was carried out at a uniform ramp rate of 2.5 °C/s for 95 °C for 10 min, 45 cycles of 95 °C for 15 s followed by 60 °C for 1 min, and a final enzyme deactivation at 98 °C for 10 min. Fluorescent signals from amplified droplets were captured individually in the QX200 droplet reader (Bio-Rad) and analyzed with QuantaSoft 1.6.6.0320 software. The target concentrations were reported as the number of copies per μL of PCR reaction after correction with the Poisson distribution.

2.2.2. Information Value

A mitochondrial *cytb* gene TaqMan MGB probe assay [25] was used. For absolute quantification, the primers and probe were fully performed in a QX200 ddPCR system. Primers and probes were purified with high-performance liquid chromatography. The preparation of the reaction mixture and optimized PCR thermal procedure is listed in the Supplementary Materials. The primers and probes are as follows:

5′-GGCCACACAAATCCTCACAG-3′;
5′-TGTGTTGGCTACTGAGGAGAAA-3′;
5′-FAM- CCTACTGGCTATGCACTACACCGCAGAC-BHQ1-3′.

2.3. Homogeneity Testing

Most CRMs are prepared as batches of bottles or vials. It is important that all units are the same within the stated uncertainty for each property value. ISO 17034 accordingly requires the assessment of the homogeneity of a certified reference material. Homogeneity includes within-unit homogeneity and between-unit homogeneity. It is always necessary to assess the between-unit homogeneity, which was assessed to ensure the equivalence between different units following the Section 7.5.2 of the ISO Guide 35:2017. Furthermore, the within-unit homogeneity is directly reflected in the minimum size of the subsample. The homogeneity of the *IL2* duck gene was evaluated by randomly selecting 15 units from the CRM candidates. Three subsamples from different positions of each unit were taken and measured by dPCR. Measurement results were used for the assessment of homogeneity. Statistical analysis of the data was performed for a 95% confidence level. The values of the copy number concentration for this *IL2* gene of the duck samples were analyzed using the F test and compared with the significance of the calculated F value of the copy number concentration and the critical $F_{0.05(14, 30)}$ value of 2.04.

An experimental assessment must consist of a determination of the minimum sample intake. It is always essential to ensure that the sample intake is sufficient. The between-unit study does not provide such assurance, so the experimental within-unit homogeneity study should be carried out.

2.4. Stability Monitoring

Stability is an important parameter for the reference material. The stability of all CRMs should be assessed. Two types of stability are relevant in the production of reference material: the long-term stability and transportation stability (short-term stability). This CRM is the DNA solution, so it needs to be repeatedly frozen and thawed, and thus the freeze-thaw stability is a concern. Stability testing was carried out to evaluate the influence of different storage temperatures and times on the CRMs using digital PCR (dPCR). The uncertainty of stability needs to include the uncertainty of long-term stability and short-term stability.

Transportation stability (short-term stability) is a property of the material referring to stability under expected transport conditions. Extreme high temperatures cannot be ruled out during transportation, so the CRM units were stored at 4 °C, 25 °C, 37 °C, and 60 °C for 1, 3, 5, 7, and 14 days. The samples were stored at −70 °C after sampling. Three tubes were randomly selected for each storage temperature and each tube was sampled 3 times. The digital PCR tested the stability of *IL2* gene of the duck samples. The *t*-test was performed on the chosen dates to evaluate the short-term stability of the CRMs. The *t*-test results showed no significant slope ($\beta1$) for the CRMs at 4 °C, 25 °C, and 37 °C at the 95% confidence level during the two weeks.

Long-term stability studies are conducted to assess stability under storage conditions specified for the lifetime of the product. The long-term stability study was extended to 6 months, the *IL2* gene of the duck CRMs was evaluated by analyzing 3 tubes stored at −20 °C for 1, 2, 4, and 6 months. The samples were stored at −70 °C after sampling. Three tubes were randomly selected for each storage temperature and each tube was sampled 3 times by the digital PCR. Because the *IL2* gene of the duck CRMs is the DNA solution, their freeze-thaw stability needs to be implemented. This batch of CRMs was 100 μL per tube and the minimum sample was 2 μL. In the experiment, 3 tubes of the *IL2* gene of the duck CRMs were taken, repeated freezing-thawing was performed 10 times, and samples were taken 3 times per tube by the digital PCR.

2.5. Collaborative Characterization

The property value of this batch of CRMs was determined by eight laboratories (Supplementary Table S1) using dPCR. The eight laboratories were engaged in DNA measurement for a long time and have a certain technical authority.

The gDNA samples of the *IL2* gene of ducks and the related ddPCR reagents were mailed to each participating laboratory in a closed box filled with dry ice. Each laboratory received two gDNA samples (label: duck1, duck2), each with a volume of 100 µL. In order to simplify the sampling procedure and reduce the deviation, primers/probes were mixed according to the proportion before sending them to each laboratory. The participating laboratories were requested to measure the copy number concentration by the operation protocol of dPCR experiments. Each sample was repeatedly measured four times, and a total of eight subsamples were required simultaneously on the same PCR plate for each participant. The eight participants need to export data and original files to the organization lab. The returned data were analyzed according to the requirements of ISO Guide 35:2017.

2.6. Evaluation of Limit of Detection of Commercial Assay Kits

In this experiment, six commercial duck-derived detection qPCR kits were purchased. There were four mitochondrial gene targets and two nuclear gene kits. Company and product information are summarized in Supplementary Table S2. The CRM was serially diluted and applied to limit of detection (LOD) probit regression analysis for six diagnostic assays [26]. Each of the concentration levels were tested with multiple replicates per concentration according to the manufacturer's instructions with blank controls. Probit regression analysis of 95% hit rates was performed with SPSS 16.0 software (SPSS Inc., Chicago, IL, USA).

3. Results

3.1. gDNA Extracted from Commercially Obtained Duck Leg Meat

Prior to testing for adulteration, we first sought to confirm the quality of gDNA extracted from leg meat samples of duck obtained from the Institute of Animal Husbandry and Veterinary Science, Zhejiang Academy of Agricultural Sciences. To this end, we first analyzed the extracted gDNA by 1% agarose gel electrophoresis and found neither a visible smear nor an RNA band, which indicated that the gDNA was intact (i.e., not degraded) and that RNA had been removed during the isolation process. In addition, the quantity and quality of each gDNA extraction were confirmed in six technical replicates using a Nanodrop 2000 spectrometer (Thermo Scientific, Wilmington, DE, USA). The A260/A280 ratio averaged 1.98, while the average A260/A230 ratio was 2.28, indicating high purity. The quantification of gDNA concentration by the Qubit 2.0 Fluorometer (Invitrogen) showed that the samples contained approximately 100 ng/µL. DNA solutions were then diluted to concentrations of 6.5 ng/µL with 0.1 × Tris-EDTA (TE) buffer and stored in a −20 °C freezer.

3.2. Bottling and Storage/Homogeneity Testing and Minimum Sample Intake

In order to establish protocols for future sample collection and preparation, we next established standardized steps for the proper storage and handling of DNA extractions following methods described before [17]. The gDNA solutions were sterilized using a 0.2 µm nylon filter and 100 µL aliquots were placed in low static, sterilized polypropylene microcentrifuge tubes, with approximately 5×10^3 copies of duck gDNA per tube. Cardboard freezer boxes accommodating 100 samples each were used for storage, and random samples were taken from each box for testing homogeneity as well as short- and long-term stability. All samples were stored in 4 °C refrigerators in the dark.

In order to establish the protocol for evaluating the duck certified reference material (CRM), we examined the homogeneity of the *IL2* gene in our gDNA extractions using a random stratified sampling method [27] to select 15 units from among 500 CRM candidate samples. Three subsamples were taken from different vial positions for each sample and analyzed by digital PCR (dPCR) to determine whether this technique was sufficiently sensitive to identify differences in *IL2* uniformity within and between samples (for raw data see Supplementary Table S3). Significant differences in the *IL2* copy number among within-vial subsamples and between different samples were then determined using an

F test to compare the $F_{calculated}$ with the $F_{critical}$ value (2.04) for a 95% confidence level. The results showed that the $F_{calculated}$ (1.53) value was lower than the $F_{critical}$ (2.04) value (Table 1), which thus indicated high homogeneity among the CRM gDNA samples in this batch.

Table 1. Results of the homogeneity analysis.

Parameter	Copy Number/Unit
Mean	5.76×10^3
Q_1	3.59×10^5
V_1	14
$S_1{}^2$	2.57×10^4
Q_2	5.02×10^4
V_2	30
$S_2{}^2$	1.67×10^4
F	1.53
$F_{0.05(14, 30)}$	2.04
Conclusion	$F < F_{0.05(14, 30)}$
μ_{bbrel}	0.0095

3.3. Stability Monitoring and Freeze-Thaw Cycles Testing

The stability of the properties of interest (in this case, the duck *IL2* copy number) represents an essential feature of any given reference material. It is well-known that long-term stability is related to storage conditions, while short-term stability is related to external factors during sample transportation. Moreover, stability studies can be categorized as either classical stability or synchronous stability studies. In this work, we incorporated data from other international studies [16,19,28] characterizing DNA standard materials in a classical stability assessment of duck *IL2* gene stability at $-70\ ^\circ$C. Under these storage conditions, we found that the *IL2* copy number of the candidate duck CRMs did not change. It warrants mention here that the copy number stability at other storage temperatures requires a synchronous stability study for comparison with storage at $-70\ ^\circ$C.

Transportation stability (short-term stability) is a property of the material that describes its stability under expected transport conditions in order to guide proper handling methods prior to evaluation. Since extremely high temperatures cannot be ruled out during transportation, gDNA CRMs were stored at $4\ ^\circ$C, $25\ ^\circ$C, and $60\ ^\circ$C for 1, 3, 5, 7, and 14 days. Three tubes were randomly selected for storage at each temperature, and each tube was tested in three technical replicates by digital PCR. The results of the *t*-test showed no significant differences in the slope ($\beta 1$) between the CRMs stored for two weeks at $4\ ^\circ$C or $25\ ^\circ$C at a 95% confidence level (Supplementary Table S4 and Figure 2). However, at $60\ ^\circ$C, the *IL2* copy number significantly decreased after a single day of storage. The relative uncertainty in the *IL2* copy number due to instability at $25\ ^\circ$C storage was calculated to be 0.035. Collectively, these results of short-term CRM stability analysis indicated that duck gDNA is stable under room temperature ($25\ ^\circ$C) storage and transport for up to 14 days, though we advise cold chain transportation to minimize the likelihood of degradation.

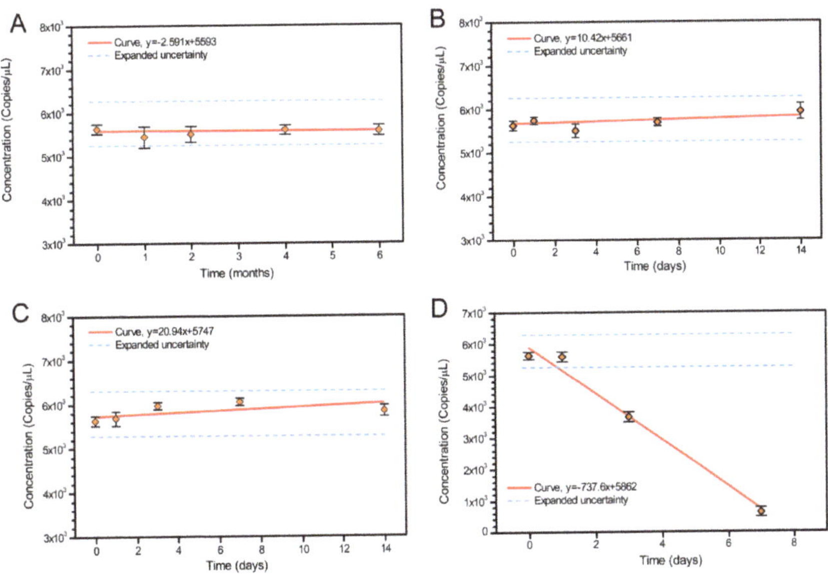

Figure 2. Trends and concentration of stability analysis for the candidate materials: (**A**) long–term stability at −20 °C and (**B–D**) short–term stability at 4 °C, 25 °C, and 60 °C. The error bars represent the standard deviation of within–bottle triplicate subsamples.

In order to determine the long-term stability of the CRMs, three randomly selected tubes were stored at −20 °C and tested for the *IL2* copy number by dPCR at 1, 2, 4, and 6 months time points. The results of this analysis showed no significant differences in the *IL2* copy number among any of the time points throughout the 6-month experiment at −20 °C ($p < 0.05$) (Supplementary Table S5 and Figure 2). We then estimated a 0.020 relative uncertainty of the copy number instability caused by long-term storage at −20 °C. Collectively, these results showed that gDNA CRMs could be stably stored at −20 °C for at least 6 months.

Finally, we investigated whether freeze-thaw cycles could adversely affect the *IL2* copy number. Since this batch of CRM samples each contained 100 µL per tube and our above results showed that 2 µL of template was appropriate for analysis, we then randomly selected three tubes of duck CRM gDNA and performed 10 repeated freezing-thaw cycles. Subsequently, dPCR was performed using each sample in three technical replicates, which showed no significant difference in the duck *IL2* gene copy number ($p > 0.05$) among samples. These results indicated that duck gDNA CRMs are stable and robust against damage due to repeated freezing and thawing (data not shown).

3.4. Characterization of Certified and Information Value

To validate our results through independent third parties, eight different laboratories each determined the *IL2* copy number for two randomly selected duck CRM samples using four technical replicates per sample in dPCR. Thus, each participating external lab returned eight dPCR results (Supplementary Table S6, Figure 3), and the copy numbers were estimated and analyzed according to ISO GUIDE 35:2017. The statistical analysis indicated that the datasets followed a normal distribution, and none of the datasets contained points outside of the uncertainty range for a 95% confidence level. Furthermore, no significant differences were found in the mean *IL2* copy numbers from each lab, and thus, the certified mean *IL2* copy number in the duck CRMs was estimated to be 5.78×10^3 copies/µL.

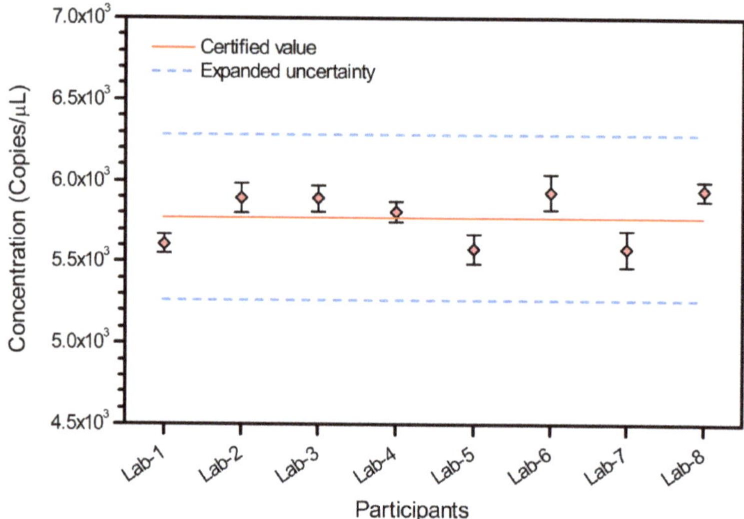

Figure 3. Results for dPCR quantification of the *IL2* gene copy number by eight independent participating laboratories. Solid line, average of participant results (♦); dotted lines, expanded uncertainty of the certified values of the copy number concentration with a 95% level of confidence.

In addition to *IL2*, mitochondrial genes can also serve as quantifiable targets for the amplification of duck gDNA in meat-based food products. Our laboratory optimized a digital PCR method for detection of the mitochondrial gene *cytb*, which revealed a concentration of 2.0×10^6 copies/µL in the CRM samples.

3.5. Statistical Estimation of Uncertainty

The uncertainty of the CRMs for the *IL2* gene in these duck CRMs was estimated from the contributions according to ISO Guide 35 and consisted of uncertainty components from characterization (u_{char}), potential between-unit heterogeneity (u_{bb}), and potential instability during long-term (u_{its}) and short-term storage (u_{sts}).

The relative standard uncertainty of characterization ($u_{char,rel}$) is estimated by Equation (1). The result is 0.012.

$$u_{char,rel} = \sqrt{u^2_{A,rel} + u^2_{B,rel}} \tag{1}$$

where $u_{A,rel}$ is type A uncertainty (random error) and $u^2_{B,rel}$ is type 2 uncertainty (systematic error).

The relative standard uncertainty of potential between-unit heterogeneity ($u_{bb,rel}$) is 0.0095, estimated as described in Table 1.

The relative standard uncertainty of the potential degradation during transport ($u_{sts,\ rel}$), and long-term storage ($u_{its,\ rel}$) is estimated by Equations (2) and (3). The results are 0.035 and 0.020, respectively.

$$u_{sts} = s(\beta_1) \times X/\bar{x} \tag{2}$$

where $s(\beta_1)$ is the standard deviation of all results of the transport time stability study (Supplementary Table S4), X is the chosen transport time (14 days at 25 °C), and $\bar{x}$ is the mean value.

$$u_{its} = s(\beta_1) \times X/\bar{x} \tag{3}$$

where $s(\beta_1)$ is the standard deviation of all the results of the long-time stability study (Table S5), X is the chosen transport time (six months at −20 °C), and $\bar{x}$ is the mean value.

The expanded relative uncertainty ($U_{CRM,rel}$) for the *IL2* gene in these duck CRMs was estimated by Equation (4), and the expanded uncertainty of the copy number concentration (U_{CRM}) was estimated to be 0.51 × 103 (coverage factor k = 2, approximate 95% confidence interval).

$$U_{CRM,rel} = k \times \sqrt{u^2_{char,rel} + u^2_{bb,rel} + u^2_{its,rel} + u^2_{sts,rel}} \qquad (4)$$

where u_{char} is the measurement of the certified value, u_{bb} is the potential between-unit heterogeneity, u_{its} and u_{sts} is the uncertainties in the long-term and short-term stability.

3.6. Results of Assay Kit Evaluations

We then investigated whether commercially available kits were sufficiently sensitive to be used for extraction of duck gDNA for comparison with CRMs characterized here. For this purpose, we used the externally validated CRM samples and accompanying data to evaluate the LODs of six diagnostic assays (Figure 4). The lot numbers and LODs claimed by each commercial diagnostic assay are listed in Supplementary Table S2. Then, using CRM samples, we tested each of these six kits and compared the results of the LOD determination using a probit regression analysis of 95% hit rates with the LOD in the manufacturer's instructions for each kit. The results showed that all of the assays met or exceeded their claimed sensitivities and are reliable for detection of *IL2* in gDNA extracted from duck meat. Moreover, the results also show that these duck gDNA CRMs are suitable for the assessment of commercial kits and allow for comparable LOD studies of various detection methodologies.

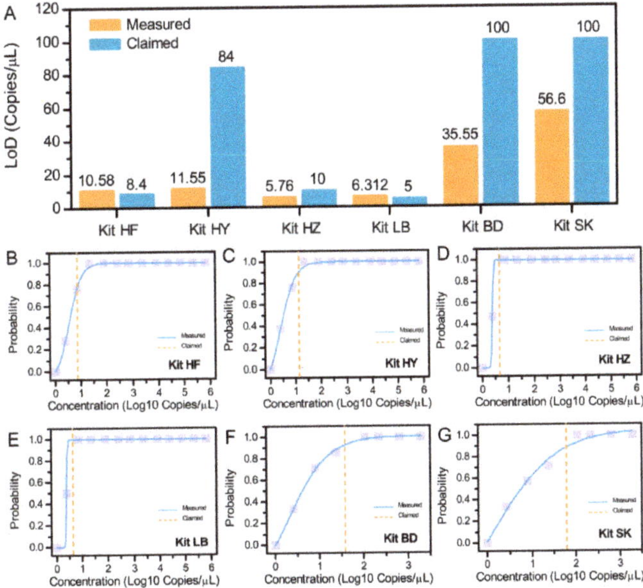

Figure 4. Evaluation of LODs of commercial assays. (**A**) Comparison of LODs claimed by the kit manufacturer using six duck DNA detection kits with the LODs determined with the CRMs. (**B–G**) Probit regression analysis of the six assays for the detection of duck *IL2* (SPSS 16.0). The probit versus *IL2* concentration ((**B–E**) mitochondrial gene, (**F,G**) nuclear gene) was obtained from 21 replicates of serial dilutions and an additional 10 replicates of a blank sample.

4. Discussion

In recent years, DNA detection methods have been an essential technique for molecular diagnosis. RM and CRM are the quality assurance of testing data, which can evaluate the measurement method and monitor the measurement process. DNA CRMs originated from the detection of genetically modified ingredients because quantitative testing is needed in this field [15]. Subsequently, a large number of molecular diagnostic standards have been developed in the field of medical testing. The international metrology field also pays great attention to DNA measurement, especially traceability to SI units [28]. As an absolute quantitative method, the digital PCR method is currently a potential primary reference measurement method for DNA target measurement. As the carrier of the measurement value, the reference material plays a vital role in the measurement traceability and the guarantee of the data quality of the measurement results.

In this study, we chose the nuclear gene *IL2* as the target gene. Compared with mitochondrial genes, nuclear genes are single-copy genes, genetically stable, and there is no difference between different tissues [6], so it is easy to perform absolute quantification. We fully optimized the *IL2* gene method on qPCR and dPCR platforms. Although the results of this study characterizing a CRM for duck meat are primarily based on dPCR, standard laboratory testing typically relies on qPCR-based analysis. Therefore, it was also necessary to investigate whether the gDNA samples contained inhibitors that could interfere with the qPCR reaction. We first performed six serial dilutions of each gDNA sample, ranging from 100 ng/μL to 0.02 ng/μL, and generated a standard curve of Ct values by qPCR. The linear regression equation between Ct (y) and log10 starting copy number (X) was Y = $-3.415X + 42.632$. The slope of the curve was -3.415, indicating a 96.3% amplification efficiency and $R^2 = 0.998$. The linear regression analysis thus confirmed that the duck gDNA samples were suitably pure for qPCR and met the performance criteria defined by the European Network of GMO Laboratories (ENGL). We next determined that the appropriate linear range for the copy number for subsequent dPCR analysis was between 20 and 14,000 copies per reaction to obtain a good correlation (i.e., $R^2 = 1$) with the expected value. Digital PCR commonly uses 20 μL reaction volumes containing 2 μL of the DNA template. The results of the dPCR showed that the *IL2* copy number in these duck CRM samples averaged 5780 copies per microliter, within the linear range, which suggested that 2 μL of the gDNA template was appropriate for analysis. We then determined the standard uncertainty of heterogeneity among technical replicates of each sample to account for variation between samples. The relative uncertainty for the *IL2* copy number was 0.0095, which corresponded with a 95% confidence level. Taken together, these results show that the samples were sufficiently homogenous for accurate evaluation as a standard CRM.

The CRM has two values to meet the detection needs of different targets. The certified value is expressed as $(5.78 \pm 0.51) \times 10^3$ copies/μL with extended uncertainty (a coverage factor $k = 2$) based on the quantification of *IL2* with a collaborative characterization of eight participants. An information value is 2.0×10^6 copies/μL based on the quantification of a mitochondrial gene. Our laboratory determined this value as an informative supplementary value for the quantification of duck meat, which may be of value to users of the CRMs. However, the replication data are currently insufficient to assess the uncertainty associated with this value rigorously.

Lots of novel detection methods of meat adulteration have been developed [29,30], while the literature on reference materials is rarely reported. Some commercial companies have developed RM, but not CRM. Therefore, it is urgent to develop animal DNA standard materials according to the ISO standard system. The identification of meat adulteration is an important part of the field of food safety. DNA-based testing is almost qualitative, and it needs to be further developed into quantitative testing. At the same time, CRM should be developed, international proficiency testing should be organized, and assay kit evaluations should be carried out to ensure accurate and consistent measurement results in the industry.

5. Conclusions

Here, we developed gDNA CRMs (GBW(E) 091060) certified by China's State Administration for Market Regulation for the analysis of duck adulteration in meat products. Eight independent diagnostic laboratories contributed to the validation and certification of these duck gDNA CRMs by digital PCR, confirming a mean of $5.78 \pm 0.51 \times 10^3$ copies/µL for the *IL2* gene. Homogeneity and stability testing demonstrated that the CRMs were homogenous and stable for at least 6 months at $-20\,^{\circ}\text{C}$ storage and for 14 days in cold chain delivery conditions. This batch of duck gDNA CRMs can serve as an essential tool for method validation and proficiency testing in the analysis of duck content in meat products. This study also provides a technical basis for development of other animal-derived reference materials.

Supplementary Materials: The following are available online at https://www.mdpi.com/article/10.3390/foods10081890/s1, Table S1: Laboratories participating in the characterization of the *IL2* gene from duck genomic DNA. Table S2: List of commercial assays assessed by the *IL2* certified reference standard. Table S3: Raw data of the homogeneity analysis. Table S4: Results of the short-term stability study. Table S5: Results of the long-term stability study. Table S6: Results of independent laboratory validation.

Author Contributions: Conceptualization, L.L. and J.X.; methodology, X.C. and X.X.; software, X.C.; validation, Y.J. and K.L.; formal analysis, Y.J. and K.L.; investigation, X.W. and C.P.; resources, L.L. and J.X.; writing—original draft preparation, X.C.; writing—review and editing, L.L. and J.X.; visualization, X.W.; supervision, C.P. and X.P. All authors have read and agreed to the published version of the manuscript.

Funding: This research was funded by Key Research and Development Program of Zhejiang (2021C02059); the Agricultural Science and Technology Innovation Program of CAAS (CAAS-ZDRW202009).

Institutional Review Board Statement: Not applicable.

Informed Consent Statement: Not applicable.

Data Availability Statement: Data are available upon request.

Acknowledgments: The authors gratefully acknowledge the financial support from the Key Research and Development Program of Zhejiang (2021C02059) and the Agricultural Science and Technology Innovation Program of CAAS (CAAS-ZDRW202009).

Conflicts of Interest: The authors declare no conflict of interest.

References

1. Köppel, R.; Daniels, M.; Felderer, N.; Brünen-Nieweler, C. Multiplex real-time PCR for the detection and quantification of DNA from duck, goose, chicken, turkey and pork. *Eur. Food Res. Technol.* **2013**, *236*, 1093–1098. [CrossRef]
2. Burns, M.; Wiseman, G.; Knight, A.; Bramley, P.; Foster, L.; Rollinson, S.; Damant, A.; Primrose, S. Measurement issues associated with quantitative molecular biology analysis of complex food matrices for the detection of food fraud. *Analyst* **2015**, *141*, 45–61. [CrossRef]
3. Kumar, A.; Kumar, R.R.; Sharma, B.D.; Gokulakrishnan, P.; Mendiratta, S.K.; Sharma, D. Identification of Species Origin of Meat and Meat Products on the DNA Basis: A Review. *Crit. Rev. Food Sci. Nutr.* **2013**, *55*, 1340–1351. [CrossRef] [PubMed]
4. Böhme, K.; Calo-Mata, P.; Barros-Velázquez, J.; Ortea, I. Review of Recent DNA-Based Methods for Main Food-Authentication Topics. *J. Agric. Food Chem.* **2019**, *67*, 3854–3864. [CrossRef] [PubMed]
5. Kitpipit, T.; Sittichan, K.; Thanakiatkrai, P. Are these food products fraudulent? Rapid and novel triplex-direct PCR assay for meat identification. *Forensic Sci. Int. Genet. Suppl. Ser.* **2013**, *4*, e33–e34. [CrossRef]
6. Li, T.; Wang, J.; Wang, Z.; Qiao, L.; Liu, R.; Li, S.; Chen, A. Quantitative determination of mutton adulteration with single-copy nuclear genes by real-time PCR. *Food Chem.* **2021**, *344*, 128622. [CrossRef] [PubMed]
7. Xu, R.; Wei, S.; Zhou, G.; Ren, J.; Liu, Z.; Tang, S.; Cheung, P.C.; Wu, X. Multiplex TaqMan locked nucleic acid real-time PCR for the differential identification of various meat and meat products. *Meat Sci.* **2018**, *137*, 41–46. [CrossRef]
8. Floren, C.; Wiedemann, I.; Brenig, B.; Schütz, E.; Beck, J. Species identification and quantification in meat and meat products using droplet digital PCR (ddPCR). *Food Chem.* **2015**, *173*, 1054–1058. [CrossRef]

9. Lin, L.; Zheng, Y.; Huang, H.; Zhuang, F.; Chen, H.; Zha, G.; Yang, P.; Wang, Z.; Kong, M.; Wei, H.; et al. A visual method to detect meat adulteration by recombinase polymerase amplification combined with lateral flow dipstick. *Food Chem.* **2021**, *354*, 129526. [CrossRef]

10. Cao, Y.; Zheng, K.; Jiang, J.; Wu, J.; Shi, F.; Song, X.; Jiang, Y. A novel method to detect meat adulteration by recombinase polymerase amplification and SYBR green I. *Food Chem.* **2018**, *266*, 73–78. [CrossRef] [PubMed]

11. Wang, J.; Wan, Y.; Chen, G.; Liang, H.; Ding, S.; Shang, K.; Li, M.; Dong, J.; Du, F.; Cui, X.; et al. Colorimetric Detection of Horse Meat Based on Loop-Mediated Isothermal Amplification (LAMP). *Food Anal. Methods* **2019**, *12*, 2535–2541. [CrossRef]

12. Hossain, M.A.M.; Ali, E.; Hamid, S.B.A.; Asing; Mustafa, S.; Desa, M.N.M.; Zaidul, I.S.M. Double Gene Targeting Multiplex Polymerase Chain Reaction–Restriction Fragment Length Polymorphism Assay Discriminates Beef, Buffalo, and Pork Substitution in Frankfurter Products. *J. Agric. Food Chem.* **2016**, *64*, 6343–6354. [CrossRef]

13. López-Oceja, A.; Nuñez, C.; Baeta, M.; Gamarra, D.; de Pancorbo, M. Species identification in meat products: A new screening method based on high resolution melting analysis of cyt b gene. *Food Chem.* **2017**, *237*, 701–706. [CrossRef]

14. ISO/TC334. *Reference Materials—Selected Terms and Definitions*; ISO Guide 30:2015; ISO: Geneva, Switzerland, 2015.

15. Yang, Y.; Li, L.; Yang, H.; Li, X.; Zhang, X.; Xu, J.; Zhang, D.; Jin, W.; Yang, L. Development of Certified Matrix-Based Reference Material as a Calibrator for Genetically Modified Rice G6H1 Analysis. *J. Agric. Food Chem.* **2018**, *66*, 3708–3715. [CrossRef] [PubMed]

16. Li, J.; Zhang, L.; Li, L.; Li, X.; Zhang, X.; Zhai, S.; Gao, H.; Li, Y.; Wu, G.; Wu, Y. Development of Genomic DNA Certified Reference Materials for Genetically Modified Rice Kefeng 6. *ACS Omega* **2020**, *5*, 21602–21609. [CrossRef]

17. Li, J.; Li, L.; Zhang, L.; Zhang, X.; Li, X.; Zhai, S.; Gao, H.; Li, Y.; Wu, G.; Wu, Y. Development of a certified genomic DNA reference material for detection and quantification of genetically modified rice KMD. *Anal. Bioanal. Chem.* **2020**, *412*, 7007–7016. [CrossRef]

18. Xu, J.; Qu, S.; Sun, N.; Zhang, W.; Zhang, J.; Song, Q.; Lin, M.; Gao, W.; Zheng, Q.; Han, M.; et al. Construction of a reference material panel for detecting KRAS/NRAS/EGFR/BRAF/MET mutations in plasma ctDNA. *J. Clin. Pathol.* **2021**, *74*, 314–320. [CrossRef] [PubMed]

19. Dong, L.; Wang, X.; Wang, S.; Du, M.; Niu, C.; Yang, J.; Li, L.; Zhang, G.; Fu, B.; Gao, Y.; et al. Interlaboratory assessment of droplet digital PCR for quantification of BRAF V600E mutation using a novel DNA reference material. *Talanta* **2020**, *207*, 120293. [CrossRef]

20. He, H.-J.; Das, B.; Cleveland, M.H.; Chen, L.; Camalier, C.E.; Liu, L.-C.; Norman, K.L.; Fellowes, A.P.; McEvoy, C.R.; Lund, S.P.; et al. Development and interlaboratory evaluation of a NIST Reference Material RM 8366 for EGFR and MET gene copy number measurements. *Clin. Chem. Lab. Med.* **2019**, *57*, 1142–1152. [CrossRef] [PubMed]

21. Vallejo, C.V.; Tere, C.P.; Calderon, M.N.; Arias, M.M.; Leguizamon, J.E. Development of a genomic DNA reference material for Salmonella enteritidis detection using polymerase chain reaction. *Mol. Cell. Probes* **2021**, *55*, 101690. [CrossRef] [PubMed]

22. Børsting, C.; Mas, C.T.; Morling, N. SNP typing of the reference materials SRM 2391b 1–10, K562, XY1, XX74, and 007 with the SNPforID multiplex. *Forensic Sci. Int. Genet.* **2011**, *5*, e81–e82. [CrossRef] [PubMed]

23. Steffen, C.R.; Kiesler, K.M.; Borsuk, L.A.; Vallone, P.M. Beyond the STRs: A comprehensive view of current forensic DNA markers characterized in the PCR-based DNA profiling standard SRM 2391D. *Forensic Sci. Int. Genet. Suppl. Ser.* **2017**, *6*, e426–e427. [CrossRef]

24. Cheng, X.; He, W.; Huang, M. Establishment and Application of Fluorescent Real-Time PCR for the Detection of Duck Meat in Foods. *Food Sci.* **2013**, *34*, 92–96. [CrossRef]

25. SAC/TC387. *Identification of Animal Ingredients from Common Livestock and Poultry Real-Time PCR*; GB/T38164; AQSIQ of China: Beijing, China, 2019.

26. Tholen, D.W.; Kallner, A.; Kennedy, J.W.; Krouver, J.S.; Meier, K. Evaluation of Precision Performance of Quantitative Measurement Methods; Approved Guideline—Second Edition. NCCLS, 2004. Available online: https://yeec.com/uploadimages1/forum/2013-2/201321114424796191.pdf (accessed on 28 May 2021).

27. ISO/Guide 35:2017(en). Reference Material-Guidance for Characterization and Assessment of Homogeneity and Stability. 2017. Available online: https://www.iso.org/obp/ui/#iso:std:iso:guide:35:ed-4:v1:en (accessed on 20 May 2021).

28. Dong, L.; Meng, Y.; Wang, J.; Liu, Y. Evaluation of droplet digital PCR for characterizing plasmid reference material used for quantifying ammonia oxidizers and denitrifiers. *Anal. Bioanal. Chem.* **2014**, *406*, 1701–1712. [CrossRef] [PubMed]

29. Kim, M.-J.; Lee, Y.-M.; Suh, S.-M.; Kim, H.-Y. Species Identification of Red Deer (*Cervus elaphus*), Roe Deer (*Capreolus capreolus*), and Water Deer (*Hydropotes inermis*) Using Capillary Electrophoresis-Based Multiplex PCR. *Foods* **2020**, *9*, 982. [CrossRef]

30. Seddaoui, N.; Amine, A. Smartphone-based competitive immunoassay for quantitative on-site detection of meat adulteration. *Talanta* **2021**, *230*, 122346. [CrossRef]

 *foods*

Article

Gas Chromatography–Mass Spectrometry-Based Metabolite Profiling for the Assessment of Freshness in Gilthead Sea Bream (*Sparus aurata*)

Athanasios Mallouchos *, Theano Mikrou and Chrysavgi Gardeli

Department of Food Science and Human Nutrition, Agricultural University of Athens, Iera Odos 75, 118 55 Athens, Greece; theano.mikrou@gmail.com (T.M.); agardeli@aua.gr (C.G.)
* Correspondence: amallouchos@aua.gr; Tel.: +30-210-529-4581

Received: 10 March 2020; Accepted: 6 April 2020; Published: 9 April 2020

Abstract: Gilthead sea bream (*Sparus aurata*) is one of the most important farmed Mediterranean fish species, and there is considerable interest for the development of suitable methods to assess its freshness. In the present work, gas chromatography–mass spectrometry-based metabolomics was employed to monitor the hydrophilic metabolites of sea bream during storage on ice for 19 days. Additionally, the quality changes were evaluated using two conventional methods: sensory evaluation according to European Union's grading scheme and K-value, the most widely used chemical index of fish spoilage. With the application of chemometrics, the fish samples were successfully classified in the freshness categories, and a partial least squares regression model was built to predict K-value. A list of differential metabolites were found, which were distinguished according to their evolution profile as potential biomarkers of freshness and spoilage. Therefore, the results support the suitability of the proposed methodology to gain information on seafood quality.

Keywords: sea bream; fish; spoilage; metabolomics; multivariate analysis; biomarkers

---

1. Introduction

Gilthead sea bream (*Sparus aurata*) is farmed intensively in Greece and accounts for over half of all production in Europe. In 2018, the volume of production reached 61,000 tons, with a value of EUR 276 million. Greece in particular is expected to double its production by 2030 in order to meet the growing demand and maintain its market position globally [1].

Fish quality is objectively the most important characteristic that affects acceptance by the consumer, and it is dependent on a wide range of factors [2]. Freshness (or degree of spoilage) is a decisive factor in assessing fish quality. Its deterioration begins immediately after slaughter and takes place through biochemical, physicochemical, and microbiological mechanisms [3]. Post-mortem changes depend on species, age, diet, slaughter method, processing, and conditions during transportation and storage, such as temperature, which is the most important factor affecting the commercial life of the product [4]. Preservation on ice is the most common method of maintaining fresh fish, which limits microbial growth, the main cause of spoilage.

The European Community has established common marketing standards for fishery products to assess freshness through organoleptic examination [5]. Thus, the fishing industry must grade the products in three freshness categories defined as Extra, A, and B. Fish not classified in any of these grades are considered unacceptable. Although organoleptic examination is still the most satisfactory way of assessing fish freshness, issues of objectivity and convenience can be claimed if compared with instrumental methods.

From an analytical point of view, several methods have been recommended in order to evaluate fish quality, which rely on the determination of chemical, microbiological, and physical parameters [2].

The K-value, one of the most widely used chemical indexes to monitor fish quality, is based on the measurement of adenosine triphosphate (ATP) and its degradation products, namely, adenosine diphosphate (ADP), adenosine monophosphate (AMP), inosine phosphate (IMP), inosine (INO), and hypoxanthine (Hx) [6]. However, the K-value is subject to large inter- and intra-species variations and is dependent on many factors [7].

The term "metabolomics" refers to the systematic study of low molecular mass metabolites, which vary under a given set of conditions in the cell, tissue, or organism [8]. In recent years, the metabolomics studies on seafood products have been steadily increasing and have focused mainly on the nutritional status of fish [9], differentiation between wild and farmed fish [10–12], and classification according to aquaculture system [13–15], but also some steps have been taken towards seafood freshness. More specifically, the changes in metabolic profiles during cold storage have been investigated on yellowtail [16], bogue [17], mussels [18], and salmon [19]. With regards to sea bream, Picone et al. [20] investigated the molecular profiles using ^{1}H-NMR only at the beginning and the end of iced storage of fish produced with different aquaculture systems. Heude and co-workers [21] proposed a method based on NMR spectroscopy for the rapid determination of K-value and trimethylamine nitrogen content on sea bream, among other fish.

In the present study, gas chromatography–mass spectrometry (GC–MS) was used to monitor the changes in the polar metabolite fraction of sea bream during storage on ice, in order to identify potential markers of freshness and spoilage. Multivariate data analysis was applied to classify fish samples in freshness categories according to EU sensory scheme, and a partial least squares regression (PLS-R) model was built to predict K-value.

2. Materials and Methods

2.1. Fish Provision, Storage, and Sampling

Gilthead sea bream samples (400–600 g, 25–30 cm) were obtained directly from a Greek fish processing plant (PLAGTON S.A., Mitikas, Aitoloakarnania, Greece). Fish were farmed in cages in the geographical area designated as Food and Agriculture Organization (FAO) 37.2.2 (Ionian Sea) and were slaughtered by immersion in ice cold water (hypothermia), packed with flaked ice into self-draining polystyrene boxes, and delivered to the laboratory within 3–4 h of harvesting. Two fish batches, each consisting of 30 ungutted whole fish, were used in the course of two independent storage trials. The fish batches were harvested in April and July of the same year. The fish samples were stored in a refrigerator, and fresh ice was added daily. The harvesting day was considered as day 0 of storage period. The sampling began the next day of storage (day 1), and afterwards continued every 2 days for a total period of 19 days. At each sampling point, three randomly chosen fish were removed from the batch and used for the subsequent analyses.

2.2. Sample Preparation for GC–MS Metabolomics

Fish were treated as described in Association of Official Agricultural Chemists (AOAC) Official Method 937.07. The heads, scales, tails, fins, guts, and inedible bones were removed and discarded. Then, fish were filleted to obtain all flesh and skin from head to tail and from top of back to belly on left side only. Each fillet (white muscle with skin) was cut quickly in small cubes and snap-frozen in liquid nitrogen to quench the metabolism. Tissue grinding was performed in a pre-cooled A11 analytical mill (IKA, Wilmington, NC, USA) to obtain a fine frozen powder. The mill was operated in pulse mode for 10–15 s per grinding batch in order to prevent the thawing of the sample. Aliquots (50 mg) of each powdered sample were accurately weighed (± 0.1 mg) into 2 mL Eppendorf tubes with O-ring screw caps (Sarstedt, Germany) and transferred to −80 °C for storage. The remaining quantity of each sample powder was stored at −80 °C in sealed bags and used for the determination of K-value. Furthermore, from each sampling point, a suitable quantity (10 g) of fish powder was pooled to obtain a single quality control (QC) sample, which was further processed similarly to unknown samples, as described below.

Tissue disruption and subsequent metabolite extraction was undertaken using a Tissuelyser LT (Qiagen, Germantown, MD, USA) according to a modified Bligh and Dyer method [22]. Pre-chilled and degassed homogenization solvent (525 μL methanol/water, 2:0.625 v/v, HPLC grade), internal standard (50 μL glycine-d5, 0.2 mg/mL in 0.1 M HCl), and two stainless-steel balls (2.5 mm diameter) were added to each Eppendorf tube and, subsequently, the fish powder was homogenized for 2 min at 20 Hz. Then, 200 μL of chloroform was added, and the homogenization was repeated for 1 min. Then, another 200 μL of chloroform was added to each tube, and the contents were mixed for 10 min using a cell shaker. During this process, the samples were always kept on ice. Finally, 200 μL HPLC-grade water was added, and the samples were vortex mixed for 15 s. To initiate phase separation, the samples were centrifuged for 2 min at 12,000 rpm. A total of 100 μL of the aqueous fraction was transferred in new Eppendorf tubes with pre-punctured screw caps and lyophilized overnight (12 h). After replacing the caps with new ones, the sample pellets were stored at -80 °C until required for analysis.

Prior to GC–MS analysis, a two-stage chemical derivatization process was carried out to impart volatility to non-volatile metabolites, while also enabling thermal stability [23,24]. The lyophilized samples were left to reach room temperature for 15 min and then 40 μL of 20 mg/mL methoxyamine solution in pyridine (Acros Organics, Geel, Belgium) was added, and they were then incubated at 30 °C for 90 min in an orbital heating block. Subsequently, 70 μL MSTFA (*N*-methyl-*N*-trimethylsilyltrifluoroacetamide—Acros Organics) was added, and the samples were incubated at 37 °C for 90 min. After cooling the samples for 5 min, 20 μL of retention index solution (*n*-alkanes C10–C24, 0.6 mg/mL in pyridine—Sigma Aldrich, Darmstadt, Germany) was added and the contents were transferred to 200 μL conical insert placed in 2 mL vial with screw cap for further GC–MS analysis.

2.3. GC–MS Analysis

GC–MS analysis was carried out using a Shimadzu GCMS QP-2010 Ultra operated with the accompanied GCMS Solution software. Helium was used as a carrier gas at a constant linear velocity of 36 cm/s. Sample injections (1 μL) were performed with AOC 20 s autosampler in split mode (split ratio 1/25). The temperature of the injection port, interface, and ion source was set at 250, 290, and 230 °C, respectively. Separation of compounds was carried out in a MEGA-5HT fused silica capillary column (30 m × 0.25 mm, 0.25 μm film thickness, MEGA S.r.l., Legnano, Italy). Oven temperature was maintained initially at 60 °C for 1 min, then programmed at 10 °C/min to 325 °C, and held for 5 min. The mass spectrometer was operated in electron ionization mode with the electron energy set at 70 eV and a scan range of 70–600 m/z. The samples (QC and blanks included) were analyzed in a predetermined order [24].

2.4. Data Processing Procedure

Raw data were processed with MS-DIAL software, which is freely available at the PRIMe website (http://prime.psc.riken.jp/) [25]. Metabolite identification was performed according to the Metabolomics Standards Initiative at four levels [26]:

- MSI level 1 (identified compounds): based on similarity of retention index (RI) and mass spectrum relative to an authentic compound analyzed under identical experimental conditions.
- MSI level 2 (putatively annotated compounds): agreement of retention index (ΔRI < 20) and mass spectrum (match > 850) coming from the publicly available libraries at PRIMe. Amdis (v. 2.72) and NIST MS Search software (v. 2.2) including NIST 14 mass spectral library were used complimentarily.
- MSI level 3 (putatively characterized compound classes): agreement of RI or mass spectrum to known compounds of a chemical class.
- MSI level 4 (unknown compounds).

The resulting output from this procedure was a retention index vs. sample data matrix with related metabolite IDs and peak heights linked to each sample injection. Subsequently, manual data curation was performed, which included the removal of metabolic features detected in < 50% of QC samples; the combination of metabolite rows that had two or more identical peaks, such as sugars; and normalization to sample mass used in extraction. Finally, the data were normalized to the QC samples using a low-order nonlinear locally estimated smoothing function (LOESS) [24], in order to correct for the signal drift within and between analytical blocks. Afterwards, metabolites with relative standard deviation (RSD) > 30% within pooled QCs were removed. The final data matrix was further processed statistically in MetaboAnalyst 4.0 web-based tool suite [27]. This included multivariate and univariate testing as detailed in the Results and Discussion section. Before statistical processing, the data were log-transformed and mean centered. Partial least squares regression (PLS-R) was performed using The Unscrambler X ver. 10.4 (CAMO Software AS, Oslo, Norway).

2.5. Freshness Assessment

The freshness rating of raw fish was performed by a panel of three trained assessors according to the European Union's grading system [5]. This system distinguishes between three freshness categories (Extra, A, B) corresponding to various levels of spoilage. Category E corresponds to the highest quality level, followed by categories A and B, whereas fish graded below B is considered unacceptable for consumption. In order to rate this evaluation, a 0–3 score scale was used (rating of categories: Extra ≥ 2.7, $2 \leq A < 2.7$, $1 \leq B < 2$, unacceptable < 1) according to [28].

2.6. ATP Breakdown Products

ATP and its degradation products (ADP, AMP, IMP, Ino, and Hx) were isolated from fish tissue according to Ryder [29]. Chromatography was performed using a JASCO HPLC system (JASCO International Co., Ltd., Tokyo, Japan) consisting of a quaternary pump (PU-2089 Plus), an autosampler (AS-1555), and a photodiode array detector (MD-910). The separation was accomplished with a Luna C18 column (250 mm × 4.6 mm i.d., 5 μm; Phenomenex, Torrance, CA, USA) using gradient elution. Mobile phase A was a 0.05 M phosphate buffer (pH 7) and mobile phase B was acetonitrile (Sigma Aldrich, Louis, MI, USA). The elution program was as follows: 0 min, 100% A; 9 min, 97% A; 15 min, 85% A; 17 min, 60% A. Final conditions were kept for 7 min and the column was equilibrated for 15 min at initial conditions. The flow rate was set at 1 mL/min and the injection volume was 20 μL. The monitoring wavelength was set at 254 nm and the molar concentration of ATP breakdown products were calculated from their corresponding calibration curves using the external standard method. The *K*-value (%) was calculated from Equation (1):

$$K(\%) = \frac{(Ino + Hx)}{(ATP + ADP + AMP + IMP + Ino + Hx)} \times 100\% \tag{1}$$

3. Results and Discussion

3.1. Freshness Assessment Using Classical Methods

Quality deterioration of fish during storage on ice was monitored using a sensory method (EU grading system), and a chemical one (*K*-value), which is based on the measurement of ATP breakdown products. The changes in sensory score and *K*-value of sea bream during 19 storage days on ice are shown in Figure 1. As expected, the sensory score decreased linearly ($y = -0.1404x + 3.0057$), showing high negative correlation with storage time ($r = -0.9880$, $p < 0.001$). Until day 1 of storage, the freshness rating of fish was evaluated as Extra. From day 3 to day 7, the freshness of fish was rated A, whereas the category B was assigned to fish stored between 9–15 days. The limit of acceptability of raw sea bream stored on ice was about 16–17 days. As the sensory quality of fish decreased, the *K*-value increased linearly ($y = 2.5033x + 3.3305$), showing a high positive correlation with storage time

($r = 0.9974$, $p < 0.001$) and a negative correlation with sensory score ($r = -0.9809$, $p < 0.001$). When the fish was considered unacceptable (day 17), the K-value was 45%. Similar findings have been reported by others authors [6,30,31]. Small variations could be attributed to the different rearing area and farming method among others.

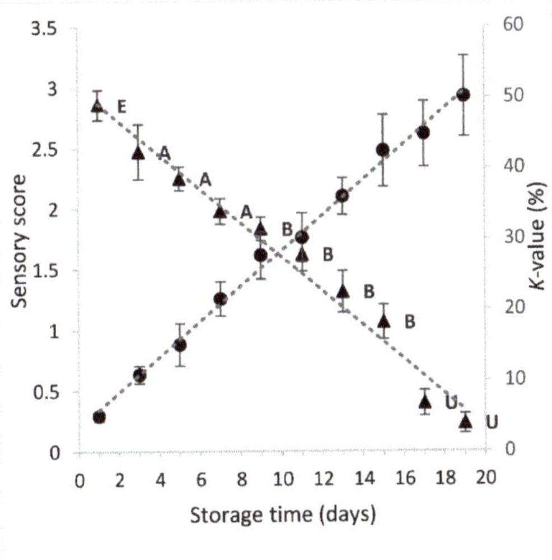

Figure 1. Changes in sensory score (-▲-) and K-value (-●-) of sea bream stored in ice. Each point represent the mean value of six replicate measurements (three fish samples x two storage experiments at each sampling point). Error bars denote standard deviation. The labels of sensory scores denote the freshness category according to EU grading system.

3.2. Freshness Assessment Using GC–MS Metabolomics

Before starting the real data elaboration, method performance was evaluated using several quality control criteria. The primary requirement was to check the relative abundance of amino acids and sugar trimethylsilyl (TMS) derivatives in QCs. A detailed description is provided elsewhere [23]. After passing the above criteria, the internal standard performance was checked. Deuterated glycine-d5 was added in every sample (including QCs and blanks) to monitor the extraction procedure. For this reason, the RSD% of peak height was calculated and the value obtained was 5.4% for QCs ($n = 15$) and 15.3% for samples ($n = 54$). As a last check, a preliminary principal component analysis (PCA) was obtained with the peak heights of the entire dataset. A tight clustering of the QCs, as well as blank samples, was observed in the score plot (Figure S1), which is a further confirmation of the robustness of the analytical procedure, not only for the internal standard but for the whole fingerprint of fish.

After data curation, the samples were grouped in 10 classes (from 0 to 9) according to the sampling sequence during fish storage on ice (class 0 represents the first day of storage, class 1 the third day, etc.). Principal component analysis (PCA) of data exhibited a significant ability to separate samples according to storage time (Figure 2a). It is evident that very fresh samples (class 0, day 1) were clearly separated from the other classes, and most distinctively from those at the later stages of storage (class 8, 9). This shows that there is quantitative information in these data, as PC1, with an explained variance of 62.5%, is the component that describes the evolution of fish spoilage. This was depicted even more clearly in the plot of PC1 scores values vs. storage time (Figure 2b). Thus, PC1 can condensate all information of the metabolite features and give a measure of the molecular quality of fish.

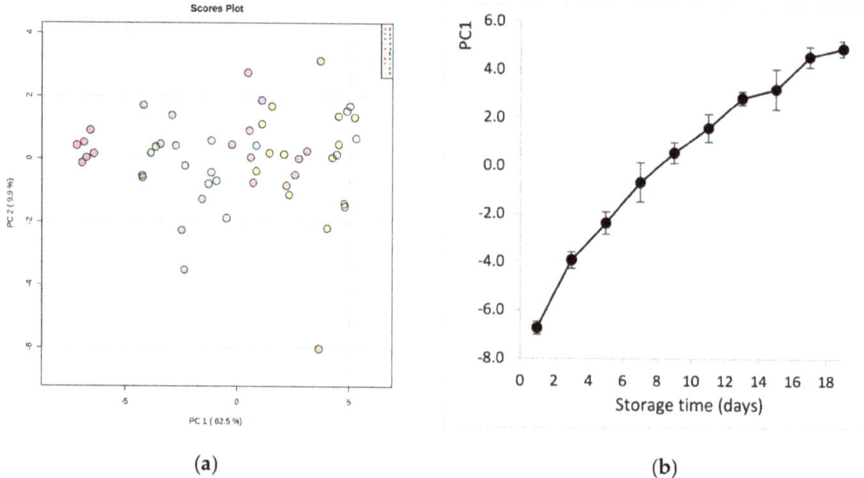

(a)   (b)

Figure 2. (**a**) Principal component analysis (PCA) score plot derived from the hydrophilic metabolites of sea bream during storage on ice. The legend indicates the sampling sequence (0-1-2-3-4-5-6-7-8-9) that corresponds to storage day (1-3-5-7-9-11-13-15-17-19), respectively; (**b**) evolution of sea bream spoilage as described by PC1 scores values vs. storage time on ice.

Studying the multivariate loadings values revealed various metabolites that either increased or decreased with fish storage. The most important loadings were established and confirmed by Kruskal–Wallis ANOVA, as well as by performing Spearman's correlation analysis. Figure 3 shows the top 25 highly correlating and significant metabolites ($p < 0.05$) that either increased (shown in red; positive R) or decreased during storage (shown in blue; negative R).

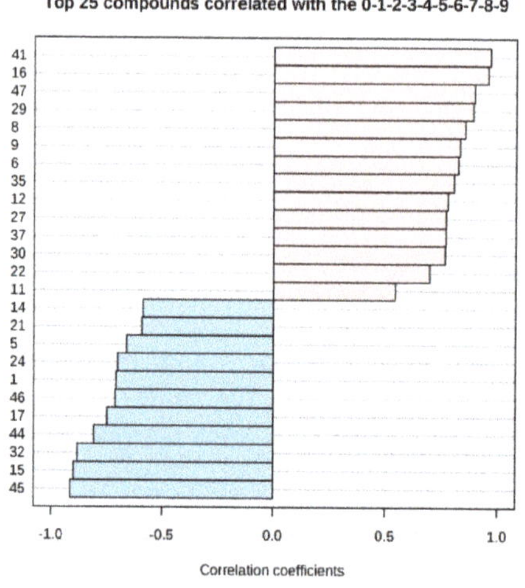

Figure 3. Pattern recognition—Spearman's correlation analysis showing the top 25 metabolite features correlated significantly with sampling sequence (0-1-2-3-4-5-6-7-8-9 is equivalent to storage day 1-3-5-7-9-11-13-15-17-19). Each row represents the most significant metabolite identified from the test ($p < 0.05$). The *x*-axis shows correlation score, whereas the *y*-axis corresponds to gas chromatography–mass spectrometry (GC–MS) peak number from peak index (see Table 1).

After these encouraging results, the data were grouped into four classes (0 to 3) representing the EU freshness grades (i.e., grade Extra: class 0; grade A: class 1; grade B: class 2; unacceptable: class 3), as described in EC no. 2396/1996. Partial least squares discriminant analysis (PLS-DA) was carried out in order to find a discriminant index of freshness. It is evident that the supervised model (Figure 4a) can clearly classify the samples into the correct freshness grade. Although there was some overlap of the confidence ellipses of grade A (class 1) and B (class 2), the grade Extra (class 0) was further apart from unacceptable samples (class 3). Similarly to the aforementioned PCA results, it seems that this separation was described mainly by PC1, which accounted for the 65.5% of model variance. The optimal number of components, as calculated by 10-fold cross-validation, was 3 (Figure S2). The predictive ability of the model (Q^2), accuracy, and coefficient of determination (R^2) were satisfactory (0.94, 0.97, and 0.76, respectively). The significance of class discrimination was verified by performing a permutation test ($p < 0.001$; 0/1000), and the performance was measured using group separation distance (B/W ratio) [32] (Figure S3). The variable importance in projection (VIP) scores were also calculated from the PLS-DA, and Figure 4b highlights the top 15 highly significant metabolites that were identified for each freshness grade.

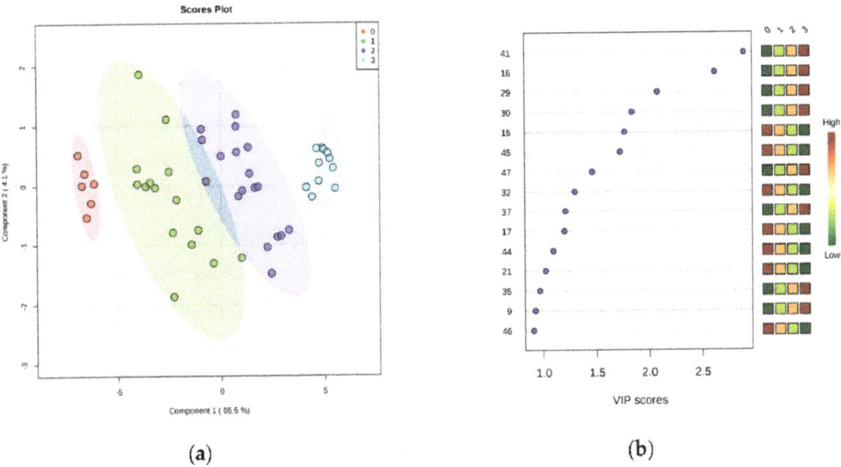

Figure 4. (a) Partial least squares discriminant analysis (PLS-DA) scores plotted for freshness grades of sea bream stored on ice. The legend indicates the four EU grades: Extra (0), A (1), B (2), unacceptable (3). (b) Top 15 metabolite features based on variable importance in projection (VIP) scores from PLS-DA. The *x*-axis shows the scores whereas the *y*-axis corresponds to GC–MS peak number from peak index (see Table 1). Color bars show median intensity of metabolite feature in the respective group.

The combined result of PCA, PLS-DA, and Kruskal–Wallis ANOVA is summarized in Table 1, which shows a list of metabolites significantly correlated with fish storage on ice. They can be distinguished in two groups—in the first group were metabolites whose relative content increased significantly during storage, whereas the second group comprised metabolites with a decreasing trend. Thus, we can infer that the first group of compounds constitute potential markers of spoilage, whereas the second group could be markers of freshness. Their respective evolution pattern during storage is summarized in Tables S1 and S2. The relative concentration of six amino acids (leucine, isoleucine, valine, phenylalanine, tyrosine, methionine) increased during storage on ice as a result of autolysis and bacterial spoilage. Similar findings were observed in bogue [17] and salmon [19]. On the contrary, the observed decrease of glycine and glutamic acid probably indicates that their degradation by microorganisms occurred in a higher rate than their release from muscle proteins. This is in contrast to the aforementioned studies, but can be rationalized by the different spoilage microorganisms developing in each fish species. The amount of inosine increased during storage, as expected, and was confirmed also by HPLC analysis of ATP breakdown products. The levels of succinic, malic, and fumaric acid, involved in the Krebs cycle, decreased during storage, thus indicating a preferential consumption for bacterial growth. In fact, organic acids or amino acids rather than glucose are the preferred carbon sources for *Pseudomonas* [33], the dominating spoilage genus in sea bream at low temperatures [34]. The increasing trend of some sugars, such as ribose and galactose, has been observed previously in other aquatic products, such as mussels and yellowtail fish [16,18]. On the contrary, the sugar phosphates, such as ribulose 5-phosphate, which is the end product of pentose phosphate pathway (oxidative branch), decreased significantly with storage, and thus may represent freshness markers. It should be noted here that the interpretation of the biological significance of metabolomics data is not always straightforward. The main difficulty arises from the nature of freshness loss, which is a process described primarily by two different phenomena—the autolysis from endogenous enzymes and the spoilage due to microbial growth. In addition, the whole picture is complicated by the evolution of microbial diversity that leads to shifts of metabolite profiles.

Table 1. Metabolites that either increased or decreased significantly[1] during storage of sea bream on ice.

Peak Number	Significant Metabolites	MSI Level	Identifier from Relevant Database
	Increasing trend		
41	Gluconic acid	2	HMDB0000625
16	Glyceric acid	2	CHEBI:32398
29	Ribose	2	CHEBI:47014
37	Galactose	1	CHEBI:4139
8	Ethanolamine	2	CHEBI:16000
47	Inosine	2	CHEBI:17596
9	Leucine	1	CHEBI:25017
6	Valine	1	CHEBI:16414
27	Phenylalanine	1	CHEBI:17295
35	Tyrosine	2	CHEBI:17895
12	Isoleucine	2	CHEBI:17191
22	Methionine	1	CHEBI:16643
11	Glycerol	1	CHEBI:17754
30	Ribitol	2	CHEBI:15963
	Decreasing trend		
45	Ribulose 5-phosphate	2	CHEBI:17363
44	Arabinose-5-phosphate	2	CHEBI:16241
46	Sugar phosphate_2381[2]	3	N/A
32	Glycerol 3-phosphate	2	CHEBI:15978
15	Succinic acid	1	CHEBI:15741
17	Fumaric acid	1	CHEBI:18012
21	Malic acid	1	HMDB0000744
2	Lactic acid	1	CHEBI:78320
5	a-Aminobutyric acid	2	CHEBI:35621
24	Creatinine	2	CHEBI:16737
1	Methylamine	2	CHEBI:16830
14	Glycine	1	CHEBI:15428
26	Glutamic acid	2	HMDB00148

[1] According to the combined results of Kruskal–Wallis ANOVA, Spearman's correlation analysis, and VIP scores from PLS-DA. [2] The number denotes the *n*-alkane retention index in MEGA HT-5 column.

The analytes of Table 1 were used further in PLS-R as input variables (predictors, X) in order to predict K-value (output variable, Y). Segmented cross validation was employed using 18 segments with three replicate samples each. Thus, each segment corresponded to a sampling point (18 segments = 9 sampling points × 2 fish batches). External validation using an independent test set was not performed due to the relatively small number of samples ($n = 54$) under study. The optimal number of factors was 3 and explained the 95% of total variance. The performance metrics and the regression line of the model are presented in Figure 5. The root mean square error (RMSE) and the coefficient of determination (R^2) at the validation stage suggested good prediction performance, with their values being 3.4710 (K-value %) and 0.9473, respectively. According to the slope of the regression line (0.9546), there was an almost perfect linear relationship between the predicted and the measured K-values.

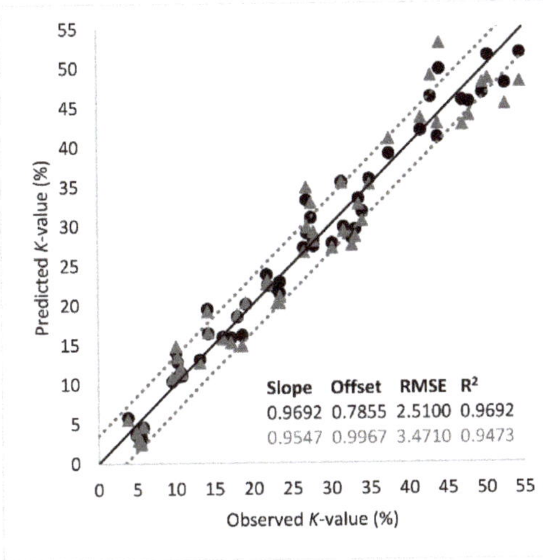

Figure 5. Comparison between the observed and predicted K-values (%) by the partial least squares regression (PLS-R) model based on important metabolites listed in Table 1. The shape of symbols denotes the calibration (-•-) and the validation (-▲-) set (black line: the ideal $y = x$ line; dotted lines: ± 3.5% K-value).

4. Conclusions

The present study demonstrated for the first time that GC–MS-based metabolomics is an efficient tool to monitor the quality loss of sea bream during storage on ice. We have clearly presented a panel of hydrophilic metabolites linked directly to storage time of fish that could be used as potential markers of freshness and spoilage. Additionally, with the application of multivariate data analysis, the samples were successfully classified to freshness grades, and the K-value was predicted using a PLS-R model. Therefore, our results support the suitability of the proposed methodology to gain information on seafood quality. However, we should note that an approach based on the hydrophilic fraction of metabolites solely, as described here, is not enough for the complete elucidation of the post-mortem changes occurring at the molecular level. Hence, future applications should include the investigation of the lipophilic metabolites using larger-scale storage experiments in combination with microbiological and sensorial data.

Supplementary Materials: The following are available online at http://www.mdpi.com/2304-8158/9/4/464/s1, Figure S1: PCA score plot for the entire data set before data curation. The legend indicates the type of samples (BL: blank, QC: quality control, S: sample). Figure S2: PLS-DA classification using different number of components. The red star indicates the best classifier based on Q2. Figure S3: PLS-DA model validation by permutation tests based on group separation distance (B/W). Table S1: Most significant metabolites with an increasing trend during storage of sea bream on ice. The y-axis represents the relative amount before and after variable transformation, respectively. The x-axis shows sampling sequence that is equivalent to storage time. Table S2: Most significant metabolites with a decreasing trend during storage of sea bream on ice. The y-axis represents the relative amount before and after variable transformation, respectively, for the bar plot and the box plot. The x-axis shows sampling sequence that is equivalent to storage time.

Author Contributions: Conceptualization, A.M.; methodology, A.M.; formal analysis, A.M.; investigation, T.M. and C.G.; data curation, A.M.; writing—original draft preparation, A.M., T.M., and C.G.; writing—review and editing, A.M., T.M., and C.G.; supervision, A.M. and C.G.; project administration, C.G. All authors have read and agreed to the published version of the manuscript.

Acknowledgments: The authors would like to thank PLAGTON S.A. (Aitoloakarnania, Greece) for the provision of fish.

Conflicts of Interest: The authors declare no conflict of interest.

References

1. Federation of Greek Maricultures. Available online: https://fgm.com.gr/uploads/file/FGM_19_ENG_WEB_spreads.pdf (accessed on 7 December 2019).

2. Botta, J.R. *Evaluation of Seafood Freshness Quality*, 1st ed.; VCH Publishers, Inc.: New York, NY, USA, 1995; ISBN 978-0-471-18580-2.

3. Dalgaard, P. FISH|Spoilage of Seafood. In *Encyclopedia of Food Sciences and Nutrition*; Caballero, B., Trugo, L., Finglas, P.M., Eds.; Academic Press: Cambridge, UK, 2003; pp. 2462–2471. ISBN 978-0-12-227055-0.

4. Olafsdottir, G.; Nesvadba, P.; Di Natale, C.; Careche, M.; Oehlenschläger, J.; Tryggvadóttir, S.V.; Schubring, R.; Kroeger, M.; Heia, K.; Esaiassen, M.; et al. Multisensor for fish quality determination. *Trends Food Sci. Technol.* **2004**, *15*, 86–93. [CrossRef]

5. Council Regulation (EC) No 2406/96 of 26 November 1996. Laying down common marketing standards for certain fishery products. *Off. J. Eur. Communities* **1996**, *L334*, 1–15.

6. Alasalvar, C.; Taylor, K.D.A.; Öksüz, A.; Garthwaite, T.; Alexis, M.N.; Grigorakis, K. Freshness assessment of cultured sea bream (Sparus aurata) by chemical, physical and sensory methods. *Food Chem.* **2001**, *72*, 33–40. [CrossRef]

7. Huidobro, A.; Pastor, A.; Tejada, M. Adenosine Triphosphate and Derivatives as Freshness Indicators of Gilthead Sea Bream (Sparus aurata). *Food Sci. Technol. Int.* **2001**, *7*, 23–30.

8. Dunn, W.B.; Ellis, D.I. Metabolomics: Current analytical platforms and methodologies. *TrAC Trends Anal. Chem.* **2005**, *24*, 285–294.

9. Gil-Solsona, R.; Nácher-Mestre, J.; Lacalle-Bergeron, L.; Sancho, J.V.; Calduch-Giner, J.A.; Hernández, F.; Pérez-Sánchez, J. Untargeted metabolomics approach for unraveling robust biomarkers of nutritional status in fasted gilthead sea bream (*Sparus aurata*). *PeerJ* **2017**, *5*, e2920. [CrossRef]

10. Melis, R.; Cappuccinelli, R.; Roggio, T.; Anedda, R. Addressing marketplace gilthead sea bream (Sparus aurata L.) differentiation by 1H NMR-based lipid fingerprinting. *Food Res. Int.* **2014**, *63*, 258–264. [CrossRef]

11. Rezzi, S.; Giani, I.; Héberger, K.; Axelson, D.E.; Moretti, V.M.; Reniero, F.; Guillou, C. Classification of Gilthead Sea Bream (*Sparus aurata*) from ^{1}H NMR Lipid Profiling Combined with Principal Component and Linear Discriminant Analysis. *J. Agric. Food Chem.* **2007**, *55*, 9963–9968. [CrossRef]

12. Mannina, L.; Sobolev, A.P.; Capitani, D.; Iaffaldano, N.; Rosato, M.P.; Ragni, P.; Reale, A.; Sorrentino, E.; D'Amico, I.; Coppola, R. NMR metabolic profiling of organic and aqueous sea bass extracts: Implications in the discrimination of wild and cultured sea bass. *Talanta* **2008**, *77*, 433–444. [CrossRef]

13. Savorani, F.; Picone, G.; Badiani, A.; Fagioli, P.; Capozzi, F.; Engelsen, S.B. Metabolic profiling and aquaculture differentiation of gilthead sea bream by 1H NMR metabonomics. *Food Chem.* **2010**, *120*, 907–914. [CrossRef]

14. Melis, R.; Anedda, R. Biometric and metabolic profiles associated to different rearing conditions in offshore farmed gilthead sea bream (*Sparus aurata* L.): General. *Electrophoresis* **2014**, *35*, 1590–1598. [CrossRef] [PubMed]

15. Del Coco, L.; Papadia, P.; De Pascali, S.; Bressani, G.; Storelli, C.; Zonno, V.; Fanizzi, F.P. Comparison among Different Gilthead Sea Bream (Sparus aurata) Farming Systems: Activity of Intestinal and Hepatic Enzymes and 13C-NMR Analysis of Lipids. *Nutrients* **2009**, *1*, 291–301. [CrossRef] [PubMed]

16. Mabuchi, R.; Adachi, M.; Ishimaru, A.; Zhao, H.; Kikutani, H.; Tanimoto, S. Changes in Metabolic Profiles of Yellowtail (Seriola quinqueradiata) Muscle during Cold Storage as a Freshness Evaluation Tool Based on GC-MS Metabolomics. *Foods* **2019**, *8*, 511. [CrossRef] [PubMed]

17. Ciampa, A.; Picone, G.; Laghi, L.; Nikzad, H.; Capozzi, F. Changes in the Amino Acid Composition of Bogue (Boops boops) Fish during Storage at Different Temperatures by 1H-NMR Spectroscopy. *Nutrients* **2012**, *4*, 542–553. [CrossRef]

18. Aru, V.; Pisano, M.B.; Savorani, F.; Engelsen, S.B.; Cosentino, S.; Cesare Marincola, F. Metabolomics analysis of shucked mussels' freshness. *Food Chem.* **2016**, *205*, 58–65. [CrossRef]

19. Shumilina, E.; Ciampa, A.; Capozzi, F.; Rustad, T.; Dikiy, A. NMR approach for monitoring post-mortem changes in Atlantic salmon fillets stored at 0 and 4°C. *Food Chem.* **2015**, *184*, 12–22. [CrossRef]

20. Picone, G.; Balling Engelsen, S.; Savorani, F.; Testi, S.; Badiani, A.; Capozzi, F. Metabolomics as a Powerful Tool for Molecular Quality Assessment of the Fish Sparus aurata. *Nutrients* **2011**, *3*, 212–227. [CrossRef]

21. Heude, C.; Lemasson, E.; Elbayed, K.; Piotto, M. Rapid Assessment of Fish Freshness and Quality by 1H HR-MAS NMR Spectroscopy. *Food Anal. Methods* **2015**, *8*, 907–915. [CrossRef]

22. Wu, H.; Southam, A.D.; Hines, A.; Viant, M.R. High-throughput tissue extraction protocol for NMR- and MS-based metabolomics. *Anal. Biochem.* **2008**, *372*, 204–212. [CrossRef]

23. Fiehn, O. Metabolomics by gas chromatography-mass spectrometry: Combined targeted and untargeted profiling. *Curr. Protoc. Mol. Biol.* **2016**, *114*, 30–34. [CrossRef]

24. The Human Serum Metabolome (HUSERMET) Consortium; Dunn, W.B.; Broadhurst, D.; Begley, P.; Zelena, E.; Francis-McIntyre, S.; Anderson, N.; Brown, M.; Knowles, J.D.; Halsall, A.; et al. Procedures for large-scale metabolic profiling of serum and plasma using gas chromatography and liquid chromatography coupled to mass spectrometry. *Nat. Protoc.* **2011**, *6*, 1060–1083. [CrossRef] [PubMed]

25. Lai, Z.; Tsugawa, H.; Wohlgemuth, G.; Mehta, S.; Mueller, M.; Zheng, Y.; Ogiwara, A.; Meissen, J.; Showalter, M.; Takeuchi, K.; et al. Identifying metabolites by integrating metabolome databases with mass spectrometry cheminformatics. *Nat. Methods* **2018**, *15*, 53–56. [CrossRef] [PubMed]

26. Sumner, L.W.; Amberg, A.; Barrett, D.; Beale, M.H.; Beger, R.; Daykin, C.A.; Fan, T.W.-M.; Fiehn, O.; Goodacre, R.; Griffin, J.L.; et al. Proposed minimum reporting standards for chemical analysis: Chemical Analysis Working Group (CAWG) Metabolomics Standards Initiative (MSI). *Metabolomics* **2007**, *3*, 211–221. [CrossRef]

27. Chong, J.; Wishart, D.S.; Xia, J. Using MetaboAnalyst 4.0 for Comprehensive and Integrative Metabolomics Data Analysis. *Curr. Protoc. Bioinforma.* **2019**, *68*, e86. [CrossRef] [PubMed]

28. Council Regulation (EEC) No 103/76 of 19 January 1976. Laying down common marketing standards for certain fresh or chilled fish. *Off. J. Eur. Communities* **1976**, *L20*, 29–34.

29. Ryder, J.M. Determination of adenosine triphosphate and its breakdown products in fish muscle by high-performance liquid chromatography. *J. Agric. Food Chem.* **1985**, *33*, 678–680. [CrossRef]

30. Lougovois, V.P.; Kyranas, E.R.; Kyrana, V.R. Comparison of selected methods of assessing freshness quality and remaining storage life of iced gilthead sea bream (Sparus aurata). *Food Res. Int.* **2003**, *36*, 551–560. [CrossRef]

31. Šimat, V.; Bogdanović, T.; Krželj, M.; Soldo, A.; Maršić-Lučić, J. Differences in chemical, physical and sensory properties during shelf life assessment of wild and farmed gilthead sea bream (Sparus aurata, L.): Shelf life parameters of wild and farmed gilthead sea bream. *J. Appl. Ichthyol.* **2012**, *28*, 95–101. [CrossRef]

32. Bijlsma, S.; Bobeldijk, I.; Verheij, E.R.; Ramaker, R.; Kochhar, S.; Macdonald, I.A.; van Ommen, B.; Smilde, A.K. Large-Scale Human Metabolomics Studies: A Strategy for Data (Pre-) Processing and Validation. *Anal. Chem.* **2006**, *78*, 567–574. [CrossRef]

33. Rojo, F. Carbon catabolite repression in Pseudomonas: Optimizing metabolic versatility and interactions with the environment. *FEMS Microbiol. Rev.* **2010**, *34*, 658–684. [CrossRef]

34. Parlapani, F.F.; Mallouchos, A.; Haroutounian, S.A.; Boziaris, I.S. Microbiological spoilage and investigation of volatile profile during storage of sea bream fillets under various conditions. *Int. J. Food Microbiol.* **2014**, *189*, 153–163. [CrossRef] [PubMed]

foods

MDPI

Article

Upstream Regulator Analysis of Wooden Breast Myopathy Proteomics in Commercial Broilers and Comparison to Feed Efficiency Proteomics in Pedigree Male Broilers

Walter G. Bottje [1,*], Kentu R. Lassiter [1], Vivek A. Kuttappan [2], Nicholas J. Hudson [3], Casey M. Owens [1], Behnam Abasht [4], Sami Dridi [1] and Byungwhi C. Kong [1]

1   Department of Poultry Science & The Center of Excellence for Poultry Science, University of Arkansas, 1260 W. Maple, Fayetteville, AR 72701, USA; klassit@uark.edu (K.R.L.); cmowens@uark.edu (C.M.O.); dridi@uark.edu (S.D.); bkong@uark.edu (B.C.K.)
2   Novus International, 20 Research Park, St. Charles, MO 63304, USA; Vivek.Kuttappan@novusint.com
3   School of Agriculture and Food Sciences, The University of Queensland, Gatton, Brisbane, QLD 4072, Australia; n.hudson@uq.edu.au
4   Department of Animal and Food Sciences, University of Delaware, Newark, DE 19716, USA; abasht@Udel.edu
*   Correspondence: wbottje@uark.edu

Citation: Bottje, W.G.; Lassiter, K.R.; Kuttappan, V.A.; Hudson, N.J.; Owens, C.M.; Abasht, B.; Dridi, S.; Kong, B.C. Upstream Regulator Analysis of Wooden Breast Myopathy Proteomics in Commercial Broilers and Comparison to Feed Efficiency Proteomics in Pedigree Male Broilers. *Foods* 2021, 10, 104. https://doi.org/10.3390/foods10010104

Received: 2 November 2020
Accepted: 23 December 2020
Published: 6 January 2021

Publisher's Note: MDPI stays neutral with regard to jurisdictional claims in published maps and institutional affiliations.

Abstract: In an effort to understand the apparent trade-off between the continual push for growth performance and the recent emergence of muscle pathologies, shotgun proteomics was conducted on breast muscle obtained at ~8 weeks from commercial broilers with wooden breast (WB) myopathy and compared with that in pedigree male (PedM) broilers exhibiting high feed efficiency (FE). Comparison of the two proteomic datasets was facilitated using the overlay function of Ingenuity Pathway Analysis (IPA) (Qiagen, CA, USA). We focused on upstream regulator analysis and disease-function analysis that provides predictions of activation or inhibition of molecules based on (a) expression of downstream target molecules, (b) the IPA scientific citation database. Angiopoeitin 2 (ANGPT2) exhibited the highest predicted activation Z-score of all molecules in the WB dataset, suggesting that the proteomic landscape of WB myopathy would promote vascularization. Overlaying the FE proteomics data on the WB ANGPT2 upstream regulator network presented no commonality of protein expression and no prediction of ANGPT2 activation. Peroxisome proliferator coactivator 1 alpha (PGC1α) was predicted to be inhibited, suggesting that mitochondrial biogenesis was suppressed in WB. PGC1α was predicted to be activated in high FE pedigree male broilers. Whereas RICTOR (rapamycin independent companion of mammalian target of rapamycin) was predicted to be inhibited in both WB and FE datasets, the predictions were based on different downstream molecules. Other transcription factors predicted to be activated in WB muscle included epidermal growth factor (EGFR), X box binding protein (XBP1), transforming growth factor beta 1 (TGFB1) and nuclear factor (erythroid-derived 2)-like 2 (NFE2L2). Inhibitions of aryl hydrocarbon receptor (AHR), AHR nuclear translocator (ARNT) and estrogen related receptor gamma (ESRRG) were also predicted in the WB muscle. These findings indicate that there are considerable differences in upstream regulators based on downstream protein expression observed in WB myopathy and in high FE PedM broilers that may provide additional insight into the etiology of WB myopathy.

Keywords: wooden breast; myopathy; proteomics; feed efficiency; upstream regulator analysis

1. Introduction

The incidence of wooden breast (WB) muscle myopathy has increased in the last several years, possibly in response to genetic selection for growth performance and a shift to heavier market weight birds [1–3]. There have been several reports recently that have investigated WB myopathy to identify genetic signatures and histological defects in the progression of the disease (e.g., [3–9]). Collectively, these studies point toward;

(a) phlebitis and inflammation, (b) oxidative stress and metabolic dysfunction, (c) myofiber degeneration, (d) lipid deposition and (e) development of hard pectoral muscle, particularly in the cranial region. All of these factors contribute to the poor muscle quality and lower shelf life of processed breast muscle fillets [10].

We have conducted a shotgun proteomic analysis on breast muscle tissue from pedigree male (PedM) broilers phenotyped for feed efficiency (FE) [11] and from commercial broilers with and without the presence of severe WB muscle myopathy [2]. It has been hypothesized that the increased incidence of WB might be related to the increased age at market weight as well as faster growth rates in broilers. Thus, there is a concern that increased selection for growth performance might be contributing to the increased incidence of WB myopathy. While WB is an area with unknowns, damaged mitochondria, low capillarity, lactic acidosis and hypoxia are all suggestive of an aerobic energy supply issue. Ingenuity Pathway Analysis (IPA) software was used to facilitate the organization and interpretation of the proteomic data in the studies by Kong et al. [11] and Kuttappan et al. [2]. A powerful feature of the IPA program is the ability to compare datasets using an overlay function in which data from one study can be projected (overlaid) onto another dataset to reveal similarities and differences between the two datasets. It can also be used to connect upstream regulators to diseases and functions. This overlay function can produce hypotheses for future studies and insight into fundamental mechanisms between these global expression datasets. Therefore, the intent of the current study is, firstly, to conduct upstream regulator analysis of WB data not previously reported in Kuttappan et al. [2], and secondly, to conduct comparisons of the WB data with the proteomic data associated with feed efficiency reported by Kong et al. [11].

2. Materials and Methods

2.1. Ethics Statement

The present study was conducted in accordance with the recommendations in the guide for the care and use of laboratory animals of the National Institutes of Health. All procedures for animal care complied with the University of Arkansas Institutional Animal Care and Use Committee (IACUC): Protocol Nos. 14,012 and 17,080.

2.2. Wooden Breast Myopathy Samples in Commercial Broilers

Breast muscle samples for wooden breast (WB) proteomics were obtained from male broilers ($n = 24$) that were randomly selected from a floor pen study at 52 d in a previous study [2]. After weighing, the birds were euthanized using carbon dioxide. The skin over the *Pectoralis major* muscle was excised and the muscle scored for woody breast [12]. Breast muscle samples (~5 g from right cranial region of the *Pectoralis major)* were collected, immediately snap frozen, and stored at $-80\ ^\circ$C. Based on the lesions scores from histological analysis conducted in the same region, the muscle samples were categorized for severity of muscle lesions. From the 24 muscle samples, five samples were chosen that exhibited normal (NORM) histology and five others were chosen with severe (SEV) amounts of lesions to be subjected to proteomic analysis [2].

2.3. Breast Muscle Samples in Pedigree Male Broilers

Shotgun proteomics was conducted on breast muscle samples obtained from Pedigree Male (PedM) broilers individually phenotyped for high or low feed efficiency (FE) ($n = 4$ per group) [11]. These samples had been obtained from the right *Pectoralis muscle* in PedM broilers between 8–9 weeks that were humanely killed as part of a larger study [13].

2.4. Protein Extraction

In Kong et al. [11], muscle was homogenized in 1.5 mL of 20 mM potassium phosphate buffer at pH 7.4 using a hand-held Tissue-Tearor (Biospec Products Inc., Bartlesville, OK, USA) at speeds varying from 5000 rpm to 32,000 rpm. Following homogenization, samples were centrifuged at $10,000\times g$, and the supernatant was collected. In Kuttappan et al. [2], protein samples were precipitated using TCA-acetone with 8 M urea, 100 mM Tris HCl (pH 8.5) with 5 mM Tris 2-carboxyethyl phosphine at room temperature. Following dissolution and reduction, a 1/20th volume of 200 mM iodoacetamide was added and alkylation carried out for 15 min in the dark at room temperature. The sample was then diluted with four volumes of 100 mM TrisHCl, and digested with trypsin overnight at 37 °C. Protein concentrations in the supernatant in both studies were determined using the Bicinchoninic Acid Protein Assay (Sigma Aldrich, St. Louis, MO, USA). NORM and SEV muscle samples were then diluted to a protein concentration of 20 µg/150 µL in phosphate buffer, and the samples were stored at −80 °C until further analysis.

2.5. Shotgun Proteomics

Individual extracted proteins were used in shotgun proteomics analysis with in-gel trypsin digestion followed by tandem mass spectrometry (MS/MS) conducted at the University of Arkansas for Medical Science proteomics core lab (UAMS, Little Rock, AR, USA) [11]. Raw data were analyzed by database searching using Masco (Matrix Science, Boston, MA, USA) and UniProtKB database (http://www.uniprot.org/help/uniprotkb) with the result compiled using the Scaffold Program (Proteome Software, Portland, OR, USA).

In Kuttappan et al. [2], digested sample homogenates were acidified with 1% formic acid, and purified by reversed phase chromatography using C18 affinity media (Omix-Agilent, Santa Clara, CA, USA). Each sample was subjected to three replicate analyses for LC-MS/MS using a hybrid-OrbitrapXL mass spectrometer (ThermoFisher Scientific, Waltham, MA, USA) according to Voruganti et al. [14]. Mass spectrometry analysis was conducted in the DNA/Protein Resource Facility (Oklahoma State University, Stillwater, OK, USA).

2.6. Upstream Regulator Analysis

Upstream regulator analysis by IPA is based on: (a) the number and degree of differential expression of downstream target molecules in the existing dataset and (b) prior knowledge of relationships between upstream transcriptional regulators and their downstream target molecules obtained in published literature citations that have been curated and stored in the IPA program. Upstream regulator analysis determines how many known targets or regulators are within the user's dataset, and compares each differentially expressed molecule to the reported relationship in the literature. If the observed direction of change is mostly consistent with either activation or inhibition of the transcriptional regulator, then a prediction is made and an activation z score generated that is also based on literature-derived regulation direction (i.e., "activating" or "inhibiting"). Activation z scores >2.0 indicate that a molecule is activated whereas activation z scores of <−2.0 indicate that a target molecule is inhibited. Qualified predictions can also be made with activation z scores between 2.0 and −2.0. The *p*-value of overlap measures whether there is a statistically significant overlap between the dataset molecules and those regulated by an upstream regulator is calculated using Fisher's Exact Test, and significance is attributed to *p*-values < 0.05.

3. Results and Discussion

The main focus of this study is on upstream regulator analysis of severe-fulminant WB myopathy. However, as parts of the results and discussion below are concerned with comparing the WB myopathy proteomic data to a dataset obtained in PedM broilers phenotyped for high and low FE, a description of proteomics methods and of the animals from which samples were obtained for the two datasets is warranted.

3.1. Phenotypic Data for Birds in Wooden Breast and Feed Efficiency Studies

Body weights, wooden breast scores for commercial broilers and growth performance during phenotyping for feed efficiency in PedM broilers from which muscle samples were obtained for proteomic studies are presented in Table 1. It can be seen that there were significant differences in body weight and wooden breast (WB) myopathy scores in commercial broilers in severe vs. normal muscle. Additionally, although there were no differences in body weights in PedM broilers, there was a significant difference in feed efficiency due to lower gain while consuming the same amount of feed in the low FE PedM phenotype.

Table 1. Body weights (BW) and wooden breast (WB) scores in commercial broilers with and without WB, and BW, weight gain (Gain), feed intake (FI) and feed efficiency (FE, gain to feed) in Pedigree Male broilers from which shotgun proteomics was conducted on breast muscle tissue [1].

| Commercial Broilers [2] | | | | Pedigree Male Broilers [3] | | |
| | Normal | Severe WB | | | High FE | Low FE |
Variable	($n = 5$)	($n = 5$)	p	Variable	($n = 4$)	($n = 4$)	p
BW Kg (52 days)	3.2 + 0.2	3.8 + 0.2	0.031	BW (49 days)	3.13 + 0.08	3.18 + 0.03	0.573
WB Score	0.2 + 0.2	1.4 + 0.2	0.003	Gain (kg/7 days)	0.64 + 0.04	0.47 + 0.04	0.022
				FI (kg/7 days)	0.99 + 0.06	1.03 + 0.09	0.689
				FE (G:F)	0.65 + 0.01	0.46 + 0.01	<0.0001

[1] Values represent the mean ± SE. [2] Kuttappan et al. [2]; [3] Bottje et al. [13]. Visual scoring values of WB (ranging 0–3) for samples selected for proteomic analysis based on histological assessment indicating normal tissue and severe WB myopathy.

3.2. Proteomic Dataset Comparison

Proteomics conducted on high and low FE PedM phenotypes in Kong et al. [11] had been obtained eight years earlier in Bottje et al. [13] than for the WB myopathy study by Kuttappan et al. [2]. The timing of sample collection is important for at least two reasons; (a) it was well before WB myopathy had been observed in commercial broilers and (b) the genetic impact of selection of the PedM broiler line would have had sufficient time in the genetic pipeline to reach the level of commercial broilers in the study by Kuttappan et al. [2]. Breast muscle samples were obtained at 52 day in the WB study and between 56 and 63 day (after FE phenotyping 6–7 weeks) in the PedM FE study. It is worth noting that the number of samples in each study are small ($n = 4$ or 5), and therefore, this paper may run the risk of not being fully representative of the two populations in each study. However, it can also be noted that the difference in phenotypes were highly significant ($p = 0.003$ and $p < 0.0001$ for the WB and FE phenotypes, respectively), suggesting that the relative differences in global expression patterns could be considered representative of a larger phenotype population.

Proteomic analysis of the two studies were conducted at different laboratories. For the commercial broiler study [2], LC-MS/MS analysis was conducted using a hybrid LTQ-OrbitrapXL mass spectrometer (ThermoFisher Scientific, Waltham, MA, USA) as described previously [14], but using 40-cm C18 columns developed over a 2-h period with 0 to 40% acetonitrile. In the PedM broiler study [11], extracted individual proteins were subjected to shotgun proteomics analysis by in-gel trypsin digestion and tandem mass spectrometry (MS/MS) at the University of Arkansas Medical Science (UAMS) Proteomics Core Lab (Little Rock, AR). Both studies used the *Gallus* reference proteome downloaded from the UniProtKB (http://www.uniprot.org/help/uniprotkb) database to identify individual proteins from the spectrometric data. The annotation of proteins upstream regulator analysis and downstream functions were conducted using Ingenuity Pathway Analysis (IPA, Qiagen, Redwood City, CA USA).

3.3. Upstream Regulators and Functional Analysis

Figure 1 provides a visual guide to assist in understanding subsequent figures and tables in this study. The predictions of activation (orange background, positive activation Z score) or inhibition (blue background, negative activation Z score) of upstream regulators (Table 2) and functions (Table 3) in WB muscle were determined using the Ingenuity

Pathway Analysis program. These predictions (activation Z scores) were based on: (1) differential protein expression in the dataset and (2) relationships of upstream regulators and downstream target molecules in the IPA scientific literature database.

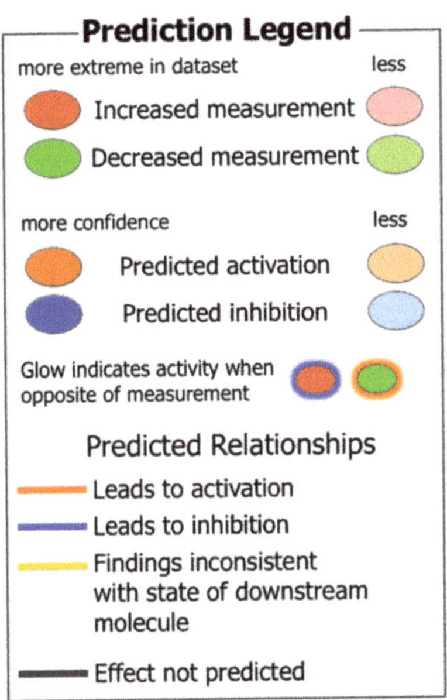

Figure 1. A prediction legend for the interpretation of protein expression and upstream regulator and function analysis in tables and figures.

Supplementary Table S1 contains all of the differentially expressed proteins in the WB myopathy datasets that will appear in tables and figures below. Proteins highlighted in green were down-regulated and those in red were upregulated in the WB myopathy muscle, respectively.

In Tables 2 and 3, proteins in red and green (underlined) lettering denote up- and down-regulation, respectively, in WB myopathy. The protein abbreviations in Tables 2 and 3 are defined in Supplementary Table S1 along with the *p* value and fold difference in expression for each protein. Using an overlay function in the IPA program, it is possible to compare the expression of proteins and predictions of upstream regulators and functions in different datasets. Thus, a qualitative comparison of the WB myopathy proteomics data in the present study can be made to the high vs. low FE proteomics data from Kong et al. [11].

Examples of functional networks in Table 2 are presented in Figure 2. In the vasculogenesis function network (Figure 2A), all up-regulated proteins, with the exception of CALR (calreticulin), contributed to the prediction of activation (activation Z score = 2.43, $p = 4.46 \times 10^{-4}$). The yellow dashed inhibitory line (indicating that the relationships of up-regulation of CALR and predicted activation of vasculogenesis are inconsistent with literature citations) for CALR indicates that enhanced vasculogenesis is typically associated with down-regulation of CALR. The down-regulation of ACP1 would contribute to the predicted activation of vasculogenesis because it would reduce the inhibitory effect on this function. On the other hand, the effects of PKM and LDHA were inconsistent with their down-regulation, indicating that increased expression of these proteins would normally be associated with increased vasculogenesis.

Table 2. Functions predicted to be activated (orange) or inhibited (blue) in wooden breast (WB) myopathy compared to normal muscle based on differentially expressed (DE) target molecules in commercial broilers.

Functions	Activation z-Score	p-Value of Overlap	Differentially Expressed Proteins [1]
Development of vasculature	2.45	1.21×10^{-3}	ACP1,ANXA2,C3,CALR, CAPNS1,CDH13,CRYAB,CTSB,FGA,FLNA,HSPA5,IGHM, LDHA, PGK1,PKM,PPIA,RHOA,VIM,YWHAZ
Vasculogenesis	2.43	4.46×10^{-4}	ACP1,ANXA2,C3,CALR, CAPNS1,CDH13,CRYAB,CTSB,FLNA,HSPA5,LDHA,PKM,PPIA,RHOA,VIM,YWHAZ
angiogenesis	2.30	1.67×10^{-3}	ACP1,ANXA2,C3,CALR, CAPNS1,CDH13,CRYAB,CTSB,FLNA,HSPA5, LDHA,PGK1,PKM,PPIA,RHOA,VIM,YWHAZ
Necrosis	−3.33	2.68×10^{-11}	ACP1,ANXA2,C3,CALR, CAPNS1,CBR1,CRYAB, CS,CTSB,DDX3X,DYNC1H1,EEF1A1,EIF3L,FDPS,FGA,FLNA, GAPDH,GLO1,HINT1,HK1,HNRNPH1,HSP90AA1,HSP90B1,HSPA5,IGHM,KLHL40, LDHA,P4HB, PARK7,PDCD6IP,PDIA3, PEBP1,PKM,POSTN,PPIA, PRDX6,PSMC2,RACK1,RAN,RHOA,RPL11,RPL6,RPLP0,RPS11,RPS14,RPS24,RPS3,Rps3a1,RPS6,RPS7,TAGLN2, UBE2V2,VIM,YWHAE,YWHAZ
Cell death	−3.05	2.01×10^{-9}	ACP1,AK1,ALDOC,ANXA2,C3,CALR, CAPNS1,CBR1,CRYAB, CS,CTSB,DDX3X,DES,DYNC1H1,EEF1A1,EIF3L, FDPS,FGA,FLNA, GAPDH,GLO1,HINT1,HK1,HNRNPH1,HSP90AA1,HSP90B1,HSPA2,HSPA5,IGHM,KLHL40, LDHA,P4HB, PARK7,PDCD6IP,PDIA3, PEBP1,PKM,POSTN,PPIA, PRDX6,PSMC2,RAB1A,RACK1,RAN,RHOA,RPL11,RPL6,RPLP0,RPS11,RPS14,RPS24,RPS3,Rps3a1,RPS6,RPS7,TAGLN2, UBE2V2,VIM,YWHAE,YWHAZ
Apoptosis	−1.98	4.61×10^{-6}	ALDOC,ANXA2,C3,CALR, CAPNS1,CRYAB, CS,CTSB,DDX3X,DES,DYNC1H1,EEF1A1,FLNA, GAPDH,GLO1,HINT1,HK1,HNRNPH1,HSP90AA1,HSP90B1,HSPA2,HSPA5,IGHM,KLHL40, LDHA,P4HB, PARK7,PDCD6IP,PDIA3, PEBP1,PKM,PPIA, PRDX6,RACK1,RHOA,RPLP0,RPS24,RPS3,RPS6,TAGLN2, UBE2V2,VIM,YWHAE,YWHAZ

[1] Target proteins in plain red type were up-regulated ($p < 0.05$, >1.3-fold difference) in WB myopathy compared to normal tissue. Target molecules in bold green and underlined type were down-regulated ($p < 0.05$, <−1.3-fold difference) in WB myopathy compared to normal tissue.

Table 3. Upstream regulators predicted to be activated (orange) or inhibited (blue) in wooden breast (WB) myopathy compared to normal muscle based on differentially expressed downstream target molecules in commercial broilers.

Upstream Regulator	Protein Name	Activation z-Score	p-Value of Overlap	Differentially Expressed Downstream Target Proteins [1]
ANGPT2	Angiopoeitin 2	3.08	7.41×10^{-9}	CALR,CRYAB,HSP90AA1,HSPA2,HSPA5,P4HB,PDIA3,PDIA6,POSTN,RPS6
EGFR	Epidermal growth factor	2.62	1.38×10^{-3}	CRYAB,HNRNPH1,HSP90B1,HSPA5,POSTN,PPIA,VIM
XBP1	X box binding protein 1	2.41	8.79×10^{-3}	CALR,HSP90B1,HSPA5,PDIA3,PDIA4,PDIA6
TGFB1	Transforming growth factor beta 1	2.21	1.95×10^{-6}	AK1, ANXA2, CAPNS1,COL7A1, CTSB, DES,EEF1A1,FLNA,FSCN1, GATD3A/GATD3B,HNT1, HNRNPH1,HSP90A1,HAPA5,IARS,IGHM,LDHA,PGK1, PLS3,POSTN,RAB1A,RACK1,RHOA,VIM,YWHAE
NFE2L2	Nuclear factor (erythroid-derived 2)-like 2	2.18	4.3×10^{-9}	ARF1, CBR1,DSTN,HSP90AA1,HSP90B1, ME1,PDIA3,PDIA4,PDIA6,PGD,RACK1,RAN,RPL18,RPLPO, TPI1
RICTOR	Regulatory-associated protein Independent of mTOR complex 2	−3.61	2.06×10^{-9}	PSMC2,RPL11,RPL18,RPL4,RPL6,RPLP0,RPS11,RPS2,RPS24,RPS3
PPARGC1A	Peroxisome proliferator activator receptor gamma-coactivator 1 α	−2.35	1.92×10^{-4}	AK1, C3, CS,GOT2,LDHA,ME1,PGAM1
AHR	Aryl hydrocarbon receptor	−2.22	2.02×10^{-3}	ABCB6,CBR1,GOT1, HSP90AA1,HSPA5,IGHM,VIM
ARNT	Aryl hydrocarbon receptor nuclear translocator	−2.20	3.35×10^{-6}	ALDOC,GAPDH, IGHM, LDHA,PGK1,TPI1, VIM
ESRRG (NR3B3)	Estrogen related receptor gamma	−2.17	1.89×10^{-6}	ALDOC,GAPDH,LDHA,PKM,TPI1

[1] Target molecules in plain red type were up-regulated ($p < 0.05$, >1.3-fold difference) in WB myopathy compared to normal tissue. Target molecules in bold green and underlined type were down-regulated ($p < 0.05$, <−1.3-fold difference) in WB myopathy compared to normal tissue. ANGPT2: Angiopoeitin 2; EGFR: epidermal growth factor; XBP1: X box binding protein; TGFB1: transforming growth factor beta 1; NFE2L2: nuclear factor (erythroid-derived 2)-like 2.

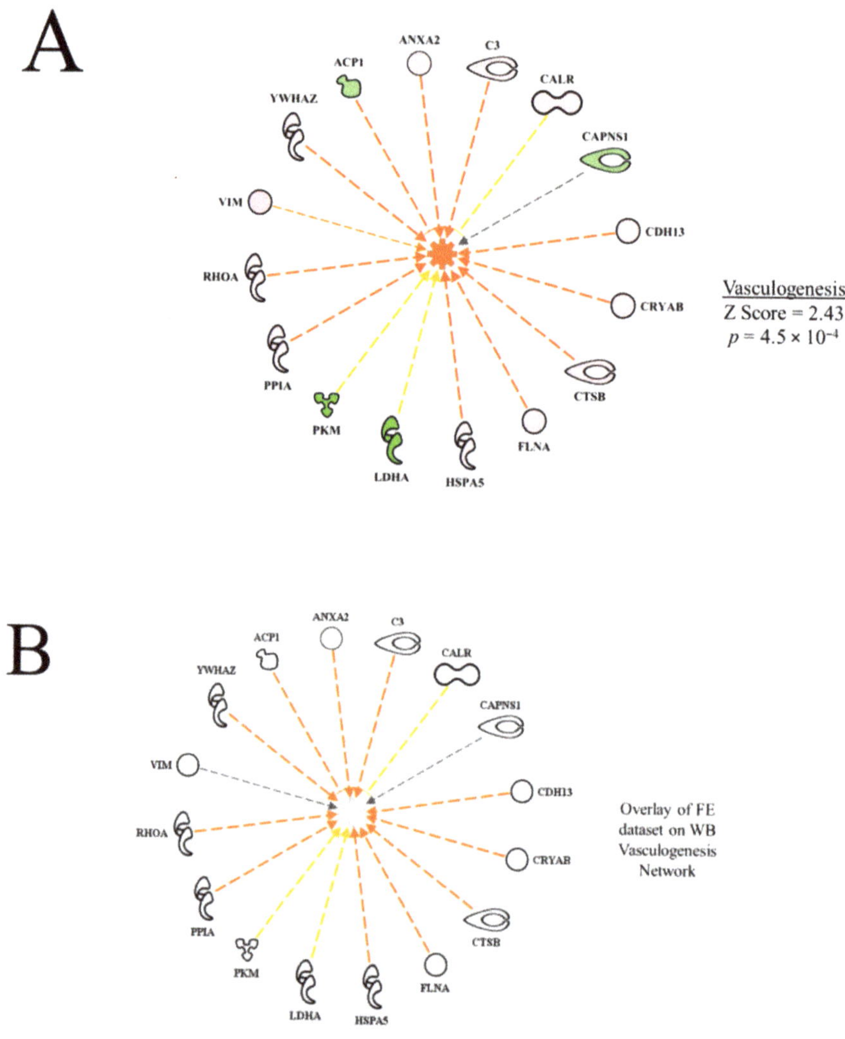

Figure 2. Network of differentially expressed proteins in the wooden breast (WB) myopathy dataset used in calculating the activation Z score for vasculogenesis (**A**). Overlay of the feed efficiency (FE) proteomic dataset from Kong et al. [11]; (**B**) indicates no commonality of differentially expressed proteins in WB vasculogenesis network Protein abbreviations are defined and differential expression of proteins are provided in Supplementary Table S1.

It is worth noting that predictions for ANGPT2 (described in more detail below) or vasculogenesis were not made in the high vs. low FE proteomic dataset; cell death of muscle cells, necrosis of muscle and apoptosis were all predicted to be inhibited in the high FE phenotype. The activation state and p value of overlap were all significant for necrosis of muscle cells (-2.19, $p = 1.52 \times 10^{-5}$), cell death of muscle cells (-2.59, $p = 1.54 \times 10^{-6}$), and apoptosis of muscles (-2.81, $p = 1.29 \times 10^{-4}$) for the high FE phenotype, which suggests that a commonality of mechanisms were presented in the high FE PedM broiler and for broilers exhibiting WB myopathy.

Using the overlay function of the IPA program, it is possible to project one dataset onto another to see commonalities and differences between the datasets. In Figure 2B, the FE proteomic dataset (from Kong et al. [11]) was projected (overlaid) onto the WB myopathy vasculogenesis functional network. The result shows all of the proteins in WB proteomics dataset were not differentially expressed (fold difference, 1.3, $p < 0.05$) in the FE dataset, i.e., vasculogenesis was not predicted to be a significant function in the high vs. low FE PedM phenotypes.

In the necrosis functional network (Figure 3), all up- and down-regulated proteins in WB myopathy contributed to the prediction of inhibition of necrosis; there were no inconsistencies in the relationships between the proteins and the downstream effect of predicted inhibition of necrosis. With the large number of predictions that were obtained in the WB myopathy dataset, the discussion of each upstream regulator may not be sufficient in detail and critical literature citations may be missed. Nonetheless, we hope it will spark hypotheses to be tested that will provide insight into this disease that can lead to preventive measures to ameliorate significant economic losses to the poultry industry.

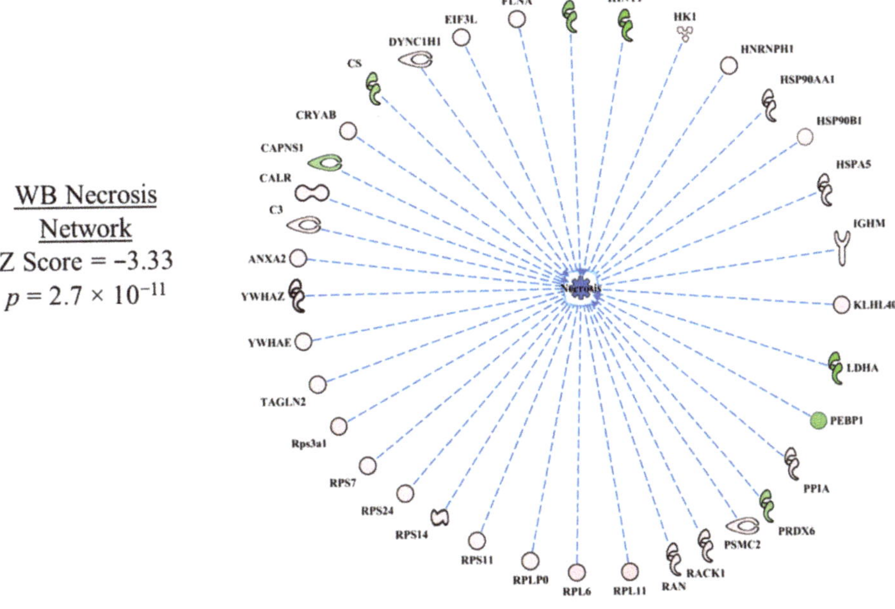

Figure 3. The network of differentially expressed proteins that was used to predict inhibition of necrosis in wooden breast (WB) myopathy. Protein abbreviations are defined and differential expression of proteins are provided in Supplementary Table S1.

Although there was no commonality between WB myopathy and FE datasets using the overlay function in IPA, three functions—muscle cell death, apoptosis of muscle cells, and necrosis of muscle cells—were predicted to be inhibited in the high FE vs. low FE PedM dataset (Supplementary Table S2). Thus, there was a commonality between WB myopathy and high FE with respect to these functions. This will be discussed in a little more detail in the summary section towards the end of this manuscript.

3.4. Upstream Regulators

3.4.1. Angiopoeitin 2

The top predicted activated upstream regulator in the WB myopathy was ANGPT2 (Table 2). Murtyn et al. [15] indicated that hypoxia and circulatory insufficiency may play a role in the development of WB myopathy. Thus, the predicted activation of ANGPT2 by IPA could be a tissue-mediated signal to increase blood vessel development in damaged tissue to enhance oxygen delivery. This hypothesis is supported by the predicted activation of functions of development of vasculature, vasculogenesis and angiogenesis shown in Table 3. Moffarri and Hossein [16] reported that ANGPT2 expression in primary skeletal muscle myocytes was responsive to hydrogen peroxide-induced oxidative stress, but not to proinflammatory cytokines. As ANGPT2 promotes skeletal myoblast survival and repair, Moffarri and Hossein [16] hypothesized that muscle-derived ANGPT2 production plays a positive role in muscle fiber repair. Conversely, ANGPT2 activation could help in the delivery of oxygen and nutrients to aid growth and muscle development as birds with WB myopathy were heavier than birds without WB myopathy (Table 1). In line with this general connection of WB to impaired aerobic energy supply, we have recently found that heavier birds (known to be pre-disposed to WB) possess a lower skeletal muscle mitochondrial content, which would reduce the capacity for aerobic ATP synthesis [17].

The ANGPT2 network of downstream molecules presented in Figure 4A indicates that all of the downstream molecules in WB myopathy were up-regulated compared to normal muscle. Each of these molecules are indicated to be up-regulated indirectly (dashed orange line and arrow) by ANGPT2, with the exception of ribosomal protein 6 (RP6), whose up-regulation was not predicted (black dashed arrow line). This relationship, i.e., an unpredicted effect, occurs when there are too few literature citations in the IPA database to establish a clear relationship between the upstream regulator and the downstream target molecule. When the FE proteomics dataset was overlaid (projected) onto the WB myopathy ANGPT2 network (Figure 4B), there were no commonalities in protein expression and therefore no prediction could be made of ANGPT2 activity.

Using the 'Grow' function of the IPA program, it is possible to link upstream regulators to downstream functions, as shown in Figure 4C. In this figure, predictions of activation of vasculogenesis and angiogenesis in WB myopathy are the result of increased VIM, CRYAB, HSPA5 and HSP90AA1 expression and the predicted activation of ANGPT2. All of these are linked by dashed orange lines with arrows. The dashed lines indicate that at least one additional step was between the upstream molecule and the downstream target. Apoptosis and necrosis were predicted to be inhibited in WB myopathy due to increased expression of CRYAB, HSPA5, HSP90AA1, CALR and HSPA2 and by the predicted activation of ANGPT2. These relationships to necrosis and apoptosis functions are indicated by dashed blue lines with a short perpendicular line at the end indicating inhibition. The lighter blue color for necrosis indicates that the prediction made for this function, based on proteins in this network, was qualitatively less than for apoptosis. The darker color (stronger qualitative prediction) for apoptosis is due in part to more proteins in the regulatory network (five up-regulated proteins plus ANGT2) compared to four proteins in the necrosis network.

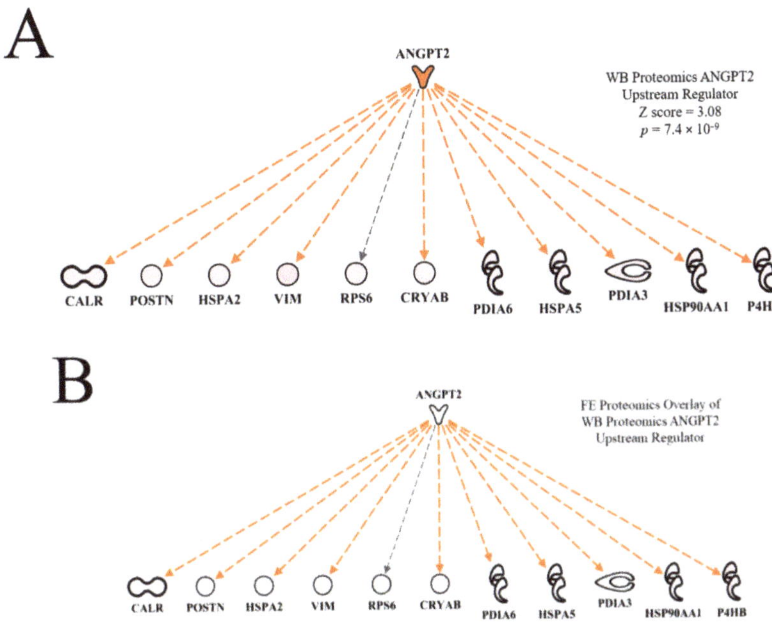

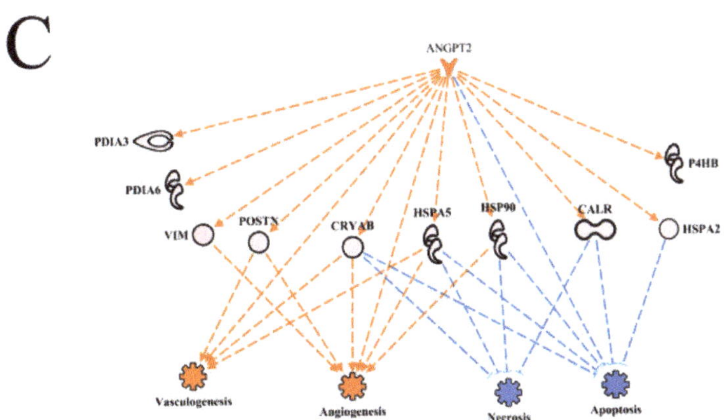

Figure 4. The upstream regulatory network of proteins displayed in hierarchical format for Angiopoeitin 2 (ANGPT2) used in the prediction of activation of this molecule (activation Z score and *p* value of overlap as shown in (**A**)). Overlay of the feed efficiency (FE) proteomics dataset from Kong et al. [11] reveals no commonality of differentially expressed proteins between the wooden breast (WB) and FE data (**B**). A regulatory network for ANGPT2 indicates target proteins in the dataset that contributed to the predictions of active vasculogenesis and angiogenesis, and inhibited necrosis and apoptosis is provided in (**C**). Protein abbreviations are defined and differential expression of proteins are provided in Supplementary Table S1.

3.4.2. Epidermal Growth Factor Receptor

The upstream regulator network of epidermal growth factor receptor (EGFR) in WB myopathy is presented in Figure 5. EGFR was predicted to be activated in WB myopathy, but the overlay of the FE proteomics dataset resulted in no prediction, indicating no commonality with this upstream regulatory network between WB myopathy and high FE. There are thousands of literature citations for EGFR involvement in carcinoma, but here we will focus on a few citations that are pertinent to EGFR in skeletal muscle development. Olwin and Hauschka [18] reported that EGFR and fibroblast growth factor receptor were permanently lost in terminal differentiation of adult mouse skeletal muscle in vitro. It was also reported that whereas EGF increases in skeletal muscle as pigs age, mRNA expression of the receptor declines [19]. Finally, Leroy et al. [20] reported that down-regulation of EGFR triggers differentiation of human myoblasts in cell culture, indicating that EGFR expression normally declines with age. Therefore, the prediction of activation of EGFR in WB myopathy might indicate an abnormal process that contributes to the pathophysiology in WB myopathy.

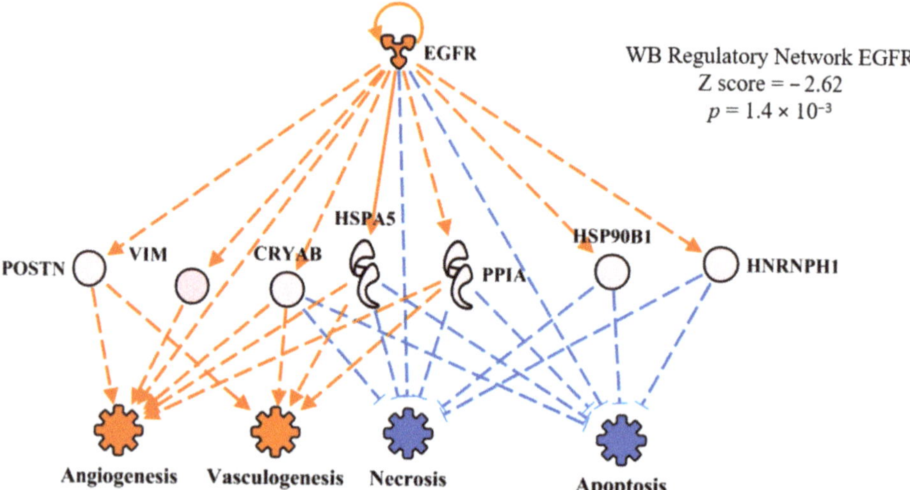

Figure 5. The regulatory network of epithelial growth factor receptor (EGFR) showing downstream target proteins contributing to the prediction of activation of vasculogenesis and inhibition of necrosis and apoptosis in wooden breast (WB) myopathy. Protein abbreviations are defined and differential expression of proteins are provided in Supplementary Table S1.

3.4.3. Transforming Growth Factor Beta 1 (TGFB1)

TGFB1 was predicted to be activated in WB myopathy proteomics (Table 2, Figure 6A) as well as in the high FE PedM phenotype (Figure 6B). While the TGFB1 networks in the two studies did not share any differentially expressed proteins, the overlay of the FE proteomics data on the WB dataset (Figure 6C) as well as the overlay of the WB proteomics data on the FE dataset (Figure 6D) resulted in predictions of the activation of TGFB1. TGFB1 is a cytokine that is involved in controlling the growth, proliferation and differentiation of cells and able to serve in an autocrine manner to regulate its own expression [21]. This autocrine stimulation is indicated by the semi-circular arrow in each of the networks in Figure 6. In the study conducted by Hubert et al. [22], TGFB1 was also identified as one of the upstream regulators of the WB myopathy. The role of TGFB1 in this myopathy may be explained in part by its effects on cell differentiation and mitochondrial function. Under normal conditions, this cytokine is involved in the regulation of cell differentiation, increased mitochondrial activity and increased oxidative phosphorylation. However, studies have shown that increased TGFB1 expression is detrimental and can produce mitochondrial dysfunction with increased oxygen radical production that compromises the mitochondrial antioxidant system [23–25]. The resulting mitochondrial perturbations have been illustrated to lead to the inhibition of myogenic differentiation and the formation of tissue fibrosis [26,27]. The potential for diminished mitochondrial biogenesis in WB myopathy is further supported by predicted inhibition of PGC1α, which is discussed in more detail below. Feed efficiency studies in cattle have also identified TGFB1 as being an upstream regulator [28,29]. This provides further evidence of the of TGFB1 involvement in genetic differences in the various feed efficiency phenotypes.

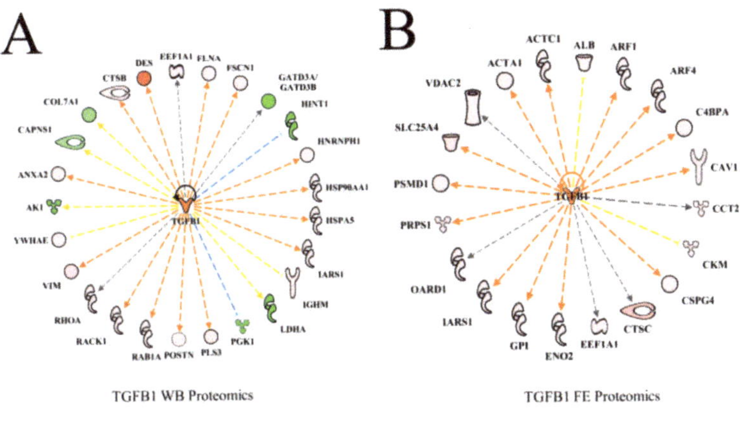

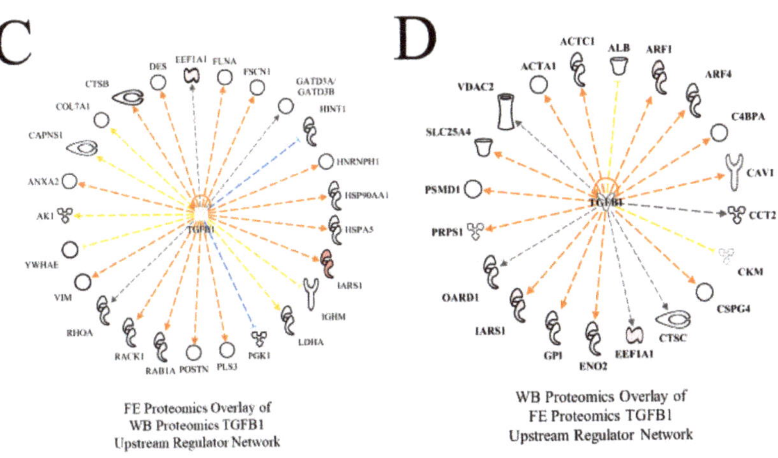

Figure 6. The upstream regulatory network of transforming growth factor beta 1 (TGFB1) for wooden breast (WB) (**A**) and feed efficiency (FE) (**B**) proteomic datasets. Downstream target proteins that were differentially expressed are shown as up-regulated in pink or red or down-regulated in WB myopathy and high FE PedM broilers. Overlay of the FE proteomic dataset on the WB myopathy network (**C**) and of the FE data on the WB network (**D**) resulted in predicted activations of TGFB1 as well. Therefore, despite dissimilar proteomic landscapes of the two datasets, TGFB1 is predicted to be active in both WB myopathy and high FE.

Protein abbreviations for WB myopathy data (A, C) are defined and differential expression of proteins are provided in Supplementary Table S1. Protein abbreviations for FE proteomics (B and D) are as follows: ACTA1 (actin, alpha 1, skeletal muscle), ACTC1(actin, alpha, cardiac muscle 1), ALB (albumin), ARF (ADP-ribosylation factor 1), ARF4 (ADP-ribosylation factor 4), C4BPA (complement component 4 binding protein, alpha), CAV1 (caveolin 1), CCT2 (cell division cycle 42), CKM (creatine kinase, muscle), CSPG4 (chondroitin sulfate proteoglycan 4), CTSC (cathepsin C), EEF1A (eukaryotic translation elongation factor 1 alpha 1), ENO2 (enolase 2, gamma neuronal), GPI (glucose 6 phosphate isomerase), IRS1 (insulin receptor S1), OARD1 (O-acyl-ADP-ribose deacylase), PRPS1 (phosphoribosyl pyrophosphate synthetase 1), PSMD1 (proteosome 26S subunit, non-ATPase, 1), SLC25A4 (solute carrier family 25 (mitochondrial carrier; adenine nucleotide translocator), member 4), VDAC2 (voltage-dependent anion channel 2).

Linkage of TGFB1 activation to downstream functions through the differentially expressed proteins is provided in Figure 7. For simplicity, only relationships that clearly predicted activation (angiogenesis, vasculogenesis) or inhibition (apoptosis, necrosis) are presented in this figure. The expression of eight proteins in this regulatory network (PGK1, CTSB, VIM, HSP90AA1, FLNA, HSPA5, ANAXA2 and RHOA) were involved with predictions of increased angiogenesis and vasculogenesis in WB myopathy. The up-regulation of all the proteins, with the exception of PGK1, which was down-regulated, were downstream targets of TGFB1 that contributed to the predictions of enhanced blood vessel-capillary formation. Eight up-regulated proteins (HSP90AAA1, FLNA, HSPA5, ANXA2, RHOA, RACK1, NFRNPH1 and DES) and one down-regulated protein (HINT1) were involved in the predictions of reduced apoptosis and necrosis in wooden breast myopathy.

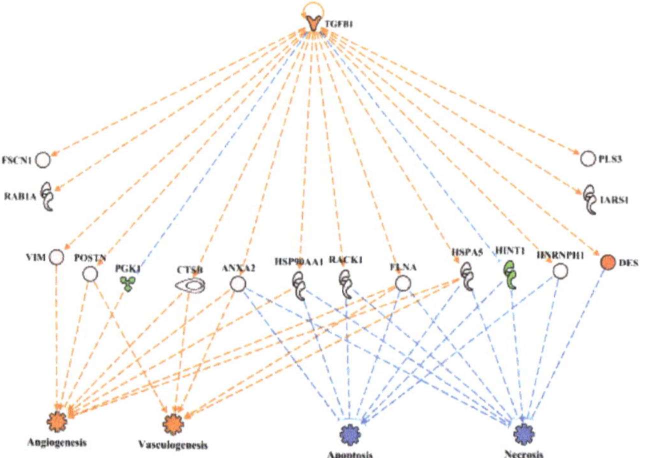

Figure 7. The regulatory network of TGFB1 in wooden breast (WB) myopathy leading to predicted activations of angiogenesis and vasculogenesis and inhibition of apoptosis and necrosis through downstream target proteins in WB myopathy. Protein abbreviations are defined and differential expression of proteins are provided in Supplementary Table S1.

3.4.4. Nuclear Factor (Erythroid-Derived 2)-Like 2 (NFE2L2)

Nuclear factor (erythroid-derived 2)-like 2 (NFE2L2) coordinates cellular response to oxidative stress [30,31]. NFE2L2 was predicted to be activated in both WB proteomics and in the FE proteomics dataset in PedM broilers (Figure 8A,B), but the predictions were based on different downstream molecules. Overlaying of each dataset on the other upstream regulatory network (Figure 6C,D) resulted in predicted activation of NFE2L2.

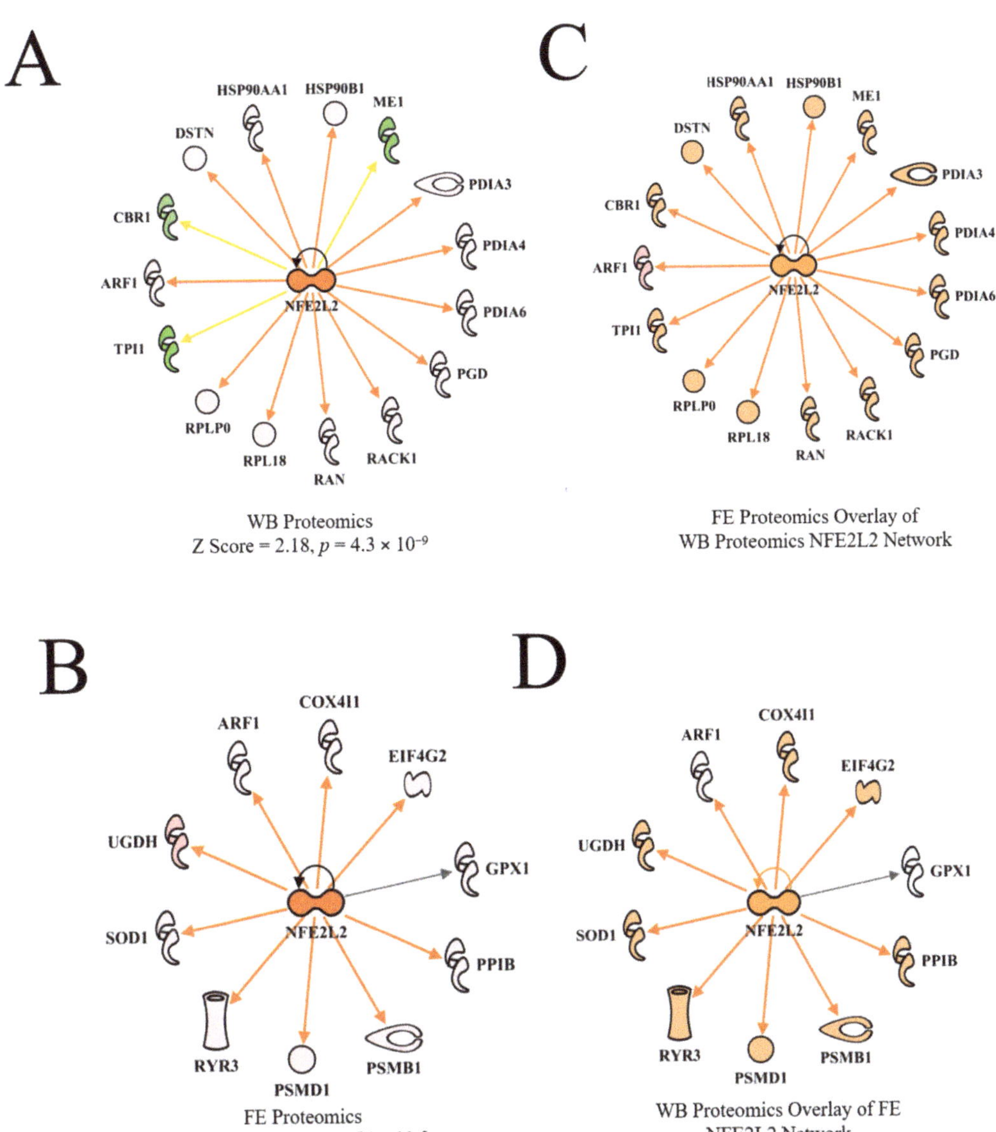

Figure 8. The upstream regulator network for NFE2L2 (nuclear factor (erythroid-derived 2)-like 2) for wooden breast (WB) myopathy (activation Z score = 2.18, 4.3×10^{-9}) compared to normal muscle (**A**) and for the high vs. low PedM broiler feed efficiency (FE) proteomic data (activation Z score = 2.38, $p = 1.54 \times 10^{-6}$) (**B**). Overlay of the FE data on the WB myopathy data (**C**) and vice versa (**D**) both resulted in predicted activations of NFE2L2. Protein abbreviations in B and D in the FE data set are as follows: ARF1 (ADP-ribosylation factor 1), COX4I1 (cytochrome oxidase subunit IV isoform 1), EIF4G2 (eukaryotic translation initiation factor 4 gamma 2), GPX1 (glutathione peroxidase 1), PPIB (peptidylprolyl isomerase B (cyclophilin B), PSMB1 (proteosome 26S subunit beta type, 1) PSMD1 (proteosome 26S subunit non-ATPase, 1), RYR3 (ryanodine receptor 3), SOD1 (superoxide dismutase 1), UGDH (UDP-glucose 6-dehydrogenase).

Under normal (non-oxidant stress) conditions, the NFE2L2 protein remains bound to Cullin3 (CUL3) and Kelch like-ECH protein 1 (KEAP1), and is then rapidly directed (within

15–20 min) to proteasomes for degradation [32]. However, in response to oxidative stress, NFE2L2 is transported to the nucleus, where it stimulates antioxidant gene expression after binding to the antioxidant response element. Recently, NFE2L2 was predicted to be activated in high feed efficiency (HFE) compared to low feed efficiency (LFE) broiler phenotypes based on downstream target molecules [11,33]. It has been shown that NFE2L2 undergoes post-transcriptional and translational modifications that can affect the amount of NFE2L2 protein produced, which, along with rapid activation or degradation, makes it an extremely labile protein [32,34]. Avian NFE2L2 appears to function the same as mammalian NFE2L2, but since it shares only 67% homology with mammalian NFE2L2 [35], it may have other roles in avian tissues besides coordination of antioxidant response. In a recent review, NFE2L2 was observed to affect mitochondrial function in many ways, including efficiency of oxidative phosphorylation, mitochondrial biogenesis and mitochondrial integrity [36].

3.4.5. X-Box Binding Protein 1

X-box binding protein 1 (XBP1) was predicted to be activated in WB myopathy (Table 2). There was no commonality of protein expression of this upstream regulator in the FE proteomic dataset. Myoblasts that were forced to differentiate in response to XBP1 overexpression produced myotubes that were shorter and less mature indicative of impaired differentiation relative to control cells [37]. Of particular potential relevance to WB myopathy is the following from this study: "Our observations suggest that skeletal muscle tissue is particularly sensitive to physiological stressors that trigger the unfolded protein response, namely glucose deprivation, anabolic stimulation, and perhaps other endoplasmic reticulum stresses, such as hypoxia, imbalances in calcium homeostasis, and ischemia. These perturbations could be triggered by normal or pathological states." The presence of hypoxia in WB myopathy has been reported [38] and will be discussed in greater detail below.

3.4.6. Rapamycin-Insensitive Companion of Mammalian Target of Rapamycin (RICTOR)

RICTOR plays an important role in actin-cytoskeletal development, as formation of this structure was impaired in a RICTOR knockout mouse model [39–41]. Ablation of RICTOR (mTORC2) did not adversely affect muscle function in mice, results that contrast dramatically with the loss of function and development of muscle dystrophy when RAPTOR (mTORC1) was removed [42].

RICTOR was predicted to be inhibited in WB myopathy with activation z scores of −3.61 (Table 2). RICTOR was also predicted to be inhibited in PedM broilers exhibiting high FE [11]. The upstream regulator networks of RICTOR in the current WB myopathy study and in the FE proteomics study are presented in Figure 9A,B, respectively. Although both studies yielded predictions of RICTOR inhibition, it is apparent that the predictions were based on different downstream target molecules in the respective datasets. The overlay of the datasets on each other yielded no prediction giving indication of no commonality in the RICTOR networks. Thus, this appears to indicate that the predictions of inhibition of RICTOR in WB myopathy and FE breast muscle proteomics are due to separate mechanisms. The connection between inhibition of RICTOR to inhibition of necrosis shown in Figure 9C is shown to be mediated by up-regulation of several ribosomal proteins. Apoptosis was predicted to be weakly activated in this network through combined inhibition of RICTOR and up-regulation of RPS3.

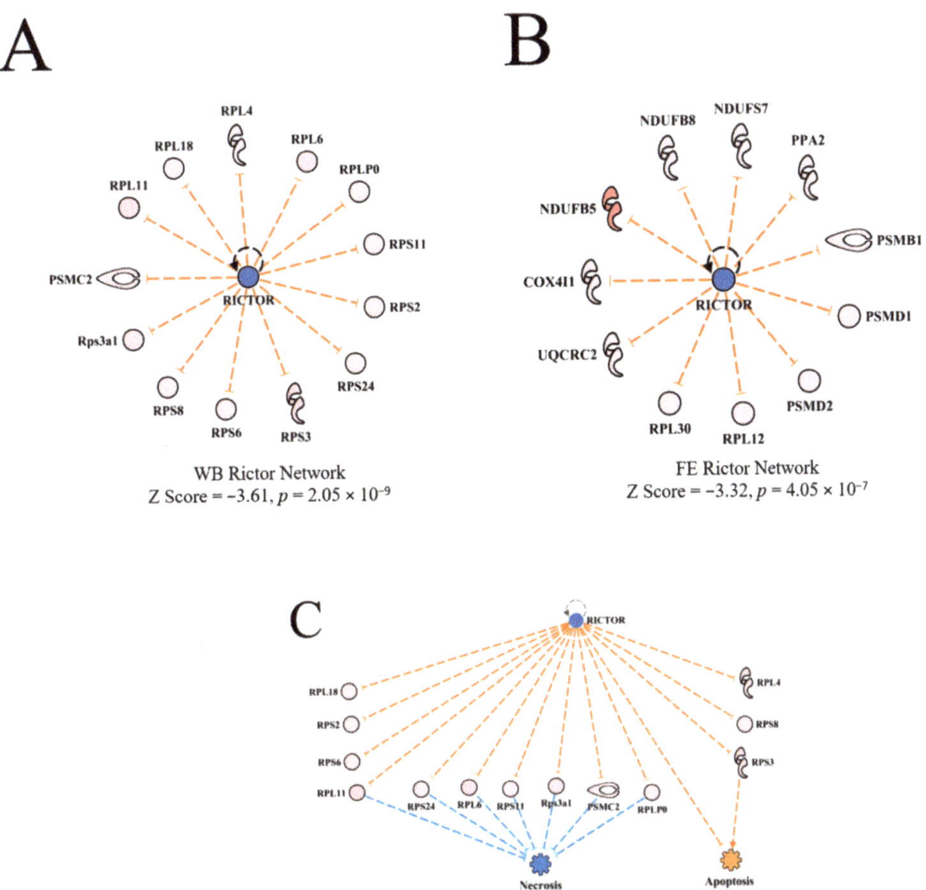

Figure 9. The upstream regulatory network for RICTOR in wooden breast (WB) myopathy (**A**) and for feed efficiency (FE) proteomics (**B**). All downstream proteins were up-regulated in both datasets, but there were no common differentially expressed proteins. The regulatory network for RICTOR linking downstream molecules to the predicted inhibition of necrosis and slight activation of apoptosis in WB myopathy (**C**). For (**A**,**C**), protein abbreviations are defined and differential expression of proteins in WB myopathy are provided in Supplementary Table S1. Proteins in (**B**) are as follows: COX4I1 (cytochrome oxidase subunit IV isoform 1), NDUFB5, B8, S7 (NADH dehydrogenase (ubiquinone) Complex I assembly factor 5, factor 8, subunit 7), PPA2 (pyrophosphate (inorganic) 2, PSMB1, D1, D2 (proteosome 26S subunit, non-ATPase beta type-1) PSMD1, PSMD2 (proteosome 26S subunit, non-ATPase, 1 & non-ATPase 2), RPL12, RPL30 (ribosomal protein L12 & L30), UQCRC2 (ubiquinol-cytochrome c-reductase core protein II).

3.4.7. Peroxisome Proliferator-Activating Gamma Coactivator 1 Alpha (PPARGC1α or PGC1α)

Because PGC1α has been described as the master regulator of mitochondrial biogenesis (Nisoli et al. [43,44], the predicted inhibition of PGC1α by upstream regulator analysis (Figure 10A, activation Z score = -2.35, $p = 1.92 \times 10^{-5}$ suggests that mitochondrial biogenesis, and possibly mitochondrial function, would be impaired in WB myopathy. In contrast, the upstream regulator analysis of PGC1α was predicted to be activated in the high FE PedM broiler phenotype (Figure 10B, activation Z score = 2.45, $p = 1.84 \times 10^{-4}$). There was no commonality of protein expression between the WB and FE datasets. The mitochondrial canonical pathways in WB and FE proteomics datasets are shown in Figure 10C,D, respectively. Whereas Complex I of the mitochondrial electron transport chain was predicted to be inhibited in WB myopathy (Figure 10C), Complex I, III and IV were predicted to be activated based on expression of proteins in the FE dataset (Figure 10D). Complex I (NADH dehydrogenase), which accepts NADH-linked energy substrates, is the largest complex in the electron transport chain, consisting of approximately 40 proteins. Thus, inhibition of this complex could have a dramatic influence on the energy production in WB tissues. The potential for diminished mitochondrial function in WB myopathy is supported by a recent report showing increased mitochondrial damage and mitophagy in histological analysis of WB myopathy [45]. Several studies provide evidence of enhanced mitochondrial function and complex activities in the high FE PedM phenotype (e.g., see review by [46]) that may be attributed to increased activity of PGC1α in the high FE phenotype shown in Figure 10B.

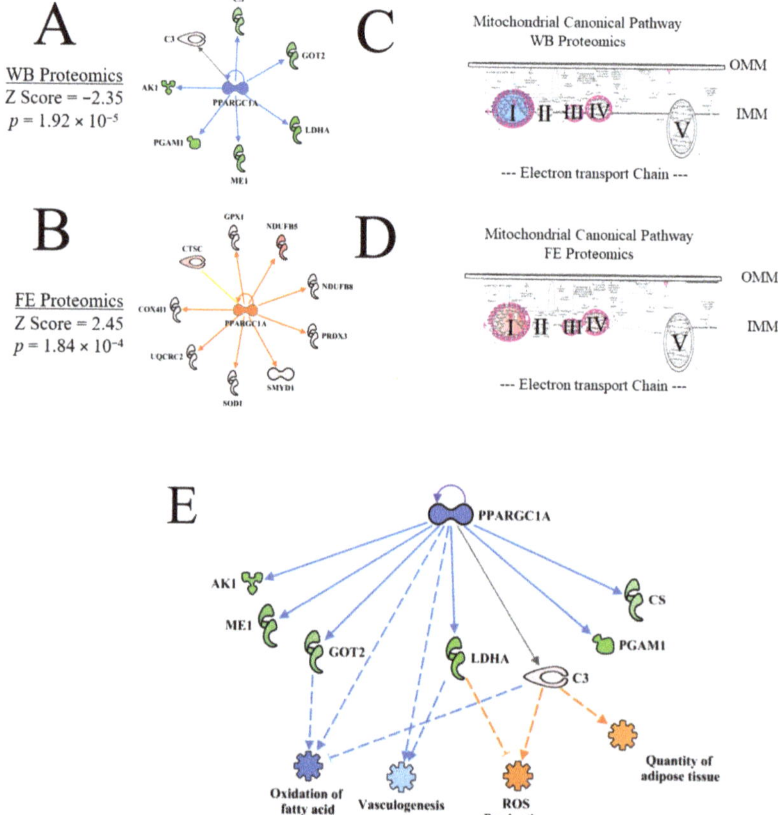

Figure 10. The upstream regulator network for PPARGC1α for WB myopathy vs. control breast muscle in commercial broilers (**A**) and high vs. low FE PedM (**B**) proteomic datasets [11]. The predicted inhibition of PGC1α in WB myopathy was associated with the predicted inhibition of Complex I of the mitochondrial electron transport chain in WB myopathy (**C**), whereas the predicted activation of PGC1α in high FE was associated with predicted activations of Complex I, III and IV of the mitochondrial electron transport chain (**D**). The regulatory network for PPARGC1A showing the downstream target molecules involved in the predictions of inhibition of vasculogenesis and oxidation of fatty acid and activation of reactive oxygen species (ROS) production and quantity of adipose tissue (**E**). Protein abbreviations in (**A,E**) are defined and differential expression of proteins are provided in Supplementary Table S1. Protein abbreviations for (**D**) are as follows: COX4I1 (cytochrome oxidase subunit IV isoform 1), CTSC (cathepsin C), GPX1 (glutathione peroxidase 1), NDUFB5 (NADH dehydrogenase (ubiquinone) Complex I assembly factor 5), NDUFB8 (NADH dehydrogenase (ubiquinone) beta subcomplex, 8), OMM (Outer Mitochondrial Membrane), IMM (Inner Mitochondrial Membrane).

The regulatory network for PGC1α presented in Figure 10E indicates that, unlike the overall prediction of vasculogenesis activation (Table 3), vasculogenesis was predicted to be inhibited by a combination of down-regulation of LDHA and predicted inhibition of PGC1α. Reactive oxygen species (ROS) production was predicted to be enhanced in WB myopathy due to a combination of LDHA down-regulation and up-regulation of C3. Enhanced ROS production may contribute to tissue damage reported in WB myopathy. In this network, oxidation of fatty acid was predicted to be inhibited in WB myopathy in the present study due to the up-regulation of C3, down-regulation of GOT2 and predicted inhibition of PGC1α and the quantity of adipose tissue was predicted to be activated in WB myopathy. These findings agree with many studies, indicating that lipid accumulation in muscle is a characteristic in WB myopathy [6,47,48].

3.4.8. Aryl Hydrocarbon Receptor Nuclear Translocator (ARNT) and Aryl Hydrocarbon Receptor Nuclear (AHR)

ARNT and AHR were predicted to be inhibited in WB myopathy (Table 2). The regulatory networks shown in Figure 11A,B, respectively. for ARNT indicate that fibrosis would be predicted to be enhanced, and glycolysis, vasculogenesis as well as apoptosis would be predicted to be inhibited based on expression of proteins within this regulatory network. In contrast, angiogenesis and necrosis were predicted to be activated in the AHR regulatory network.

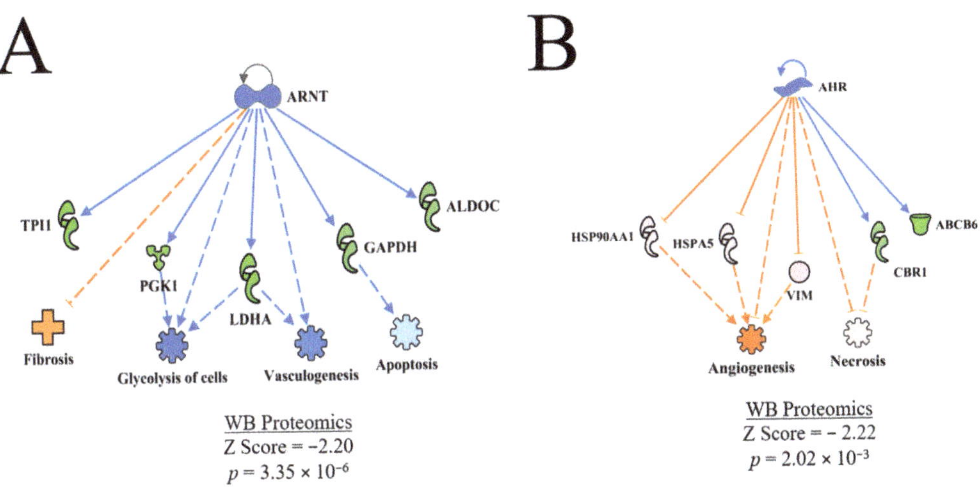

Figure 11. The upstream regulatory networks for ARNT (**A**) and AHR (**B**). The predicted inhibition of ARNT was based on the downstream expression of proteins that are shown and was associated with predicted inhibition of apoptosis, vasculogenesis and glycolysis and prediction of activation of fibrosis. Conversely, the expression of downstream target molecules of involved in predicting inhibition of AHR were associated with predictions of enhanced angiogenesis and necrosis. The discrepancy in the prediction of active necrosis in the AHR and prediction of inhibition of necrosis in the entire dataset shown in Table 2 is based on the smaller subset of expression data for AHR compared to the overall dataset. Protein abbreviations are defined and differential expression of proteins are provided in Supplementary Table S1.

Murtyn et al. [15] provided a sound argument for the presence of hypoxic conditions playing a role in WB myopathy that was based on the differential expression of several downstream molecules that are responsive to hypoxia inducible factor 1 (HIF1 and HIF-dependent expression). However, they also indicated that transcripts of the two subunits of HIF-1, HIF-1α and ARNT were not differentially expressed between normal and WB myopathy breast muscle. Since both AHR and ARNT are involved in other cellular processes (e.g., xenobiotic metabolism), the predictions of inhibition in the present study (and lack of difference in gene expression in Murtyn et al. [15]) may have nothing to do with tissue hypoxia, but rather, are playing roles in other cellular processes. For example, AHR knockout in aortic smooth muscle cells resulted in the deregulation of components of TGF-β signaling and could directly affect the metabolism of toxic substances [49]. Recent reviews indicated that there are numerous functions that AHR may play in the cell [50] and immunity [51]. Greene et al. [38] measured lower oxygen levels in breast muscle tissue exhibiting WB myopathy compared to normal tissue. Furthermore, several genes associated with hemoglobin in red blood cells and HIF-1α in breast muscle were down-regulated in WB compared to normal breast muscle tissue [38].

Some discussion is warranted with regard to the predicted inhibition of AHR in the present study. First, predictions of activation or inhibition of an upstream molecule generated in the IPA program are based on the expression of downstream molecules specifically associated with the dataset being examined. In the case of the WB proteomic dataset in the present study, there were far fewer differentially expressed proteins (~130) that were detected compared to approximately 1500 differentially expressed transcripts detected by Murtyn et al. (2015) [15]. Secondly, denaturing gels used in separating proteins prior to protein expression analysis can result in a loss of many proteins with large hydrophobic regions (e.g., membrane-bound or membrane associated proteins), which in turn could influence the predictions generated by the IPA program. Third, the lack of differential expression of ARNT and HIF-1α m-RNA reported by Murtyn et al. [15] does not mean that the proteins would also not be differentially expressed. Furthermore, these proteins may also undergo post-translational modifications that would be required for activation of downstream target molecules. The predicted activation of ANGPT2 in WB myopathy could be a response to tissue hypoxia to increase vascularization caused by tissue damage and/or hypoxia. Thus, further examination AHR and ARNT signaling in WB myopathy is warranted.

Although no predictions were made regarding AHR and ARNT in the FE proteomics study by Kong et al. [11], ESR1 (estrogen receptor 1) was predicted to be activated (activation Z score = 1.97, 3.93 × 10^{-3}). In the IPA gene view in the Qiagen software, ESR1 is indicated to be a member of AHR-aryl hydrocarbon-Arnt-ESR1. Thus, this would appear a major difference between the WB and FE proteomics datasets.

3.4.9. Estrogen Regulator Receptor Gamma (ERSSγ)

ERSSγ was predicted to be inhibited in WB myopathy (Table 2). The regulatory network of ERSSγ shown in Figure 12 indicates that vasculogenesis would be inhibited by the down-regulation of LDHA and PKM, whereas down-regulation of GAPDH caused a prediction of enhanced necroptosis in WB myopathy. Again, both of these predictions are opposite to functions outlined in Table 3, with the discrepancy attributed to the larger number of proteins involved in the overall predictions shown in Table 3.

Fan et al. [52] reported that ESRRγ helps in reducing muscle damage and improves muscle function in a double PGC1α/β knockout mouse model. The loss of PGC1α/β resulted in a decrease in expression of myoglobin and a lighter color of muscle consistent with a change from oxidative to glycolytic fibers, which were largely restored when ESRRγ transgenic mice were crossed with the PGC1α/β knockout mice [52]. Defects in mitochondrial function from the PGC1 α/β knockout were also restored by crossing with the ESRRγ transgenic mice. Furthermore, ESRRγ also increased the expression of genes associated with angiogenesis (e.g., vascular endothelial growth factor A and fibroblast growth factor 1) and vascular density (CD31).

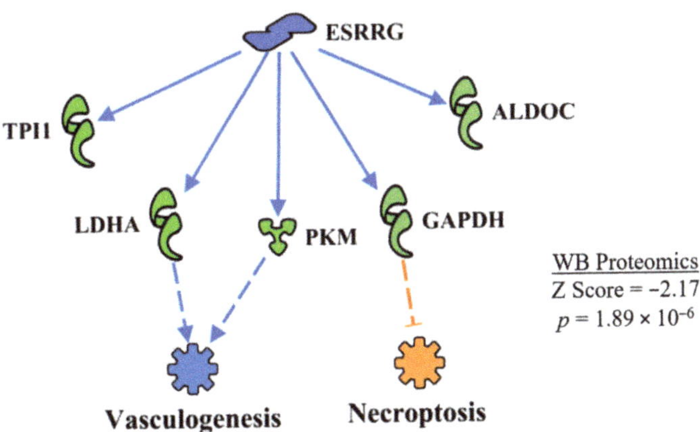

Figure 12. ESRRGamma (ESRRG or ESSRγ) was predicted to be inhibited in WB myopathy based on the expression of downstream molecules shown in this regulatory network. The regulatory network resulted in the prediction of the inhibition of vasculogenesis and activation of necroptosis. As in Figure 9, discrepancies in the predictions of these functions and those in Table 2 are due to a smaller subset of proteins in the ERRG upstream regulatory network, compared to the larger set of proteins listed for functions in Table 2. Protein abbreviations are defined and differential expression of proteins are provided in Supplementary Table S1.

3.5. Summary and Synopsis

There were two major goals of this study; (1) to present an upstream regulatory factor analysis associated with wooden breast myopathy to expand our understanding of this disease and (2) to determine where contributions to WB myopathy might be linked to selection for performance (high FE) in PedM broilers. Table 4 summarizes the results of comparison of upstream regulator analysis between the WB proteomic data to the FE proteomic data reported by [11]. With the exception of TGFB1 and NFE2L2, there were no commonalities for upstream regulators between the WB and FE proteomics data. While PPARGC1α was predicted to be inhibited in WB myopathy, this upstream regulator was predicted to be activated (based on a completely different set of differentially expressed proteins) in muscle of the high FE compared to the low FE PedM broiler. From these findings, we hypothesize that mitochondrial biogenesis would be inhibited in WB myopathy and activated in the high FE PedM broiler [11].

Table 4. Upstream regulators that were predicted to be activated (+) or inhibited (–) in wooden breast myopathy (WB) proteomics (from Kuttappan et al. [2]) with the resulting prediction when the FE proteomics dataset from high and low feed efficiency (FE) Pedigree Broilers males [11] was overlaid on the WB upstream regulator network. The designation 'no' indicates that no prediction was made with the FE overlay (projection of FE dataset onto WB myopathy dataset).

Upstream Regulator	WB	FE Overlay	Upstream Regulator	WB	FE Overlay
ANGPT2	+	no	RICTOR	–	no
EGFR	+	no	PPARGC1α	–	no
XBP1	+	no	AHR	–	no
TGFB1	+	+	ARNT	–	no
NFE2L2	+	+	ESRRG	–	no

Figure 13 provides a graphical presentation of the upstream regulators that were predicted to be activated (orange) or inhibited (blue) in WB myopathy (13A) or FE proteomics (13B) and predicted interactions with other upstream regulators described in this study.

This figure was generated by uploading the upstream regulator networks in Table 2, and then removing differentially expressed downstream target proteins to reveal predicted interactions between the upstream regulators. Major functions/processes, including the prediction of the activation of angiogenesis and the inhibition of cell death (apoptosis and necrosis), were discussed previously. The predicted activation of TGFB1 is indicated as having an indirect (through one or more mediators) inhibitory effect on PPARGC1A and predicted inhibition of PPARGC1A would directly inhibit EGFR. Predictions of reduced activity of AHR and ARNT both contribute to the prediction of lower ESRRG activity. Similarly, predicted inhibition of ARNT and AHR would inhibit ESRRG as well. Because PPARGC1A and ESRRG enhance mitochondrial biogenesis, their predicted inhibition could be hypothesized to lower mitochondrial biogenesis. Elevated NFE2L2 activity is predicted to result in enhanced activation of TGFB1. The predicted inhibition of AHR would function to enhance TGFB1 activity, since AHR has been shown to inhibit TGFB1. NFE2L2 activation would contribute to the predicted activation of TGFB1. Effects of ANGPT2 and TGFB1 would enhance EGFR and their combination would lead to the prediction of enhanced angiogenesis in WB myopathy.

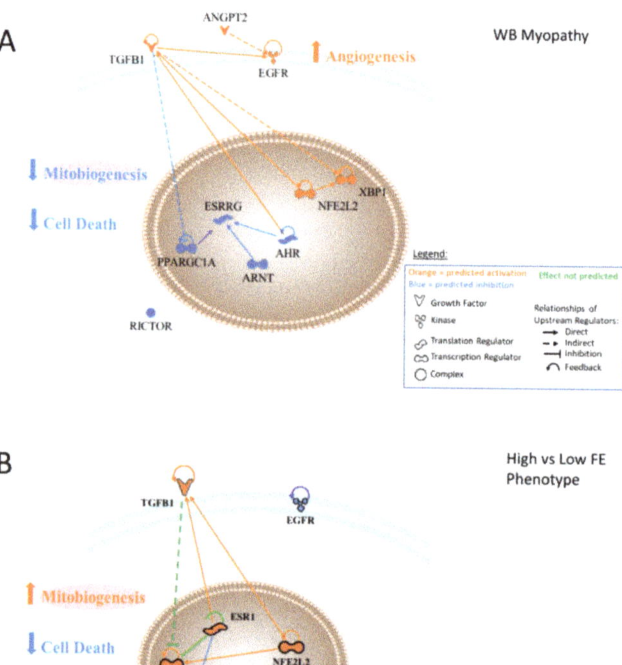

Figure 13. (**A**) Upstream regulator interactions contributing to the overall prediction of increased vasculogenesis and decreased cell death in wooden breast muscle myopathy in commercial broilers. The combination of the predicted activities of these upstream regulators would support the prediction of increased angiogenesis and inhibition of cell death in wooden breast myopathy. (**B**) Arrangement of similar upstream regulators in the high vs. low feed efficiency PedM phenotype dataset revealed a scenario in which angiogenesis was not predicted to be active or inhibited while predictions were made of inhibition cell death, which is similar to that for WB myopathy.

Figure 13B presents upstream regulators from the FE proteomics data. No prediction was made concerning ANGPT2, and unlike WB myopathy, EGFR was predicted to be inhibited in the high FE phenotype. No prediction was made in high vs. low FE data concerning angiogenesis/vasculogenesis. TGFB1 was predicted to be activated in high FE, as were NFE2L2 and PPARGC1A, but as indicated in the text, these were based on different downstream target molecules compared to the WB myopathy data. No predictions were made for AHR, ESSRG or ARNT, but estrogen receptor 1 (ESR1), a member or the ARN family, was predicted to be activated in the high FE phenotype. The combination of increased PPARGC1A and ESR1 would act to enhanced mitochondrial biogenesis in the high FE phenotype.

All of the predictions presented in this study are just that—predictions. Each of these predictions of upstream regulator activity represents hypotheses and mechanistic studies that need to be conducted in order to determine if these upstream regulators contribute to the development of WB muscle myopathy. We hope that they will help shed light on the understanding of WB myopathy that will lead to methods, whether nutritional or chemical, that can be used to ameliorate WB muscle myopathy in commercial broilers.

Supplementary Materials: The following are available online at https://www.mdpi.com/2304-815 8/10/1/104/s1, Supplementary Table S1. Abbreviations and names with P value and fold change of proteins presented in Tables 2 and 3 listed alphabetically. Fold Differences in green (negative values) were down-regulated in wooden breast (WB) myopathy whereas fold differences in red (positive values) were up-regulated in WB myopathy breast muscle relative to values expressed in normal breast muscle tissue. Supplementary Table S2. Functions predicted to be inhibited (blue) in breast muscle of Pedigree Male Broilers exhibiting a high compared to low feed efficiency phenotype (previously unpublished data from Kong et al. [11]).

Author Contributions: Design of studies: W.G.B., V.A.K., C.M.O. and B.C.K.; bioinformatics: V.A.K. and B.C.K.; interpretation of data—pathway analysis: W.G.B., B.C.K., N.J.H. and V.A.K.; drafting and critical review of manuscript: W.G.B., K.R.L., N.J.H., V.A.K., B.A., S.D., C.M.O. and B.C.K. All authors have read and agreed to the published version of the manuscript.

Funding: This research was provided by the USDA-NIFA (#2013-01953) to W.B., B.K. and N.J.H., USDA-NIFA SAS (#2019 69012-29905) to W.B., S.D., and B.K. and the Arkansas Biosciences Institute to W.B., B.K. and S.D.

Institutional Review Board Statement: The present study was conducted in accordance with the recommendations in the guide for the care and use of laboratory animals of the National Institutes of Health. All procedures for animal care complied with the University of Arkansas Institutional Animal Care and Use Committee (IACUC): Protocol Nos. 14,012 and 17,080.

Informed Consent Statement: Not applicable.

Data Availability Statement: No new data were created in this study. Data sharing is not applicable to this article. Support for this research was provided by the USDA-NIFA (#2013-01953) to W.B., B.K. and N.H., USDA-NIFA SAS (#2019 69012-29905) to W.B., S.D., B.K. and N.H. and the Arkansas.

Acknowledgments: Support was provided by USDA-NIFA (#2013-01953), Chancellors Challenge Fund (University of Arkansas), USDA-NIFA Sustainable Agriculture Systems (#2019-69012-29905), and Division of Agriculture, University of Arkansas.

Conflicts of Interest: The authors declare no conflict of interest. The funding sponsors had no role in the design of the study; in the collection, analyses or interpretation of data; in the writing of the manuscript, and in the decision to publish the results.

References

1. Velleman, S.G. Relationship of Skeletal Muscle Development and Growth to Breast Muscle Myopathies: A Review. *Avian Dis.* **2015**, *59*, 525–531. [CrossRef]
2. Kuttappan, V.A.; Bottje, W.; Ramnathan, R.; Hartson, S.D.; Coon, C.N.; Kong, B.-W.; Owens, C.M.; Vazquez-Añon, M.; Hargis, B.M. Proteomic analysis reveals changes in carbohydrate and protein metabolism associated with broiler breast myopathy. *Poult. Sci.* **2017**, *96*, 2992–2999. [CrossRef] [PubMed]

3. Lake, J.A.; Abasht, B. Glucolipotoxicity: A Proposed Etiology for Wooden Breast and Related Myopathies in Commercial Broiler Chickens. *Front. Physiol.* **2020**, *11*, 1–13. [CrossRef] [PubMed]

4. Velleman, S.G.; Clark, K.L. Histopatholoic and myogenic gene expression changes associated with wooden breast in broiler breast muscle. *Avian Dis.* **2015**, *59*, 410–418. [CrossRef] [PubMed]

5. Abasht, B.; Mutryn, M.F.; Mickalek, R.D.; Lee, W.R. Oxidative stress and metabolic perturbations in wooden breast disorder in chickens. *PLoS ONE* **2016**, *11*, e0153750. [CrossRef]

6. Papah, M.B.; Brannick, E.M.; Schmidt, C.J.; Abasht, B. Evidence and role of phlebitis and lipid infiltration in the onset and pathogenesis of Wooden Breast Disease in modern broiler chickens. *Avian Pathol.* **2017**, *46*, 623–643. [CrossRef]

7. Sihvo, H.-K.; Lindén, J.; Airas, N.; Immonen, K.; Valaja, J.; Puolanne, E. Wooden Breast Myodegeneration of Pectoralis Major Muscle over the Growth Period in Broilers. *Vet. Pathol.* **2017**, *54*, 119–128. [CrossRef]

8. Brothers, B.; Zhuo, Z.; Papah, M.B.; Abasht, B. RNA-Seq Analysis Reveals Spatial and Sex Differences in Pectoralis Major Muscle of Broiler Chickens Contributing to Difference in Susceptibility to Wooden Breast Disease. *Front. Physiol.* **2019**, *10*, 764. [CrossRef]

9. Petracci, M.; Soglia, F.; Madruga, M.; Carvalho, L.; Ida, E.; Estévez, M. Wooden-Breast, White Striping, and Spaghetti Meat: Causes, Consequences and Consumer Perception of Emerging Broiler Meat Abnormalities. *Compr. Rev. Food Sci. Food Saf.* **2019**, *18*, 565–583. [CrossRef]

10. Gratta, F.; Fasolato, L.; Birolo, M.; Zomeño, C.; Novelli, E.; Massimiliano, P.; Pascual, A.; Xiccato, G.; Trocino, A. Effect of breast myopathies on quality and microbial shelf life of broiler meat. *Poult. Sci.* **2019**, *98*, 2641–2651. [CrossRef]

11. Kong, B.; Lassiter, K.; Piekarski-Welsher, A.; Dridi, S.; Reverter-Gomez, A.; Hudson, N.J.; Bottje, W.G. Proteomics of breast muscle tissue associated with the phenotypic expression of feed efficiency within a pedigree male broiler line: I. Highlight on mitochondria. *PLoS ONE* **2016**, *11*, e0155679. [CrossRef]

12. Tijare, V.; Yang, V.; Kuttappan, V.; Alvarado, C.; Coon, C.; Owens, C. Meat quality of broiler breast fillets with white string and woody breast muscle myopathies. *Poult. Sci.* **2016**, *95*, 2167–2173. [CrossRef] [PubMed]

13. Bottje, W.; Brand, M.D.; Ojano-Dirain, C.; Lassiter, K.; Toyomizu, M.; Wing, T. Mitochondrial proton leak kinetics and relationship with feed efficiency within a single genetic line of male broilers. *Poult. Sci.* **2009**, *88*, 1683–1693. [CrossRef] [PubMed]

14. Voruganti, S.; Lacroix, J.C.; Rogers, C.N.; Rogers, J.; Matts, R.L.; Hartson, S. The Anticancer Drug AUY922 Generates a Proteomics Fingerprint That Is Highly Conserved among Structurally Diverse Hsp90 Inhibitors. *J. Proteome Res.* **2013**, *12*, 3697–3706. [CrossRef] [PubMed]

15. Mutryn, M.F.; Brannick, E.M.; Fu, W.; Lee, W.R.; Abasht, B. Characterization of a novel chicken muscle disorder through differential gene expression and pathway analysis using RNA-sequencing. *BMC Genom.* **2015**, *16*, 1–19. [CrossRef] [PubMed]

16. Mofarrahi, M.; Husain, S.N.A. Expression and Functional Roles of Angiopoietin-2 in Skeletal Muscles. *PLoS ONE* **2011**, *6*, e22882. [CrossRef]

17. Reverter, A.; Okimoto, R.; Sapp, R.; Bottje, W.G.; Hawken, R.; Hudson, N.J. Chicken muscle mitochondrial content appears co-ordinately regulated and is associated with performance phenotypes. *Biol. Open* **2017**, *6*, 50–58. [CrossRef]

18. Olwin, B.B.; Haushka, S.D. Cell surface fibroblast growth factor and epidermal growth factor receptors are permanently lost during skeletal muscle terminal differentiation in culture. *J. Cell Biol.* **1988**, *107*, 761–769. [CrossRef]

19. Peng, M.; Palin, M.-F.; Véronneau, S.; Lebel, D.; Pelletier, G. Ontogeny of epidermal growth factor (EGF), EGF receptor (EGFR) and basic fibroblast growth factor (bFGF) mRNA levels in pancreas, liver, kidney, and skeletal muscle of pig. *Domest. Anim. Endocrinol.* **1997**, *14*, 286–294. [CrossRef]

20. Leroy, M.C.; Perroud, J.; Darbellay, B.; Bernheim, L.; Konig, S. Epidermal Growth Factor Receptor Down-Regulation Triggers Human Myoblast Differentiation. *PLoS ONE* **2013**, *8*, e71770. [CrossRef]

21. Van Obberghen-Schilling, E.; Roche, N.S.; Flanders, K.C.; Sporn, M.B.; Roberts, A.B. Transforming growth factor B1 positively regulates its own expression in normal and transformed cells. *J. Biol. Chem.* **1988**, *263*, 7741–7746. [CrossRef]

22. Hubert, S.M.; Williams, T.J.; Athrey, G. Insights into the molecular basis of wooden breast based on comparative analysis of fast-and slow-growth broilers. *bioRxiv* **2018**, 356683. [CrossRef]

23. Casalena, G.; Daehn, I.; Bottinger, E.P. Transforming Growth Factor-β, Bioenergetics, and Mitochondria in Renal Disease. *Semin. Nephrol.* **2012**, *32*, 295–303. [CrossRef] [PubMed]

24. Abe, Y.; Sakairi, T.; Beeson, C.; Kopp, J.B. TGF-beta1 stimulates mitochondrial oxidative phosphorylation and generation of reactive oxygen species in cultured mouse podocytes, mediated in part by the mTOR pathway. *Am. J. Physiol. Ren. Physiol.* **2013**, *305*, F1477–F1490. [CrossRef] [PubMed]

25. Jain, M.; Rivera, S.; Monclus, E.A.; Synenki, L.; Zirk, A.; Eisenbart, J.; Feghali-Bostwick, C.; Mutlu, G.M.; Budinger, G.R.; Chandel, N.S. Mitochondrial reactive oxygen species regulate transforming growth factor-beta signaling. *J. Biol. Chem.* **2013**, *288*, 770–777. [CrossRef]

26. Liu, D.; Black, B.L.; Derynck, R. TGF-β inhibits muscle differentiation through functional repression of myogenic transcription factors by Smad3. *Genes Dev.* **2001**, *15*, 2950–2966. [CrossRef]

27. Negmadjanov, U.; Godic, Z.; Rizvi, F.; Emelyanova, L.; Ross, G.; Richards, J.; Holmuhamedov, E.; Jahangir, A. TGF-β1-Mediated Differentiation of Fibroblasts Is Associated with Increased Mitochondrial Content and Cellular Respiration. *PLoS ONE* **2015**, *10*, e0123046. [CrossRef]

28. Widmann, P.; Reverter, A.; Weikard, R.; Suhre, K.; Hammon, H.M.; Albrecht, E.; Kühn, C. Systems Biology Analysis Merging Phenotype, Metabolomic and Genomic Data Identifies Non-SMC Condensin I Complex, Subunit G (NCAPG) and Cellular Maintenance Processes as Major Contributors to Genetic Variability in Bovine Feed Efficiency. *PLoS ONE* **2015**, *10*, e0124574. [CrossRef] [PubMed]

29. Alexandre, P.A.; Naval-Sanchez, M.; Porto-Neto, L.R.; Ferraz, J.B.S.; Reverter, A.; Fukumasu, H. Systems Biology Reveals NR2F6 and TGFB1 as Key Regulators of Feed Efficiency in Beef Cattle. *Front. Genet.* **2019**, *10*, 230. [CrossRef]

30. Shelton, L.M.; Park, B.K.; Copple, I.M. Role of Nrf2 in protection against acute kidney injury. *Kidney Int.* **2013**, *84*, 1090–1095. [CrossRef]

31. Chen, Q.M.; Maltagliati, A.J. Nrf2 at the heart of oxidative stress and cardiac protection. *Physiol. Genom.* **2018**, *50*, 77–97. [CrossRef] [PubMed]

32. Kobayashi, A.; Kang, M.-I.; Watai, Y.; Tong, K.I.; Shibata, T.; Uchida, K.; Yamamoto, M. Oxidative and Electrophilic Stresses Activate Nrf2 through Inhibition of Ubiquitination Activity of Keap1. *Mol. Cell. Biol.* **2006**, *26*, 221–229. [CrossRef] [PubMed]

33. Zhou, N.; Lee, W.R.; Abasht, B. Messenger RNA sequencing and pathway analysis provide novel insights into the biological basis of chickens' feed efficiency. *BMC Genom.* **2015**, *16*, 195. [CrossRef] [PubMed]

34. Furukawa, M.; Xiong, Y. BTB Protein Keap1 Targets Antioxidant Transcription Factor Nrf2 for Ubiquitination by the Cullin 3-Roc1 Ligase. *Mol. Cell. Biol.* **2005**, *25*, 162–171. [CrossRef]

35. Maher, J.; Yamamoto, M. The rise of antioxidant signaling—The evolution and hormetic actions of Nrf2. *Toxicol. Appl. Pharmacol.* **2010**, *244*, 4–15. [CrossRef]

36. Dinkova-Kostova, A.T.; Abramov, A.Y. The emerging role of Nrf2 in mitochondrial function. *Free. Radic. Biol. Med.* **2015**, *88*, 179–188. [CrossRef]

37. Acosta-Alvear, D.; Zhou, Y.; Blais, A.; Tsikitis, M.; Lents, N.H.; Arias, C.; Lennon, C.J.; Kluger, Y.; Dynlacht, B.D. XBP1 Controls Diverse Cell Type- and Condition-Specific Transcriptional Regulatory Networks. *Mol. Cell* **2007**, *27*, 53–66. [CrossRef]

38. Greene, E.; Flees, J.; Dadgar, S.; Mallman, B.; Orlowski, S.; Dhamad, A.; Rochell, S.; Kidd, M.; Laurendon, C.; Whitfield, H.; et al. Quantum blue reduces the severity of muscle myopathy via modulation of oxygen homeostasis-related genes in broiler chickens. *Front. Physiol.* **2019**, *10*, 1251. [CrossRef]

39. Jacinto, E.; Loewith, R.; Schmidt, A.; Lin, S.; Rüegg, M.A.; Hall, A.; Hall, M.N. Mammalian TOR complex 2 controls the actin cytoskeleton and is rapamycin insensitive. *Nat. Cell Biol.* **2004**, *6*, 1122–1128. [CrossRef]

40. Kim, D.-H.; Sarbassov, D.D.; Ali, S.M.; King, J.E.; Latek, R.R.; Erdjument-Bromage, H.; Tempst, P.; Sabatini, D.M. mTOR Interacts with Raptor to Form a Nutrient-Sensitive Complex that Signals to the Cell Growth Machinery. *Cell* **2002**, *110*, 163–175. [CrossRef]

41. Sarbassov, D.D.; Ali, S.M.; Kim, D.-H.; Guertin, D.A.; Latek, R.R.; Erdjument-Bromage, H.; Tempst, P.; Sabatini, D.M. Rictor, a Novel Binding Partner of mTOR, Defines a Rapamycin-Insensitive and Raptor-Independent Pathway that Regulates the Cytoskeleton. *Curr. Biol.* **2004**, *14*, 1296–1302. [CrossRef]

42. Bentzinger, C.F.; Romanino, K.; Cloëtta, D.; Lin, S.; Mascarenhas, J.B.; Oliveri, F.; Xia, J.; Casanova, E.; Costa, C.F.; Brink, M.; et al. Skeletal muscle-specific ablation of raptor, but not of rictor, causes metabolic changes and results in muscle dystrophy. *Cell Metab.* **2008**, *8*, 411–424. [CrossRef]

43. Nisoli, E.; Clementi, E.; Moncada, S.; Carruba, M.O. Mitochondrial biogenesis as a cellular signaling framework. *Biochem. Pharmacol.* **2004**, *67*, 1–15. [CrossRef]

44. Nisoli, E.; Clementi, E.; Paolucci, C.; Cozzi, V.; Tonello, C.; Sciorati, C.; Bracale, R.; Valerio, A.; Francolini, M.; Moncada, S.; et al. Mitochondrial Biogenesis in Mammals: The Role of Endogenous Nitric Oxide. *Science* **2003**, *299*, 896–899. [CrossRef]

45. Hosotani, M.; Kawasaki, T.; Hasegawa, Y.; Wakasa, Y.; Hoshino, M.; Takahashi, N.; Ueda, H.; Takaya, T.; Iwasaki, T.; Watanabe, T. Physiological and Pathological Mitochondrial Clearance Is Related to Pectoralis Major Muscle Pathogenesis in Broilers with Wooden Breast Syndrome. *Front. Physiol.* **2020**, *11*. [CrossRef]

46. Bottje, W.; Carstens, G.E. Association of mitochondrial function and feed efficiency in poultry and livestock species1. *J. Anim. Sci.* **2009**, *87*, E48–E63. [CrossRef]

47. Papah, M.B.; Brannick, E.M.; Schmidt, C.J.; Abasht, B. Gene expression profiling of the early pathogenesis of wooden breast disease in commercial broiler chickens using RNA-sequencing. *PLoS ONE* **2018**, *13*, e0207346. [CrossRef]

48. Lake, J.A.; Papah, M.B.; Abasht, B. Increased Expression of Lipid Metabolism Genes in Early Stages of Wooden Breast Links Myopathy of Broilers to Metabolic Syndrome in Humans. *Genes* **2019**, *10*, 746. [CrossRef]

49. Guo, J.; Sartor, M.; Karyala, S.; Medvedovic, M.; Kann, S.; Puga, A.; Ryan, P.; Tomlinson, C.R. Expression of genes in the TGF-B signaling pathway is significantly deregulated in smooth muscle cells from aorta of aryl hydrocarbon receptor knockout mice. *Toxicol. Appl. Pharmacol.* **2004**, *194*, 76–89. [CrossRef]

50. Larigot, L.; Juricek, L.; Dairou, J.; Coumoul, X. AhR signaling pathways and regulatory functions. *Biochim. Open* **2018**, *7*, 1–9. [CrossRef]

51. Gutierrez-Vasquez, C.; Quintana, F. Regulation of the immune response by the aryl hydrocarbon receptor. *Immun. Cell Biol.* **2018**, *48*, 19–29. [CrossRef] [PubMed]

52. Fan, W.; He, N.; Lin, C.S.; Wei, Z.; Hah, N.; Waizenegger, W.; Ming-Xiao, J.; Liddle, C.; Yu, R.T.; Atkins, A.R.; et al. Errg promotes angiogenesis, mitochondrial biogenesis, and oxidative remodeling in PGC1a/b-deficient mice. *Cell Rep.* **2018**, *22*, 2521–2529. [CrossRef] [PubMed]